Representation Theory, Dynamical Systems, and Asymptotic Combinatorics

Courtesy of Boris Shvartsman

Anatoly Vershik

American Mathematical Society

TRANSLATIONS

Series 2 • Volume 217

Advances in the Mathematical Sciences—58

(*Formerly Advances in Soviet Mathematics*)

Representation Theory, Dynamical Systems, and Asymptotic Combinatorics

V. Kaimanovich
A. Lodkin
Editors

American Mathematical Society
Providence, Rhode Island

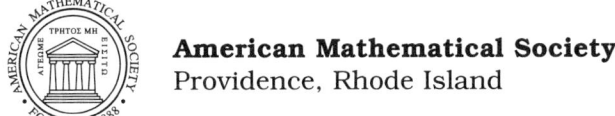

2000 *Mathematics Subject Classification.* Primary 00B15, 05Exx, 22Dxx, 37–XX.

Library of Congress Card Number 91-640741
ISBN-13: 978-0-8218-4208-9
ISBN-10: 0-8218-4208-0
ISSN 0065-9290

∞ The paper used in this book is acid-free and falls within the guidelines
established to ensure permanence and durability.
Visit the AMS home page at http://www.ams.org/

10 9 8 7 6 5 4 3 2 1   11 10 09 08 07 06

Contents

Preface

In 2004 two conferences on the occasion of the 70th birthday of Professor Anatoly Vershik were held: "Geometry and Analysis on Random Structures", Lille, May 25–28, and "Representation Theory, Dynamical Systems, and Asymptotic Combinatorics", St. Petersburg, June 8–13[1]. This volume is mainly based on the contributions of the participants of the St. Petersburg conference. Another volume appeared in 2005[2].

The titles of the conferences encompass the main range of interests of Professor Vershik. With no hope of just listing more or less completely all of his favorite topics, we can only try to mention the most important ones.

Vershik's early work in the theory of linear topological spaces (the late 1950s) was inspired by his first teacher, Gleb Akilov. In the beginning of the 1960s his research interests moved to the theory of measures on infinitely dimensional spaces and Gaussian dynamical systems: the latter topic was suggested by Vladimir Rokhlin, who had strongly influenced him at this stage.

One of Vershik's remarkable achievements in ergodic theory and measure theory is the discovery, in the late 1960s, of *non-standard filtrations* (non-standard decreasing sequences of sigma-fields) of a measure space. This result together with his lacunary theorem laid the foundation of orbit theory in measure spaces developed by Vershik and related to the classical Dye theorem. In the framework of this theory he defined the entropy of a filtration and new invariants of measure-preserving actions of groups (the *scale*) which led to the notion of the so-called *Vershik equivalence*. The theory of filtrations which was initiated by him has now become quite popular in probability theory and its applications (see the talk of M. Émery at the Bourbaki Seminar in 2000).

In the beginning of the 1970s the main interests of Vershik moved from dynamical systems and ergodic theory to representation theory (especially of infinite-dimensional groups) and to the theory of C^*-algebras. A series of pioneering papers with I. M. Gelfand and M. I. Graev on representations of current groups completely changed the field. The main contribution of these papers was a construction of the *integral of representations* for some semi-simple groups of rank one based on the so-called *special representation* with non-trivial first cohomology. Later this construction was generalized to smooth current groups, groups of diffeomorphisms, etc.

---

[1]For a report on the conferences see *Russian Math. Surveys* **60** (2005), no. 5.

[2]*Zapiski Nauchn. Sem. POMI* **326** (`http://www.pdmi.ras.ru/znsl/2005/v326.html`). An English translation will appear in *J. Math. Sci.*

Vershik called one of his main projects *Asymptotic Representation Theory*: it brought together the ideas of classical representation theory, probability theory, and dynamical systems (*ergodic method*). In its origin was the paper (in collaboration with A. Schmidt) on the limit distribution of the length of cycles in random permutations (the Poisson–Dirichlet measure); it was continued (jointly with S. Kerov) by a new proof of Thoma's theorem on the characters of the infinite symmetric group based on the consideration of limits of Young diagrams and by a description of the limit shape of Young diagrams with respect to Plancherel measure.

In particular, the latter result gave a solution of the longstanding combinatorial Ulam problem on longest increasing subsequences of a sequence of independent identically distributed random variables, which later became a starting point for many combinatorial and probabilistic papers. It also initiated a large cycle of works on representations of the infinite symmetric and unitary groups, one of the most important results being the approximation theory of the representations of infinite symmetric groups which leads to a complete harmonic analysis for this group. This line of research is closely connected with the study of AF-algebras and locally finite groups, their K-functor, semi-finite traces, etc. Moreover, this point of view also prompted a new approach to the classical theory of representations of finite Coxeter groups. This "revised" theory was elaborated upon in a joint paper with A. Okounkov and is currently being continued by Vershik and his collaborators.

It was in the process of the study (in collaboration with S. Kerov) of AF-algebras related to infinite groups that the notion of the *adic transformation* of the Cantor compactum (nowadays often called *Vershik transformation*) was invented. It was based on a new order on the path space of a Bratteli diagram (the resulting structure is now called *Bratteli-Vershik diagram*). It turned out to be very useful for approximation of transformations and for constructing interesting examples of adic transformations (Pascal transformation associated with the Pascal graph, Young transformation, etc.). Vershik proved that an arbitrary measure-preserving transformation can be realized in the adic model. The adic transformation also played a crucial role in the construction of a uniform simultaneous finite-dimensional approximation of the multiplier and of the shift operator.

A long series of papers by Vershik is devoted to the general subject of *limit shape problems* for configurations, partitions (involving ideas from statistical mechanics), diagrams, etc.

A recent subject of Vershik's interests is universality and randomness. In particular, he proved that universal Urysohn space is generic and that a random Polish space is the Urysohn space with probability one.

The current active development of many other topics was also strongly influenced by Vershik's contributions. Let us just mention

- linear programming and optimization, duality (in the framework of the classical papers of one of his teachers, Leonid Kantorovich);
- random walks on groups, entropy and Poisson boundaries (together with V. Kaimanovich);
- non-holonomic mechanics (with L. Faddeev) and non-holonomic dynamics;
- non-Fock factorizations (with B. Tsirelson);

- ergodic theory of polymorphisms (multivalued Markov maps);
- arithmetic coding of hyperbolic automorphisms and substitutions.

Dozens of former students of Professor Vershik are active mathematicians throughout the world, and his famous seminar in St. Petersburg has been well known for about 40 years. His extraordinary erudition, profound knowledge, and active position made him an established authority in the mathematical community. Since 1998 he has been president of the St. Petersburg Mathematical Society.

The authors and the editors of the present volume dedicate it, with admiration, to Anatoly Moiseevich Vershik and wish him further success in all areas of his activity.

V. Kaimanovich and A. Lodkin

Amer. Math. Soc. Transl.
(2) Vol. **217**, 2006

Statistics of the Symmetric Group Representations as a Natural Science Question on Asymptotics of Young Diagrams

Vladimir Arnold

ABSTRACT. The main content of the paper is a description of some unsolved problems related to the Vershik–Kerov–Plancherel statistics of large Young diagrams. It also describes some relations of these problems to other branches of mathematics and to natural sciences, such as the symmetries statistics for the eigenfunctions of the Laplacians on Riemannian manifolds with symmetrical metrics, Coxeter groups representation theory and arithmetical turbulence in number theory.

Evidence is diminished by proofs.

M. T. Cicero

I first heard about the remarkable works of Vershik on Young diagrams from Yu. V. Linnik, who had clearly explained to me the content and the perspectives of these studies. Yurii Vladimirovich had added, quoting Bertrand Russell, that "an exposition may be either clear or strict, both qualities together being incompatible."

Unfortunately, he was right: trying later to understand the talks and papers by Vershik and his collaborators, like [**VK**], and even continuing these studies, as in [**A2**], I was unable to find in the published texts the simple and fundamental statements explained to me by Linnik: I only found proofs of some complicated inequalities instead. These papers do contain, perhaps, the proofs of the simple statements that I had been interested in. However, even the most elaborate generalizations of the Stirling formula do not explain the fundamental importance and the generality of the Gaussian normal probability distribution. Similarly, the articles on Vershik's theory provide insufficient explanation of its deep meaning. "Write it laconically and vaguely," suggested Napoleon to the authors of the Constitution.

I therefore decided to present here a short and informal review of the astonishing facts, basic for the remarkable Vershik theory, which lead to some special kind of intuition deserving to be included into the mathematical luggage both of

---

2000 *Mathematics Subject Classification.* Primary 20C30; Secondary 05E10.

Key words and phrases. Catalan numbers, reflection groups, Weyl groups, frequent representations, Frobenius numbers, weak asymptotics, Cesaro averaging, astroidal triangle, spectral representation of Laplace operators on Riemannian manifolds with symmetries.

Partially supported by the RFBR, grant 02-01-00655.

theoreticians and of experts in applied mathematics (mentioning below only a small part of the applied domains).

The basic subject studied in this theory is so general, as is, for instance, that of the Pascal triangle or of the cycloid curve or of the Lobachevski plane, taking into account their applications to such sciences as combinatorics and probability theory, or variational calculus and mechanics (called "mathematical paradise" by Leonardo da Vinci) or applications to the geometry of De Sitter's world and to relativity theory.

The intuition, allowing one to replace a small angle measure by its sine value, is more important than the division of epsilon by 3 (and not by 2) in the search of delta, and I will try to communicate below this basic intuition rather than the formal proofs.

It is well known (going back to Frobenius) that the irreducible representations of the group $S(n)$ of all the $n!$ permutations of n elements are characterized by their *Young diagrams* of area n.

Such a diagram describes a partition into natural summands

$$n = a_1 + a_2 + \ldots + a_s.$$

It is usually represented by the following set of n unit squares in the plane. First, order the summands in the nonincreasing way:

$$a_1 \geq a_2 \geq \ldots \geq a_s.$$

The first line of the diagram ($0 \leq x \leq a_1, \quad -1 \leq y \leq 0$) consists of a_1 unit squares.

The second line (lying lower) consists of a_2 squares ($0 \leq x \leq a_2, \quad -2 \leq y \leq -1$), and so on, down to the last line ($0 \leq x \leq a_s, \quad -s \leq y \leq -s-1$).

The resulting set of n unit squares is the Young diagram of the partition.

Vershik's theory studies the asymptotical properties of the large Young diagrams as their area n tends to infinity. Perhaps the most striking result is the *universality theorem* claiming that *"most" Young diagrams of large area look alike when seen from a distance.*

Namely, there exists a special function f of one variable, growing from $f(0) = 0$ to $f(1) = 1$ (and calculated implicitly by Vershik and Kerov), defining an astroid-like curve $f(x) + f(y) = 1$ (in the square $0 \leq x \leq 1$, $0 \leq y \leq 1$), such that a typical Young diagram of large area n is well approximated by the triangle obtained by a rigid motion and a homothety from the curved triangular domain below the astroidal curve

$$f(x) + f(y) \leq 1, \qquad x \geq 0, \quad y \geq 0.$$

The explicit equation of the "astroid-like" curve is presented by Vershik in terms of the coordinates $x - y$ and $x + y$ (the function f can be calculated by integration of his function representing the dependence of $x + y$ on $x - y$).

Of course, the area of the curved triangular domain described above is a quantity of order of 1, while the area of the Young diagram is n. Therefore, the homothety of the Young diagram to the universal triangle should be a contraction (one should contract the Young diagram approximately $\sqrt{n}$ times to reduce its area n to the quantity of order of 1).

Such a contraction would reduce the Young diagram to a domain bounded by a broken line with short edges. The universality theorem claims that *this broken line* *lives (for large n) in a small neighborhood of the universal astroidal curve defined by the equation $f(x) + f(y) = 1$.*

Of course, in order to transform this description into a formal theorem, one should explain the words "a typical diagram", "for most diagrams", "small neighborhood" defining the convergence of the broken line to the universal curve as $n \to \infty$.

All these details are presented in the papers by Vershik and his collaborators, the convergence character being different in the interior part of the curve (where both x and y are far from 0 and from 1) and near the vertices ($x = 0$, $y = 1$ and $x = 1$, $y = 0$) of the curve. Namely, near the vertices one observes some kind of the "Gibbs phenomenon" (also similar to the large deviation phenomena in the limit theorems of probability theory).

A very astonishing part of Vershik's theory is the special averaging used by him to define the "typical diagrams" and the "majority of the diagrams D of large area n".

Namely, let the integer $a(D)$ be the (complex) dimension of the unitary irreducible representation (in the Hermitian space $\mathbb{C}^{a(D)}$) of the symmetric group $S(n)$ that is characterized by the diagram D in Frobenius theory.

The average is then the weighting accordingly to the "Plancherel measure", *the weights w of the diagrams being proportional to the squares of the dimensions of the corresponding representations* (as in the decomposition of the regular representation into the irreducible ones):

$$w(D) = Ca^2(D), \qquad C = \left(\sum a^2(D)\right)^{-1}.$$

It is not easy to guess why the squares are present in this choice of weights, but the theorems proved by Vershik and his collaborators show their necessity.

Vershik's theory contains, besides the universality theorem for the Plancherel measure described above, many other remarkable results, of which I shall discuss only a small part.

For different diagrams D of fixed area n, the dimensions $a(D)$ of representations have different values, from 1 up to some *maximal dimension* $\bar{a}(n)$.

An explicit combinatorial description of the integer $a(D)$ (which is an extension of the *Catalan number*) represents it as *the number of monotonic fillings* of the n squares of the Young diagram with n natural numbers $(1, 2, \ldots, n)$. The monotonicity condition claims the decreasing of the numbers along each line (as x grows) and along each column (as $|y|$ is growing in our description of the diagram).

Thus, for the partition $5 = 3 + 2$ of area $n = 5$ with lines of lengths 3 and 2, the monotonic fillings are

$$\begin{pmatrix} 5 & 4 & 3 \\ 2 & 1 & \end{pmatrix}, \quad \begin{pmatrix} 5 & 4 & 1 \\ 3 & 2 & \end{pmatrix}, \quad \begin{pmatrix} 5 & 4 & 2 \\ 3 & 1 & \end{pmatrix}, \quad \begin{pmatrix} 5 & 3 & 2 \\ 4 & 1 & \end{pmatrix}, \quad \begin{pmatrix} 5 & 3 & 1 \\ 4 & 2 & \end{pmatrix};$$

therefore, $a(D) = 5$. The maximal dimension for $n = 5$ is $\bar{a}(5) = 6$; it is attained at the partition $5 = 3 + 1 + 1$.

A convenient formula representing $a(D)$ in terms of the easily computable combination of factorials is given in article [**A2**], where one can also find some remarkable relations between the combinatorics of these extended Catalan numbers $a(D)$, singularity theory, and the Euler–Jacobi residues formula of complex analysis.[1]

An astonishing result of Vershik and Kerov [**VK**] is the following theorem:

The maximal dimension (for a given area n of D) of the representation, $a(D)$, behaves asymptotically (for large N) in the same way as the mean dimension (if the averaging, to obtain the mean, is done according to the Plancherel measure on the set of the Young diagrams of area n).

This claim means that, besides the diagram D providing the maximal value $a(D) = \bar{a}(n)$, there exist extremely many other diagrams for which the dimensions of their representations are close to the maximal value.

Perhaps the proof of Vershik and Kerov also provides the information about the quantity of these "almost maximal-dimensional" diagrams. However, I have not succeeded to extract the answer to this natural science question from their paper [**VK**].

As W. Blake said, the truth does not allow to be understood: it needs to be believed.

The study of the *histogram of dimensions*, that is, of the distribution of the numbers $a(D)$ corresponding to different Young diagrams D of area n, in the neighborhood of the maximal point $\bar{a}(n)$ of this set of dimensions, is still waiting for a study (even for the study of its asymptotical behavior for large values of n, which is perhaps easier).

Yu. V. Linnik also described to me some other striking applications of the same theory (which I was later unable to reconstruct from Vershik's articles). According to Linnik, the main point is to understand which of the logarithms in the asymptotics might be considered as being "constant" coefficients of the powers containing asymptotics terms.

Let N be a large natural number. Consider its decomposition into the prime factors,

$$N = 2^{a_2} 3^{a_3} 5^{a_5} \dots .$$

The problem is *to find the behavior of the multiplicities a_p of different prime factors p (in the average, for a random choice of a large number N).*

Yu. V. Linnik told me that the answer to this natural science question is provided by Vershik's theory of the asymptotical statistic of the Young diagrams (with a suitable non-Plancherel probability distribution on the set of Young diagrams) in the following way.

Consider the partition

$$\ln N = a_2 \ln 2 + a_3 \ln 3 + \dots$$

as a *generalized Young diagram* with a_2 lines of lengths $\ln 2$, a_3 lines of lengths $\ln 3$, etc.

The claim is that the formal application of Vershik's statistics of the Young diagrams to these generalized diagrams provides (at least in the average) the correct large N asymptotics.

---

[1] This formula has rich history. See, for instance, J. S. Frame, G. de B. Robinson, R. M. Thrall, *The hook graphs of the symmetric groups,* Canadian J. Math. **6** (1954), 316–324 (*Editors' note*).

For instance, *the diagram's shape* (after a normalizing contraction) *becomes in most cases close to some universal astroidal curve*, in spite of the fact that the logarithms are rather irrational numbers (the lines of Young diagrams in the preceding Vershik theory had integral lengths).

Therefore, the asymptotics of Young diagrams (for suitable Vershik statistics) provide the answers to the natural science questions on the factors of the large random integers. I would compare these answers to the Dirichlet theorem claiming that the number $\varphi(n)$ of the integers smaller than n and relatively prime to n grows (averaged) like cn. The constant $c \approx 2/3$ in this asymptotical expansion represents the probability of the irreducibility of a fraction p/q. Euler calculated c to be $6/\pi^2$. It was for this goal that he introduced his zeta-function formula

$$\zeta(s) = \prod \left(1 - \frac{1}{p^s}\right)^{-1} = \sum_{n=1}^{\infty} \frac{1}{n^s}$$

(the product along all the prime numbers $p = 2, 3, 5, \ldots$). Namely, his formula provides the value $1/\zeta(s)$ for the probability of the absence of integer points on the interval connecting 0 to a random point in $\mathbb{Z}^s$. To calculate $\zeta(2) = \pi^2/6$, Euler used the "Fourier" series.

One more application of Vershik's statistic studies the following problem.

Consider a segment of a straight line. Break it at a random point decomposing it into two segments. Iterating this breaking with each part (for instance, distributing the breaking points uniformly along the segments and independently each of the others), we get after n stages 2^n parts of different lengths.

The problem is to study the statistics of the resulting partition of the length of the initial segment into 2^n *summands.*

The claim is that *the corresponding Young diagrams are in most cases asymptotically standard*, the shape of the (homothetically normalized) diagram approaching some universal curve for n tending to infinity.

I shall not formulate the corresponding limit theorems in detail: they are perhaps *proved*, rather then *formulated*, in the literature (in spite of their evident natural science importance).

Now I shall add some evident missing generalizations, which are also waiting for courageous researchers. One usually finds, together with a result on the symmetric groups, similar results for other *Coxeter groups* (at least for the four classical series, A_k, B_k, C_k, D_k, one might study the asymptotics as $k \to \infty$; there are also many questions to study for the other Weyl and Coxeter groups, E_6, E_7, E_8, F_4, G_2, $I_2(p)$, H_3, H_4).

Unfortunately, I did not find results resembling Vershik's theory in the literature (neither for the theorems proved in the symmetric case A_k, nor for those which are left open even in this case, like the question on the histogram of the distribution of the dimensions $a(D)$ near the maximal value $\bar{a}(n)$ of these dimensions for the Young diagrams of large area n).

I mention the applications of these questions about the dimensions of the irreducible unitary representations in the study of the *frequent representations* [**A2**].

A unitary representation of a group in the Hermitian space $\mathbb{C}^N$ is called *frequent* if *the dimension of the variety of those unitary representations in the same space that are unitarily equivalent to it is maximal* (compared to the other representations in $\mathbb{C}^N$).

It is proved in [**A2**] that *the multiplicities of the irreducible summands of the frequent representations (of any finite group) in $\mathbb{C}^N$ are asymptotically (for $N \to \infty$) proportional to the dimensions of these irreducible components.*

Therefore, the problem arises about the histogram of the distribution of the numbers of the unitary equivalence classes of those (reducible in general) representations T in $\mathbb{C}^N$ for which the dimension of the variety of the representations unitarily equivalent to T is close to the maximal such dimension.

Even in the case of symmetric groups A_n, this question is open. It is also open in the asymptotical versions for large n and large N (where the question is easier).

The missing information is, first of all, that on the histograms of the distribution of dimensions of irreducible representations (near the maximal dimension) discussed above.

Moreover, to pass from the statistic of irreducible representations to the statistic of reducible representations, one should study the behavior of the so-called *Frobenius constant* $K(a_1, \ldots, a_s)$ *that is the smallest integer such that it as well as all the bigger integers belong to the additive semigroup of integers generated by the integers* $a_1, \ldots, a_s$ (with no common factor). This semigroup consists of all the linear combinations $k_1 a_1 + \ldots + k_s a_s$ with nonnegative integral coefficients k_i.

For $s = 2$, the number $K(a, b)$ was calculated by Sylvester [**S**], who proved it to be $(a-1)(b-1)$ (like $K(3,5) = 8$). In this case the semigroup fills one half of the segment $(0, K)$.

Using the natural science (heuristical) arguments on forest projection shapes, I suggested for the Frobenius number the hypothetical weak asymptotics [**A1**]:

$$K(a) \sim \left(C \prod a(i)\right)^{1/(s-1)}, \quad \text{where } C = (s-1)!\,.$$

The weakness means in [**A1**] the averaging along a neighborhood of the vector a (hypothetically, the relation is approximately valid for most vectors a).

But the mathematical theorems on the behavior of $K(a)$ (even in the averaged version) are still unproved, as well as similar conjectures on the asymptotical density of the elements of the distribution of the additive subgroup $\{(ka)\}$ along the segment $(0, K(a))$. These conjectures suggest that the density (asymptotically) behaves like t^{s-1} at the point tK; therefore, the semigroup must fill $1/s$ part of the segment $(0, K)$.

The theory of unitary frequent representations developed in [**A2**] has undoubtedly both the *orthogonal version* and the (*compact*) *symplectic one*, and both versions should be accompanied by the corresponding versions of Vershik's asymptotics.

Unfortunately, except for my 1972 examples, I know of no published results in this direction (as well as no results extending the theory of frequent representations of finite groups [**A2**] to the case of infinite discrete groups and to the case of compact Lie groups, where one should also study the Plancherel measure whose weights are proportional to the dimensions of the spaces where the representation is a multiple of an irreducible one).

One may hope to find such extensions both for the discrete generalizations of finite groups (with sums replaced by series) and for the continuous case (replacing sums by integrals).

I have in mind first of all the regular representation (in the space of functions on the group) whose dimension is equal to the number of the elements of the finite group and to the sum of squares of the dimensions of irreducible representations.

Replacing sums by series or by integrals, one might perhaps obtain useful generalizations, including even the generalizations of Vershik's Plancherel measure and its applications both to the irreducible representations and to the frequent ones: the regular representation is well defined for infinite groups, and one might decompose it.

Of course, the extensions of Vershik's theory discussed above yet need to be constructed.

However,

> Every science must be founded on concrete theorems discovered directly from the observations of the particular experiences. Believing in these facts one is led by them to the acceptance of the general laws of nature.
>
> U. Eco, "The name of the Rose", Day 3, Hour 9

References

[A1] V. I. Arnold, *Weak asymptotics of the solutions numbers of Diophantine problems,* Func. Anal. Appl. **33** (1999), no. 3, 65–66.

[A2] V. I. Arnold, *Frequent representations,* Moscow Math. J. **3** (2003), 1209–1221.

[A3] V. I Arnold, *Number-theoretical turbulence in Fermat–Euler arithmetics and large Young diagrams geometry statistics,* Journal of Mathematical Fluid Mechanics **7** (2005), 4–50.

[S] J. S. Sylvester, *Mathematical questions with their solutions,* Educational Times **41** (1884), 21.

[VK] A. M. Vershik, S. V. Kerov, *Asymptotics of maximal and of typical dimension of irreducible representations of symmetric groups,* Func. Anal. Appl. **19** (1985), no. 1, 25–39.

Steklov Mathematical Institute, Gubkina St. 8, Moscow 119991, GSP-1, Russia

Amer. Math. Soc. Transl.
(2) Vol. **217**, 2006

Stochastic Dynamics Related to Plancherel Measure on Partitions

Alexei Borodin and Grigori Olshanski

Abstract. Consider the standard Poisson process in the first quadrant of the Euclidean plane, and for any point (u, v) of this quadrant take the Young diagram obtained by applying the Robinson–Schensted correspondence to the intersection of the Poisson point configuration with the rectangle with vertices $(0,0)$, $(u,0)$, (u,v), $(0,v)$. It is known that the distribution of the random Young diagram thus obtained is the poissonized Plancherel measure with parameter uv.

We show that for (u, v) moving along any southeast-directed curve C in the quadrant, these Young diagrams form a Markov process Λ_C with continuous time. We also describe Λ_C in terms of jump rates.

Our main result is the computation of the dynamical correlation functions of such Markov processes and their bulk and edge scaling limits.

Introduction

For any $n = 1, 2, \ldots$, consider a measure on the set of all partitions of n which assigns to a partition λ the square of the dimension of the corresponding irreducible representation of the symmetric group $S(n)$, divided by $|S(n)| = n!$. The classical Burnside formula implies that this is a probability measure (the sum of all weights equals one). It is usually referred to as the nth *Plancherel measure*.

As was independently shown by Logan and Shepp [**LS**] and Vershik and Kerov [**VK1, VK3**][1], for large n the random partitions distributed according to the nth Plancherel measure have a typical (limit) shape. More detailed information about local behavior of the random partitions in different regions of the limit shape was later obtained in [**BDJ1, BDJ2**], [**O**], [**BOO**], and [**J1**] for the "edge" of the limit shape and in [**BOO**] for the "bulk" of the limit shape.

2000 *Mathematics Subject Classification.* Primary 60G55; Secondary 60J27, 60C05.

Key words and phrases. Plancherel measure, determinantal point processes, Poisson process, Markov processes on partitions, dynamical correlation functions, Robinson–Schensted correspondence, discrete Bessel kernel, discrete sine kernel, Airy kernel.

This research was partially conducted during the period the first author served as a Clay Mathematics Institute Research Fellow. He was also partially supported by the NSF grant DMS-0402047.

The second author was supported by CRDF grant RM1-2543-MO-03.

[1]A different proof was later found by Kerov; see [**IO**].

One key observation that allowed performing such a detailed analysis was that the mixture of the Plancherel measures with different n's by a Poisson distribution, the so-called *poissonized* Plancherel measure, has a nice algebraic structure: it defines a *determinantal point process*[2] [**BOO, J1**].

In this work we construct stationary Markov processes on the set of all partitions which have the poissonized Plancherel measures as their invariant distributions. We prove that for any finite number of time moments, the corresponding joint distribution of the same number of random partitions defines a determinantal point process, and we compute its correlation kernel.

As in the "static" case, there are two limit transitions, at the edge and in the bulk of the limit shape. The corresponding limit of the correlation kernel at the edge turns out to be the well-known *extended Airy kernel*, while the limiting kernel in the bulk appears to be a new one.

As a matter of fact, we obtain these results for more general, *nonstationary* Markov processes on partitions, which change the value of the poissonization parameter of the poissonized Plancherel measure with time. This allows us to interpret the results in terms of the Poisson process in a quadrant and its projection by the Robinson–Schensted correspondence.

Our results also extend to more general measures on partitions, the so-called *z-measures* [**BO1**]. Markov processes related to these measures are studied in detail in our paper [**BO2**]. The Plancherel measures may be viewed as appropriate limits of the z-measures, and we use this connection extensively in our proofs: the results of §3 are obtained from the similar results of [**BO2**] by a degeneration.

Our work was largely inspired by previous papers by Okounkov and Reshetikhin [**OR**], Prähofer and Spohn [**PS**], and Johansson [**J2**]. Thus, some of the results below may already be known to experts. In particular, the results of §3 can be obtained using the formalism of Schur processes [**OR**] (this is not true for the z-measures, however), and for certain special cases of our Markov processes the determinantal structure of the correlation functions and the edge scaling limit were obtained by Prähofer and Spohn [**PS**] in their work on polynuclear growth processes.

1. Construction of Markov processes

As in Macdonald [**M**] we identify partitions and Young diagrams. By $\mathbb{Y}_n$ we denote the set of partitions of a natural number n or, equivalently, the set of Young diagrams with n boxes. By $\mathbb{Y}$ we denote the set of all Young diagrams, that is, the disjoint union of the finite sets $\mathbb{Y}_n$, where $n = 0, 1, 2, \dots$ (by convention, $\mathbb{Y}_0$ consists of a single element, the empty diagram $\varnothing$). Given $\lambda \in \mathbb{Y}$, let $|\lambda|$ denote the number of boxes of λ (so that $\lambda \in \mathbb{Y}_{|\lambda|}$), and let $\ell(\lambda)$ be the number of nonzero rows in λ (the length of the partition).

For two Young diagrams λ and μ we write $\mu \nearrow \lambda$ (equivalently, $\lambda \searrow \mu$) if $\mu \subset \lambda$ and $|\mu| = |\lambda| - 1$, or, in other words, μ is obtained from λ by removing one box.

Let $\dim \lambda$, the *dimension* of λ, be the number of all standard tableaux of shape λ. Equivalently, $\dim \lambda$ is the dimension of the irreducible representation of the

[2] This means that its correlation functions can be written as minors of a suitable matrix called *the correlation kernel*.

symmetric group $S(|\lambda|)$ labelled by λ. A convenient explicit formula for $\dim\lambda$ is

$$\dim\lambda = \frac{n!}{\prod_{i=1}^{N}(\lambda_i+N-i)!}\prod_{1\le i<j\le N}(\lambda_i-i-\lambda_j+j), \qquad \lambda\in\mathbb{Y}_n,$$

where N is an arbitrary integer $\ge \ell(\lambda)$ (the above expression is stable in N).

For $\lambda\in\mathbb{Y}_n$, $\mu\in\mathbb{Y}_{n-1}$ set

$$p^{\downarrow}(n,\lambda;n-1,\mu) = \begin{cases} \dfrac{\dim\mu}{\dim\lambda}, & \mu\nearrow\lambda,\\ 0, & \text{otherwise,}\end{cases}$$

and for $\lambda\in\mathbb{Y}_n$, $\nu\in\mathbb{Y}_{n+1}$ set

$$p^{\uparrow}(n,\lambda;n+1,\nu) = \begin{cases} \dfrac{\dim\nu}{\dim\lambda\,(n+1)}, & \lambda\nearrow\nu,\\ 0, & \text{otherwise.}\end{cases}$$

Then we have (see [**VK2**])

$$\sum_{\mu\in\mathbb{Y}_{n-1}} p^{\downarrow}(n,\lambda;n-1,\mu)=1, \qquad \sum_{\nu\in\mathbb{Y}_{n+1}} p^{\uparrow}(n,\lambda;n+1,\nu)=1.$$

The nth *Plancherel measure* is a probability measure on the finite set $\mathbb{Y}_n$ which is defined by

$$M^{(n)}(\lambda)=\frac{(\dim\lambda)^2}{n!}, \qquad \lambda\in\mathbb{Y}_n\,; \tag{1.1}$$

see [**VK2**]. The Plancherel measures with various indices n are related to each other by means of the "down probabilities" $p^{\downarrow}(n,\lambda;n-1,\mu)$ and the "up probabilities" $p^{\uparrow}(n,\lambda;n+1,\nu)$, as follows (see [**VK2**]):

$$\begin{aligned} M^{(n-1)}(\mu) &= \sum_{\lambda\in\mathbb{Y}_n} M^{(n)}(\lambda)p^{\downarrow}(n,\lambda;n-1,\mu),\\ M^{(n+1)}(\nu) &= \sum_{\lambda\in\mathbb{Y}_n} M^{(n)}(\lambda)p^{\uparrow}(n,\lambda;n+1,\nu). \end{aligned} \tag{1.2}$$

Note also the following relation:

$$M^{(n)}(\lambda)p^{\uparrow}(n,\lambda;n+1,\nu) = M^{(n+1)}(\nu)p^{\downarrow}(n+1,\nu;n,\lambda). \tag{1.3}$$

Consider the Poisson distribution on the set $\mathbb{Z}_+=\{0,1,2,\dots\}$, with parameter $\theta>0$:

$$Poisson_\theta(n)=e^{-\theta}\,\frac{\theta^n}{n!}, \qquad n\in\mathbb{Z}_+\,. \tag{1.4}$$

Mixing all measures $M^{(n)}$ together by means of the Poisson distribution (1.4), we obtain a probability measure on the set $\mathbb{Y}$. We denote it by M_θ and call it the *poissonized Plancherel measure* with parameter θ:

$$M_\theta(\lambda)=e^{-\theta}\,\theta^n\left(\frac{\dim\lambda}{n!}\right)^2, \qquad n=|\lambda|. \tag{1.5}$$

We are going to define a stationary Markov process $\Lambda_\theta=\Lambda_\theta(t)$ with discrete state space $\mathbb{Y}$ and continuous time $t\in\mathbb{R}$ and such that M_θ is an invariant measure of Λ_θ. Moreover, Λ_θ is reversible with respect to M_θ.

The trajectories of Λ_θ are step functions (in other words, piecewise constant functions) $\Lambda(t)$ of variable $t\in\mathbb{R}$, with values in $\mathbb{Y}$. We say that a trajectory $\Lambda(t)$

makes a *jump* at a moment t if the left limit $\Lambda(t^-) = \lim_{t'\uparrow t}\Lambda(t')$ differs from the right limit $\Lambda(t^+) = \lim_{t'\downarrow t}\Lambda(t')$. We reserve the notation $\Lambda_\theta(t)$ to denote the *random* trajectory.

DEFINITION 1.1 (Jump rates of Λ_θ). For any $t\in\mathbb{R}$ and any $\lambda\in\mathbb{Y}_n$ we have by definition: conditional on $\Lambda_\theta(t^-)=\lambda$, the probability that $\Lambda_\theta(\,\cdot\,)$ makes a jump to a diagram $\mu\in\mathbb{Y}_{n-1}$ in the time interval $[t,t+dt]$ is equal to $R^\downarrow(n,\lambda;n-1,\mu)dt+o(dt)$, where

$$R^\downarrow(n,\lambda;n-1,\mu) = n\,p^\downarrow(n,\lambda;n-1,\mu). \tag{1.6}$$

Likewise, the (conditional) probability of jumping to a diagram $\nu\in\mathbb{Y}_{n+1}$ in the time interval $[t,t+dt]$ is equal to $R^\uparrow(n,\lambda;n+1,\nu)dt+o(dt)$, where

$$R^\uparrow(n,\lambda;n+1,\nu) = \theta\,p^\uparrow(n,\lambda;n+1,\nu). \tag{1.7}$$

Finally, any other jumps in $[t,t+dt]$ are excluded (with probability $1-o(dt)$). For obvious reasons we refer to (1.6) and (1.7) as to the *jump rates*.

The knowledge of the jump rates makes it possible (in our concrete case) to define uniquely a *transition function*

$$P_{\Lambda_\theta}(t,\lambda;s,\kappa) = \mathrm{Prob}\{\Lambda_\theta(s)=\kappa \mid \Lambda_\theta(t)=\lambda\}, \qquad s>t, \quad \lambda,\kappa\in\mathbb{Y},$$

which depends only on $s-t$. The poissonized Plancherel measure is compatible with the transition function,

$$\sum_{\lambda\in\mathbb{Y}} M_\theta(\lambda)P_{\Lambda_\theta}(t,\lambda;s,\kappa) = M_\theta(\kappa),$$

which allows us to define the Markov process in question. Moreover we can define the process not only on a half-line $[t_0,+\infty)\subset\mathbb{R}$ but on the whole real line; that is, we can construct a probability measure on the set $\{\Lambda(t)\}$ of $\mathbb{Y}$-valued step functions $\Lambda(t)$ defined for all $t\in\mathbb{R}$.[3] Since the transition function is translation invariant in time, the process is *stationary* (that is, the above measure on the set $\{\Lambda(t)\}$ is invariant under shifts of time, $t\to t+\text{const}$). Finally, the process turns out to be *reversible* (that is, the measure on $\{\Lambda(t)\}$ is also invariant under the time reversion $t\to -t$).

REMARK 1.2. The above definition of the Markov process Λ_θ can be rephrased as follows. Introduce an auxiliary Markov process N_θ: a birth-death process on $\mathbb{Z}_+=\{0,1,2,\dots\}$, which can be defined by the "down" and "up" jump rates

$$R^\downarrow(n;n-1)=n, \qquad R^\uparrow(n;n+1)\equiv\theta. \tag{1.8}$$

The key property of N_θ is that it has the Poisson distribution (1.4) as the invariant measure.[4] The birth-death process N_θ governs the jumps of Λ_θ: each moment $N_\theta(\,\cdot\,)$ jumps down, say, from n to $n-1$, the trajectory $\Lambda_\theta(\,\cdot\,)$ makes a jump from $\mathbb{Y}_n$ to $\mathbb{Y}_{n-1}$ (and the target Young diagram $\mu\in\mathbb{Y}_{n-1}$ is chosen according to the "down" probabilities $p^\downarrow(n,\lambda;n-1,\mu)$, where λ stands for the preceding state). Likewise, when $N_\theta(\,\cdot\,)$ jumps up, say, from n to $n+1$, the trajectory $\Lambda_\theta(\,\cdot\,)$ makes a jump from λ to a random diagram $\nu\in\mathbb{Y}_{n+1}$ chosen according to the "up" probabilities

[3]Let us note that the relevant step functions have only finitely many jumps on finite time intervals $[t,s]\subset\mathbb{R}$. Markov processes with such a property (finitely many jumps in finite time) are sometimes called *regular*.

[4]Recall that this distribution is precisely the "mixing" measure used in the definition of the poissonized Plancherel measure M_θ.

$p^{\uparrow}(n,\lambda;n+1,\nu)$. The fact that M_θ is the invariant measure is deduced from the fact that the Poisson distribution is the invariant measure of N_θ and from relations (1.2). The reversibility property [5] of Λ_θ is deduced from the reversibility of N_θ and relation (1.3).

The above construction of the Markov process can be generalized. The idea is to use more general birth-death processes, with time-dependent jump rates.

Let $\mathbb{R}^2_{>0}$ denote the open first quadrant of the Euclidean plane $\mathbb{R}^2$ with coordinates $u>0, v>0$. Consider a parameterized curve $C=(u(t),v(t))$ in $\mathbb{R}^2_{>0}$ subject to the following conditions: the functions $u(t)>0$, $v(t)>0$ are continuous and piecewise continuously differentiable; the curve is directed southeast, that is, $\dot u(t)\ge 0$, $\dot v(t)\le 0$, and $\dot u(t)$ and $\dot v(t)$ do not vanish simultaneously (here the dot means derivative with respect to t, and t is interpreted as time). Such curves C will be called *admissible.*

We modify formulas (1.8) as follows:

$$R^{\downarrow}(n;n-1;t)=-n\,\frac{\dot v(t)}{v(t)},\qquad R^{\uparrow}(n;n+1;t)=\dot u(t)v(t). \tag{1.9}$$

Note that if C is the hyperbola $uv=\theta$ parameterized by $t=\ln u$, then (1.9) reduces to (1.8). In the general case we set

$$\theta(t)=u(t)v(t). \tag{1.10}$$

There exists a birth-death process determined by the jump rates (1.9); we denote it by N_C. Let

$$P_{N_C}(t,n;s,m)=\operatorname{Prob}\{N_C(s)=m\mid N_C(t)=n\} \tag{1.11}$$

be the transition function of N_C (here $s>t$ and $n,m\in\mathbb{Z}_+$). The process N_C is no longer stationary in time, and instead of a single Poisson distribution with fixed parameter θ (see (1.4)) we deal with the whole family of such distributions indexed by the varying parameter $\theta(t)$ given by (1.10). These distributions are compatible with the transition function (1.11):

$$\sum_{n\in\mathbb{Z}_+} e^{-\theta(t)}\,\frac{(\theta(t))^n}{n!}P_{N_C}(t,n;s,m)=e^{-\theta(s)}\,\frac{(\theta(s))^m}{m!},\qquad s>t,\quad m\in\mathbb{Z}_+.$$

Therefore, we can make the assumption that

$$\operatorname{Prob}\{N_C(t)=n\}=e^{-\theta(t)}\,\frac{(\theta(t))^n}{n!}$$

for any t.

Now we can construct a Markov process Λ_C with state space $\mathbb{Y}$ precisely as in Remark 1.2. It should be added that we assume that conditional on $N_C(t)=n$, the distribution of $\Lambda_C(t)$ coincides with the nth Plancherel measure. This implies that

$$\operatorname{Prob}\{\Lambda_C(t)=\lambda\}=M_{\theta(t)}(\lambda)$$

for any $\lambda\in\mathbb{Y}$.

---

[5] Any birth-death process is reversible.

REMARK 1.3. One can show that the Markov process obtained from Λ_C by the reversion of time $t \to -t$ has a similar form, $\Lambda_{\widehat{C}}$, where the curve $\widehat{C}$ is the image of C under the transposition of coordinate axes. That is, $\widehat{C}$ is given by

$$\widehat{u}(t) = v(-t), \quad \widehat{v}(t) = u(-t).$$

REMARK 1.4. A reparametrization of the curve C (which leaves it admissible) leads simply to a reparametrization of time in the corresponding Markov process Λ_C. There is a distinguished parametrization of C which is unique within an additive constant:

$$t = \frac{1}{2}(\ln u - \ln v) + \text{const}, \qquad (u,v) \in C. \tag{1.12}$$

We call t the *interior time* along the curve. If C is a hyperbola $uv = \theta$, then the interior time coincides with the natural time in the stationary Markov process Λ_θ.

REMARK 1.5. There are two particular cases of nonstationary Markov processes Λ_C which can be called the *descending* and the *ascending* ones. By definition, they are obtained when we take as C a vertical or horizontal line, respectively. These processes correspond to pure death or pure birth processes, respectively. More generally, one considers broken lines C with alternating vertical and horizontal segments. Note that any admissible curve can be approximated by such broken lines, which suggests the idea that a general Markov process Λ_C can be approximated, in an appropriate sense, by processes with alternating descending and ascending fragments.

2. Interpretation of Markov processes via Poisson process in quadrant

In this section we describe a nice interpretation of the Markov processes Λ_C. As in §1, we are dealing with the quadrant $\mathbb{R}^2_{>0} \subset \mathbb{R}^2$ with coordinates $u > 0, v > 0$.

DEFINITION 2.1. Let π be an n-point configuration in $\mathbb{R}^2_{>0}$ such that no two points lie on the same vertical or horizontal line. We assign to π a permutation σ in the symmetric group $S(n)$ and then a Young diagram $\lambda \in \mathbb{Y}_n$, as follows. Let $u_1 < \cdots < u_n$ and $v_1 < \cdots < v_n$ be the u- and v-coordinates of the points in π. By definition, the permutation σ determines a matching of these coordinates. That is, the points in π are of the form $(u_i, v_{\sigma(i)})$, where $i = 1, \dots, n$. Next, to obtain λ, we apply the Robinson–Schensted algorithm (RS for short). Recall (see, e.g., Sagan's book [**S**, §3.3 and §3.8]) that RS establishes an explicit bijection between permutations $\sigma \in S(n)$ and pairs $(\mathcal{P}, \mathcal{Q})$ of standard Young tableaux of the same shape $\lambda \in \mathbb{Y}_n$, and we just take this diagram λ.

Consider the Poisson process Π in the quadrant $\mathbb{R}^2_{>0}$ with constant density 1. We also denote by Π the random point configuration in $\mathbb{R}^2_{>0}$ produced by the process. We can assume that no two points in Π lie on the same vertical or horizontal line, because this condition holds for almost all configurations Π.

DEFINITION 2.2. To any point $(u,v) \in \mathbb{R}^2_{>0}$ we assign a random permutation $\sigma_\Pi(x,v)$ and a random Young diagram $\lambda_\Pi(u,v)$, both depending on a realization Π of the Poisson process, as follows. Let $\square(u,v)$ denote the rectangle with vertices (u,v), $(u,0)$, $(0,v)$, $(0,0)$, and let $\Pi(u,v) = \Pi \cap \square(u,v)$ be the random point configuration in this rectangle. Then we set $\pi = \Pi(u,v)$, $n = |\pi|$, and apply Definition 2.1.

The construction of Definition 2.2 is well known. It was widely used in the literature since Hammersley's paper [**H**]. Note that for a fixed point (u, v), the number $n = |\Pi(u, v)|$ has Poisson distribution (1.4) with parameter $\theta = uv$ and that $\lambda_\Pi(u, v)$ is distributed according to the poissonized Plancherel measure M_θ. Now we let (u, v) vary.

THEOREM 2.3. *Let C be an admissible curve in the sense of §1 and let a point $(u, v) = (u(t), v(t))$ move along C. Let, as above, Π be the random Poisson point configuration in $\mathbb{R}^2_{>0}$, and consider the $\mathbb{Y}$-valued stochastic process $\widetilde{\Lambda}_C$ with random trajectories $\lambda_\Pi(u(t), v(t))$, where the random Young diagram $\lambda_\Pi(u(t), v(t))$ is afforded by Definition 2.2.*

The process $\widetilde{\Lambda}_C$ is a Markov process, equivalent to the Markov process Λ_C of §1.

By Theorem 2.3, each Markov process Λ_C can be interpreted as a certain projection of the Poisson process in the quadrant. Actually, a more precise result holds. Assume that the curve C satisfies the conditions

$$\lim_{t\to-\infty} u(t) = 0, \qquad \lim_{t\to+\infty} v(t) = 0, \tag{2.1}$$

and let $\mathcal{D} \subset \mathbb{R}^2_{>0}$ denote the subgraph of C, that is, the part of the quadrant which is below and on the left of C. For instance, in the case of the stationary process Λ_θ, one has $\mathcal{D} = \{(u, v) \in \mathbb{R}^2_{>0} \mid uv < \theta\}$.

THEOREM 2.4. *In the situation of Theorem 2.3, assume additionally that condition (2.1) is satisfied. Then the construction of Theorem 2.3 provides a measure space isomorphism between the realizations of the Poisson process in the domain $\mathcal{D}$ and the trajectories of $\widetilde{\Lambda}_C$.*

Clearly, any trajectory $\{\lambda_\Pi(u(t), v(t))\}_{t\in\mathbb{R}}$ of $\widetilde{\Lambda}_C$ depends only on the restriction $\Pi|_\mathcal{D}$ of the corresponding Poisson configuration Π to the domain $\mathcal{D}$. It turns out that, conversely, $\Pi|_\mathcal{D}$ can be reconstructed from $\{\lambda_\Pi(u(t), v(t))\}_{t\in\mathbb{R}}$. This implies the theorem.

REMARK 2.5. Let, as in Theorem 2.3, a point $(u, v) = (u(t), v(t))$ move along an admissible curve C and replace $\lambda_\Pi(u, v)$ by $\sigma_\Pi(u, v)$; see Definition 2.2. Then we obtain a random process taking values in permutations $\sigma \in S(n)$ with varying n. One can show that this is again a Markov process. Clearly, it "covers" the Markov process $\widetilde{\Lambda}_C$ of Theorem 2.3 (the latter is a projection of the former). On the other hand, Theorem 2.4 implies a somewhat paradoxical claim that the projection $\sigma \mapsto \lambda$ given by the algorithm RS defines a measure space isomorphism between the trajectories of both processes.

REMARK 2.6. In case the curve C is a straight line $u + v = \text{const}$, the corresponding Markov process $\Lambda_C = \widetilde{\Lambda}_C$ was earlier described in very different terms by Prähofer and Spohn [**PS**]; see also Remarks 3.4 and 4.5.

REMARK 2.7. The assumption that the curve C goes in a southeast direction is crucial for the Markov property in Theorem 2.3. This can be demonstrated in the following simple example. Consider three points in the quadrant: $a = (1, 1)$, $b = (2, 1)$, and $c = (2, 2)$. Then, conditional on $\lambda_\Pi(b)$ is the one-box diagram, the random diagrams $\lambda_\Pi(a)$ and $\lambda_\Pi(c)$ are not independent. This shows that on the broken line going northeast, from a to b to c, the Markov property does not hold.

3. Dynamical correlation functions

Consider the lattice of (proper) half-integers

$$\mathbb{Z}' = \mathbb{Z} + \tfrac{1}{2} = \{\dots, -\tfrac{5}{2}, -\tfrac{3}{2}, -\tfrac{1}{2}, \tfrac{1}{2}, \tfrac{3}{2}, \tfrac{5}{2}, \dots\}.$$

We can write $\mathbb{Z}' = \mathbb{Z}'_- \cup \mathbb{Z}'_+$, where $\mathbb{Z}'_-$ consists of all negative half-integers and $\mathbb{Z}'_+$ consists of all positive half-integers. For any $\lambda \in \mathbb{Y}$ we set

$$\mathcal{L}(\lambda) = \{\lambda_i - i + \tfrac{1}{2} \mid i = 1, 2, \dots\} \subset \mathbb{Z}'.$$

For instance, $\mathcal{L}(\varnothing) = \mathbb{Z}'_-$. The correspondence $\lambda \mapsto \mathcal{L}(\lambda)$ is a bijection between the Young diagrams λ and those (infinite) subsets $\mathcal{L} \subset \mathbb{Z}'$ for which the symmetric difference $\mathcal{L} \triangle \mathbb{Z}'_-$ is a finite set with equally many points in $\mathbb{Z}'_+$ and $\mathbb{Z}'_-$.

We regard $\mathcal{L}(\lambda)$ as a point configuration on the lattice $\mathbb{Z}'$. Assume we are given a probability measure M on $\mathbb{Y}$. Then we can speak about the random diagram λ and hence about the random point configuration $\mathcal{L}(\lambda)$. The nth *correlation function* of M is defined as follows

$$\rho_n(x_1, \dots, x_n) = \mathrm{Prob}\{x_1, \dots, x_n \in \mathcal{L}(\lambda)\},$$

where $n = 1, 2, \dots$ and $x_1, \dots, x_n$ are *pairwise distinct* points of $\mathbb{Z}'$. In other words, the correlation functions tell us what is the probability that the random point configuration $\mathcal{L}(\lambda)$ contains a given finite set of points. The collection of all correlation functions determines the initial probability measure M uniquely.

For the poissonized Plancherel measure M_θ defined in (1.5) the correlation functions were found independently by Borodin, Okounkov, and Olshanski [**BOO**] and Johansson [**J1**]:

THEOREM 3.1.

(i) *The correlation functions of the measure M_θ have determinantal form*

$$\rho_n(x_1, \dots, x_n) = \det_{1 \le i,j \le n} [K(x_i, x_j)],$$

where $n = 1, 2, \dots$ and

$$K(x, y) = \sqrt{\theta}\, \frac{J_{x-\frac{1}{2}}(2\sqrt{\theta}) J_{y+\frac{1}{2}}(2\sqrt{\theta}) - J_{y-\frac{1}{2}}(2\sqrt{\theta}) J_{x+\frac{1}{2}}(2\sqrt{\theta})}{x - y}. \tag{3.1}$$

Here $J_z(\,\cdot\,)$ denotes the Bessel function which is regarded as a function of its index z. When $x = y$, the indeterminacy in formula (3.1) *is resolved by l'Hôpital's rule.*

(ii) *The kernel can also be written in the form*

$$K(x, y) = \sum_{a \in \mathbb{Z}'_+} J_{x+a}(2\sqrt{\theta}) J_{y+a}(2\sqrt{\theta}). \tag{3.2}$$

The function $K(x, y)$ on $\mathbb{Z}' \times \mathbb{Z}'$ is called the *discrete Bessel kernel*.

We are now going to state an analog of Theorem 3.1 for the Markov processes Λ_C. The correspondence $\lambda \mapsto \mathcal{L}(\lambda)$ allows us to regard each trajectory $\Lambda(t)$ in $\mathbb{Y}$ as a varying in time point configuration $\mathcal{L}(\Lambda(t))$. In this picture, a jump consists of shifting one of the points of the configuration by ± 1.

DEFINITION 3.2. Let $\Lambda(t)$ stand for the random trajectory of a $\mathbb{Y}$-valued stochastic process with time parameter t. By the nth *dynamical correlation function* of the process we mean the function

$$\rho_n(t_1, x_1; \ldots; t_n, x_n) = \operatorname{Prob}\{x_i \in \mathcal{L}(\Lambda(t_i)), i = 1, \ldots, n\}$$

where $(t_1, x_1), \ldots, (t_n, x_n)$ are pairwise distinct elements of $\mathbb{R} \times \mathbb{Z}'$.

In other words, the dynamical correlation functions tell us what are the probabilities of the following events: for each t in a given finite subset of $\mathbb{R}$, the point configuration $\mathcal{L}(\Lambda(t))$ contains a given finite subset $X(t) \subset \mathbb{Z}'$. Note that this definition somewhat resembles that of the finite-dimensional distribution, because both definitions describe the behavior of the process at a finite number of moments of time.

THEOREM 3.3. *Let $C = (u(t), v(t))$ be an admissible curve in $\mathbb{R}^2_{>0}$ with its canonical parametrization* (1.12) *given by interior time t, let $\theta(t) = u(t)v(t)$ as in* (1.10)*, and let Λ_C be the corresponding Markov process.*

(i) *The dynamical correlation functions of Λ_C have determinantal form*

$$\rho_n(t_1, x_1; \ldots; t_n, x_n) = \det_{1 \le i,j \le n} [K_C(t_i, x_i; t_j, x_j)],$$

where $n = 1, 2, \ldots$ and $K_C(s, x; t, y)$ is a kernel on $(\mathbb{R} \times \mathbb{Z}') \times (\mathbb{R} \times \mathbb{Z}')$ which can be written as a double contour integral

$$K_C(s, x; t, y) \tag{3.3}$$
$$= \frac{e^{\frac{1}{2}(s-t)}}{(2\pi i)^2} \int_{\{\omega_1\}} \int_{\{\omega_2\}} \frac{e^{\sqrt{\theta(s)}(\omega_1 - \omega_1^{-1}) + \sqrt{\theta(t)}(\omega_2 - \omega_2^{-1})}}{e^{s-t}\omega_1\omega_2 - 1} \omega_1^{-x-\frac{1}{2}} \omega_2^{-y-\frac{1}{2}} \, d\omega_1 \, d\omega_2$$

where $\{\omega_1\}$ and $\{\omega_2\}$ are any two contours which go around 0 in the positive direction and satisfy the following:

- *if $s \ge t$, so that $e^{s-t} \ge 1$, then the contour $\{\omega_1\}$ must contain the contour $\{e^{t-s}\omega_2^{-1}\}$;*
- *if $s < t$, so that $e^{s-t} < 1$, then, on the contrary, the contour $\{\omega_1\}$ must be contained in the contour $\{e^{t-s}\omega_2^{-1}\}$.*

(ii) *The kernel can also be written in the form*

$$K_C(s, x; t, y) = \pm \sum_{a \in \mathbb{Z}'_+} e^{-a|s-t|} J_{x \pm a}(2\sqrt{\theta(s)}) J_{y \pm a}(2\sqrt{\theta(t)}) \tag{3.4}$$

where the plus sign is chosen if $s \ge t$ whereas the minus sign is chosen if $s < t$.

One can verify that if $s = t$, then (3.3) can be reduced to (3.1), and it is immediate that (3.4) becomes (3.2).

Note the asymmetry between the conditions $s \ge t$ and $s < t$. One can show that

$$\lim_{s \to t^+} K_C(s, x; t; y) = K_C(t, x; t, y) = \lim_{s \to t^-} K_C(s, x; t; y) + \delta_{xy}.$$

This agrees with the fact that

$$\sum_{a \in \mathbb{Z}'} J_{x+a}(2\sqrt{\theta}) J_{y+a}(2\sqrt{\theta}) = \delta_{xy}\,.$$

REMARK 3.4. For the curves $u + v = \text{const}$, formula (3.4) was earlier proved by Prähofer and Spohn [**PS**, (3.52)].

4. Scaling limits

The study of the asymptotic behavior of the random Young diagrams distributed according to the nth Plancherel measure (1.1) as $n \to \infty$, or according to the poissonized Plancherel measure (1.5) as $\theta \to \infty$, is an important and nontrivial problem which has many different aspects. We refer the reader to [**BOO, IO**] for a general discussion and relevant references, and we restrict ourselves here to considering two types of the asymptotics: in the middle (bulk) and at the edge of the Young diagrams distributed according to M_θ and Markov processes introduced in §1.

4.1. Bulk. We start with recalling the following statement.

THEOREM 4.1 ([**BOO**, Section 3]). *The correlation functions of the measure M_θ have the following limit as $\theta \to \infty$: Fix $c \in (-2, 2)$ and let $x_0(\theta) \in \mathbb{Z}'$ be such that*

$$x_0(\theta) = c \cdot \sqrt{\theta} + o(\sqrt{\theta}), \quad \theta \to \infty.$$

Furthermore, for any $n = 1, 2 \dots$ and arbitrary $x_1, \dots, x_n \in \mathbb{Z}$, set

$$x_i(\theta) = x_0(\theta) + x_i, \qquad i = 1, \dots, n.$$

Then

$$\lim_{\theta \to \infty} \rho_n(x_1(\theta), \dots, x_n(\theta)) = \det_{1 \le i,j \le n} [S_c(x_i - x_j)],$$

where

$$S_c(r) = \frac{\sin\bigl(\arccos(c/2) \cdot r\bigr)}{\pi r}. \tag{4.1}$$

The function $S_c(x - y)$ on $\mathbb{Z} \times \mathbb{Z}$ is called the *discrete sine kernel.*

In Theorem 4.1, the pairwise distances $x_i(\theta) - x_j(\theta)$ do not vary as $\theta \to +\infty$; if, instead of this, one supposes that for some i and j, the distance between two points $x_i(\theta)$ and $x_j(\theta)$ from the bulk[6] tends to infinity together with θ, then the events of finding particles (= elements of $\mathcal{L}(\lambda)$) at these locations become asymptotically independent.

Our goal is to extend Theorem 4.1 to the Markov processes Λ_C.

Let us return for a moment to the interpretation of Λ_C via the Poisson process in the first quadrant; see §2. Theorem 4.1 deals with Young diagrams $\lambda_\Pi(u, v)$ sitting at the points (u, v) such that $uv = \theta \to \infty$. Now let us assume that such a point (u, v) sits on an admissible curve.

It turns out that the bulk of $\lambda_\Pi(u, v)$ will have nontrivial correlations with the bulks of the Young diagrams corresponding to other points of the curve if the distance to these points in the (u, v)-plane remains finite as $\theta \to \infty$. If we assume that we are far enough from the boundary of the quadrant (u/v is bounded away from zero and infinity), then the interior time change between such points has to be of order $1/\sqrt{\theta}$.

---

[6]Meaning that $x_i(\theta)/\sqrt{\theta}$ and $x_j(\theta)/\sqrt{\theta}$ are bounded away from the edges -2 and 2.

THEOREM 4.2. *Let $C_\theta = (u_\theta(t), v_\theta(t))$ be a family of admissible curves in $\mathbb{R}^2_{>0}$ with their canonical parametrizations given by the interior time (1.12). Here $\theta > 0$ is a parameter, and we assume that there exists a constant $T \in \mathbb{R}$ such that $u_\theta(T)v_\theta(T) = \theta$. For instance, we may take as the curves C_θ the hyperbolas $uv = \theta$.*

The dynamical correlation functions of Λ_{C_θ} have the following limit as $\theta \to +\infty$. Fix an arbitrary $c \in (-2, 2)$ and let $x_0(\theta) \in \mathbb{Z}'$ be such that

$$x_0(\theta) = c \cdot \sqrt{\theta} + o(\sqrt{\theta}), \qquad \theta \to \infty.$$

Let $n = 1, 2, \ldots$, and let $\tau_1, \ldots, \tau_n \in \mathbb{R}$ and $x_1, \ldots, x_n \in \mathbb{Z}$ be arbitrary. Furthermore, assume that

$$t_i(\theta) = T + \tau_i/\sqrt{\theta} + o(1/\sqrt{\theta}).$$

Then

$$\lim_{\theta\to\infty} \rho_n(t_1(\theta), x_0(\theta) + x_1; \ldots; t_n(\theta), x_0(\theta) + x_n) = \det_{1\le i,j\le n} \left[S_c(\tau_i - \tau_j; x_i - x_j)\right],$$

where

$$S_c(h; r) = \frac{1}{2\pi i} \int_{\{\omega\}} e^{-h(\omega+\omega^{-1}-c)} \frac{d\omega}{\omega^{r+1}}, \tag{4.2}$$

and $\{\omega\}$ is a contour in $\mathbb{C}$ going from the point $e^{-i\phi}$ to the point $e^{i\phi}$ with $\phi =$ $\arccos(c/2)$, in such a way that it passes to the right of the origin if $h \ge 0$ and to the left of the origin if $h < 0$.

Note that the limit correlations do not depend on the choice of the curves C_θ. It is readily verified that when $h = 0$, (4.2) coincides with (4.1).

Another, somewhat similar, extension of the discrete sine kernel called the *incomplete beta kernel* was obtained by Okounkov and Reshetikhin [**OR**, Section 3].

An infinite-dimensional family of extensions of the discrete sine kernel which includes the kernel of Theorem 4.2 and the incomplete beta kernel was constructed in [**B**].

4.2. Edge. We now concentrate our attention on the asymptotics of the correlation functions at the edge of the Young diagrams, where $x_i(\theta) \sim \pm 2\sqrt{\theta}$. Due to the symmetry of our measures with respect to transposition of Young diagrams, which also swaps the edges $2\sqrt{\theta}$ and $-2\sqrt{\theta}$, it suffices to consider one of the edges.

THEOREM 4.3 ([**BOO**, Section 4], [**J1**]). *The correlation functions of M_θ have the following scaling limit as $\theta \to \infty$: For any $n = 1, 2, \ldots$ and $x_1, \ldots, x_n \in \mathbb{R}$, let $x_1(\theta), \ldots, x_n(\theta) \in \mathbb{Z}'$ be such that*

$$x_i(\theta) = 2\sqrt{\theta} + x_1\theta^{\frac{1}{6}} + o\left(\theta^{\frac{1}{6}}\right), \qquad i = 1, \ldots, n.$$

Then

$$\lim_{\theta\to\infty} (\theta^{\frac{1}{6}})^n \rho_n(x_1(\theta), \ldots, x_n(\theta)) = \det_{1\le i,j\le n} [\mathcal{A}(x_i, x_j)],$$

where

$$\mathcal{A}(x, y) = \frac{Ai(x)Ai'(y) - Ai'(x)Ai(y)}{x - y},$$

and $Ai(x)$ is the Airy function.

Note that the factor $(\theta^{\frac{1}{6}})^n$ comes from the scaling of the space variable x.

The function $\mathcal{A}(x,y)$ on $\mathbb{R}^2$ is called the *Airy kernel*. Another useful formula for the Airy kernel is

$$\mathcal{A}(x,y)=\int_0^{\infty} Ai(x+a)Ai(y+a)da.$$

Once again, let us return to the interpretation of Λ_C through the Poisson process. It turns out that the edges of the Young diagrams sitting on a curve $(u(t),v(t))$ with $uv=\theta\to\infty$ will have nontrivial correlations if the distance between the points in the (u,v)-plane grows as $\theta^{\frac{1}{3}}$. This means that if we are away from the coordinate axes, then the interior time change is of order $\theta^{-\frac{1}{6}}$.

THEOREM 4.4. *Let $C_\theta=(u_\theta(t),v_\theta(t))$ be a family of admissible curves in $\mathbb{R}^2_{>0}$ with their canonical parametrizations given by the interior time* (1.12). *Here $\theta>0$ is a parameter, and we assume that there exists a constant $T\subset\mathbb{R}$ such that $u_\theta(T)v_\theta(T)=\theta$. For instance, we may take as the curves C_θ the hyperbolas $uv=\theta$.*

The dynamical correlation functions of Λ_{C_θ} have the following limit as $\theta\to+\infty$. For arbitrary $\{\tau_i\}_{i=1}^n\subset\mathbb{R}$, choose $t_i(\theta)$ such that

$$t_i(\theta)=T+\tau_i\theta^{-\frac{1}{6}}+o\left(\theta^{-\frac{1}{6}}\right),\qquad i=1,\dots,n.$$

Furthermore, for arbitrary $x_1,\dots,x_n\in\mathbb{R}$, choose $x_1(\theta),\dots,x_n(\theta)\in\mathbb{Z}'$ such that

$$x_i(\theta)=2\sqrt{u_\theta(t_i(\theta))v_\theta(t_i(\theta))}+x_i\theta^{\frac{1}{6}}+o\left(\theta^{\frac{1}{6}}\right),\qquad i=1,\dots,n.$$

Then

$$\lim_{\theta\to\infty}(\theta^{\frac{1}{6}})^n\rho_n(t_1(\theta),x_1(\theta)\,;\,\dots\,;\,t_n(\theta),x_n(\theta))=\det_{1\le i,j\le n}\left[\mathcal{A}(\tau_i-\tau_j;x_i,x_j)\right],$$

where

$$\mathcal{A}(\tau;x,y)=\begin{cases}\int_0^{\infty}e^{-\tau a}Ai(x+a)Ai(y+a)da, & \tau\ge 0,\\ -\int_0^{+\infty}e^{-|\tau|a}Ai(x-a)Ai(y-a)e^{-\tau a}da, & \tau<0.\end{cases}$$

The kernel $\mathcal{A}(\tau;x,y)$ is called the *extended Airy kernel*. It is stationary in time.

REMARK 4.5. A special case of Theorem 4.4 with C_θ being the lines $u+v=$ const was proved in Prähofer and Spohn [**PS**].

The next statement uses the notion of the *Airy process*, introduced in [**PS**]; see also Johansson [**J2**].

COROLLARY 4.6. *Let $\{(u_\theta(t),v_\theta(t))\mid t\in(T-\epsilon,T+\epsilon)\}$ be a family of admissible curves in $\mathbb{R}^2_{>0}$ with their canonical parametrizations, and let $u_\theta(T)v_\theta(T)=\theta$. Denote by $l(t,\theta)$ the length of the first row of the random Young diagram $\lambda_\Pi(u_\theta(t),v_\theta(t))$ and set $t(\tau)=T+\tau\theta^{-\frac{1}{6}}$. Then as $\theta\to\infty$, the random variable*

$$L(\tau)=\frac{l(t(\tau),\theta)-2\sqrt{u_\theta(t(\tau))\,v_\theta(t(\tau))}}{\theta^{\frac{1}{6}}}$$

converges, as a function of τ, to the Airy process.

References

[BDJ1] J. Baik, P. Deift, K. Johansson, *On the distribution of the length of the longest increasing subsequence of random permutations,* J. Amer. Math. Soc. **12** (1999), 1119–1178; arXiv:math.CO/9810105.

[BDJ2] J. Baik, P. Deift, K. Johansson, *On the distribution of the length of the second row of a Young diagram under Plancherel measure,* Geom. Funct. Anal. **10** (2000), 702–731, 1606–1607; arXiv:math.CO/9901118.

[B] A. Borodin, *Periodic Schur process and cylindric partitions,* preprint, 2006; arXiv:math.CO/0601019.

[BOO] A. Borodin, A. Okounkov, G. Olshanski, *Asymptotics of Plancherel measures for symmetric groups,* J. Amer. Math. Soc. **13** (2000), 481–515; arXiv:math.CO/9905032.

[BO1] A. Borodin, G. Olshanski, *Distributions on partitions, point processes, and the hypergeometric kernel,* Commun. Math. Phys. **211** (2000), 335–358; arXiv:math.RT/9904010.

[BO2] A. Borodin, G. Olshanski, *Markov processes on partitions,* preprint, 2004; arXiv:math-ph/0409075.

[H] J. M. Hammersley, *A few seedlings for research,* Proc. Sixth Berkeley Symposium on Mathematical Statistics and Probability, vol. 1, Univ. of California Press, 1972, pp. 345–394.

[IO] V. Ivanov, G. Olshanski, *Kerov's central limit theorem for the Plancherel measure on Young diagrams,* Symmetric functions 2001. Surveys of developments and perspectives. Proc. NATO Advanced Study Institute (S. Fomin, editor), Kluwer, 2002, pp. 93–151; arXiv:math.CO/0304010.

[J1] K. Johansson, *Discrete orthogonal polynomial ensembles and the Plancherel measure,* Ann. Math. (2) **153** (2001), 259–296; arXiv:math.CO/9906120.

[J2] K. Johansson, *Discrete polynuclear growth and determinantal processes,* Comm. Math. Phys. **242** (2003), 277–329; arXiv:math.PR/0206208.

[LS] B. F. Logan, L. A. Shepp, *A variational problem for random Young tableaux,* Advances in Math. **26** (1977), 206–222.

[M] I. G. Macdonald, *Symmetric functions and Hall polynomials,* 2nd edition, Oxford University Press, 1995.

[O] A. Okounkov, *Random matrices and random permutations,* Intern. Math. Res. Notices **20** (2000), 1043–1095; arXiv:math.CO/9903176.

[OR] A. Okounkov, N. Reshetikhin, *Correlation functions of Schur process with applications to local geometry of a random 3-dimensional Young diagram,* J. Amer. Math. Soc. **16** (2003), 581–603; arXiv:math.CO/0107056.

[PS] M. Prähofer, H. Spohn, *Scale invariance of the PNG droplet and the Airy process,* J. Stat. Phys. **108** (2002), 1071–1106; arXiv:math.PR/0105240.

[S] B. E. Sagan, *The symmetric group. Representations, combinatorial algorithms, and symmetric functions,* Wadsworth and Brooks/Cole, Pacific Grove, CA, 1991.

[VK1] A. M. Vershik, S. V. Kerov, *Asymptotics of the Plancherel measure of the symmetric group and the limiting form of Young tableaux,* Doklady Acad. Nauk SSSR **233**:6 (1977), 1024–1027; English translation: Soviet Mathematics Doklady **18** (1977), 527–531.

[VK2] A. M. Vershik, S. V. Kerov, *Asymptotic theory of characters of the symmetric group,* Function. Anal. i Prilozhen. **15**:4 (1981), 15–27; English translation: Funct. Anal. Appl. **15** (1985), 246–255.

[VK3] A. M. Vershik, S. V. Kerov, *Asymptotics of the largest and the typical dimensions of irreducible representations of a symmetric group,* Funktsional. Anal. i Prilozhen. **19**:1 (1985), 25–36; English translation: Funct. Anal. Appl. **19** (1985), 21–31.

MATHEMATICS 253-37, CALTECH, PASADENA, CALIFORNIA 91125, USA
E-mail address: borodin@caltech.edu

DOBRUSHIN MATHEMATICS LABORATORY, INSTITUTE FOR INFORMATION TRANSMISSION PROBLEMS, BOLSHOY KARETNY 19, 127994 MOSCOW GSP-4, RUSSIA
E-mail address: olsh@online.ru

Amer. Math. Soc. Transl.
(2) Vol. **217**, 2006

A Condition for Continuous Spectrum of an Interval Exchange Transformation

Alexander Bufetov, Yakov G. Sinai, and Corinna Ulcigrai

ABSTRACT. We consider interval exchange transformations (which are piecewise isometries of an interval) and give a condition for an interval exchange to have continuous spectrum. The values that spectral measures assign to points are studied, using the renormalization algorithm for interval exchange transformations introduced by Rauzy and Veech.

I had the pleasure and the privilege to know A. M. Vershik for more than forty years, even when the other authors of this paper were not born. We shared our common love for ergodic theory and discussed many problems of mutual interest. It is a pity that our contacts now are less regular.

Ya. G. Sinai

1. Introduction

We consider interval exchange transformations (IET) of κ intervals. Any such IET is given by a partition of $I = [0,1]$ onto κ subintervals, $I_1, I_2, \dots, I_\kappa$, defined by a length vector, $\lambda = (\lambda_1, \lambda_2, \dots, \lambda_\kappa)$, $\lambda_i > 0$, $\sum_{i=1}^{\kappa} \lambda_i = 1$, where $\lambda_i = |I_i|$, and by a permutation $\pi \in S_\kappa$; after the transformation I_i becomes the $\pi(i)^{\text{th}}$ interval, so the order of the subintervals after applying T is $I_{\pi^{-1}(1)}, I_{\pi^{-1}(2)}, \dots, I_{\pi^{-1}(\kappa)}$. We shall often use the notation $T = (\lambda, \pi)$.

It is clear that T preserves the Lebesgue measure. Veech [**V3**] and Masur [**M**] proved that, for a.e. λ with respect to the Lebesgue measure, T is ergodic, and Katok [**Ka**] proved that any T is not mixing.

In the case $\kappa = 3$ Katok and Stepin proved in [**KS**] that, for Lebesgue a.e. λ, the spectrum of (λ, π) is continuous.

2000 *Mathematics Subject Classification.* Primary 37A05.

Key words and phrases. Interval exchange transformations, spectral measures, weak mixing.

The second author would like to express thanks for financial support from NSF grant DMS-0070698 and RFFI grant 99-01-00314.

Nogueira and Rudolph proved that a typical T (not isomorphic to a rotation) does not admit continuous eigenfunctions [**NR**]. Weak mixing for typical IET under some additional assumption on the combinatorics of the permutation was obtained by Veech [**V4**].

While writing this paper, we learned about a proof by Avila and Forni of weak mixing for a typical IET not isomorphic to a rotation of the circle (see [**AF**]).

Rauzy, Veech, and Zorich [**R, V2, Z**] developed a renormalization algorithm for IET which is a multi-dimensional generalization of the continued fraction algorithm for rotations. This algorithm is the main tool in proving some of the former results.

In this paper we use the Rauzy–Veech algorithm to derive a sufficient condition under which a given IET has continuous spectrum. For typical interval exchange transformations we also state a formula for the values of spectral measures of points.

In [**SU**] we use the condition for continuous spectrum derived in this paper to construct explicit examples of weakly mixing interval exchange transformations.

1.1. Outline. In Section 2 we recall the definition of the Rauzy–Veech algorithm and of the associated cocycle (see 2.1), as well as the definition of Rauzy classes and of the renormalized map on the space of IET (see 2.1.1). The original IET can be viewed as piecewise suspension over the IET obtained at any step n of the algorithm, given by κ towers, denoted by $Z_j^{(n)}$, $1 \leq j \leq \kappa$ (see 2.2). The recursive structure of these towers can be studied with the help of the measurable partitions defined in 2.2.1.

In Section 3 the condition under which T has continuous spectrum (Proposition 3.2) is proved. It is based on a condition of continuity of the spectral measures of the characteristic functions χ_A, where A are the floors $Z_{j,l}^{(n)}$ of the towers defined in 2.2 (Proposition 3.3). The spectral measure ρ_A of χ_A is related to the recursive structure of the towers.

In 3.1 the spectral measure ρ_A of a point ω is expressed as a limit of a sum over elements of the partitions introduced in 2.2.1. The terms appearing in the sum can be rewritten using the functions $\varphi_\omega^\pm(Z_j^{(n)})$ defined in 3.1.2, which depend on the structure of the towers $Z_j^{(n)}$. In 3.1.4 it is proved that $\max_{1\leq j\leq\kappa} |\varphi_\omega^\pm(Z_j^{(n)})|$ converges as n grows to a limit $\psi_{\omega,A}$. The sufficient condition in Proposition 3.2 is formulated in terms of $\psi_{\omega,A}$ and is proven in 3.2. Moreover, an explicit expression for $\rho_A(\omega)$ in terms of $\psi_{\omega,A}$ is stated in 3.2.1 and shows that for a typical IET the condition is also necessary.

2. Algorithm, towers, and their partitions

Given (λ, π) where $\lambda \in \mathbb{R}_+^\kappa$ and $\pi \in S_\kappa$, let $\beta_0 = 0$ and $\beta_j \doteqdot \sum_{i=1}^j \lambda_i$ for $j = 1, \dots, \kappa$; the exchanged interval I_j is $[\beta_{j-1}, \beta_j[$. Let $\beta_0^\pi = 0$ and $\beta_j^\pi \doteqdot \sum_{i=1}^j \lambda_{\pi^{-1}i}$ for $j = 1, \dots, \kappa$. The interval exchange transformation is defined by

$$T(x) = x + \beta_{j-1}^\pi - \beta_{j-1} \quad \text{for} \quad x \in I_j.$$

We say that T satisfies the infinite distinct orbit condition (IDOC) introduced by Keane in [**Ke**] if the T-orbits $\mathcal{O}(\beta_j)$ of the discontinuities of T, $1 \leq j \leq \kappa - 1$, are infinite and distinct, i.e., $\mathcal{O}(\beta_j) \cap \mathcal{O}(\beta_i) = \emptyset$, $1 \leq j \neq i \leq \kappa - 1$. We will always assume that T satisfies the (IDOC).

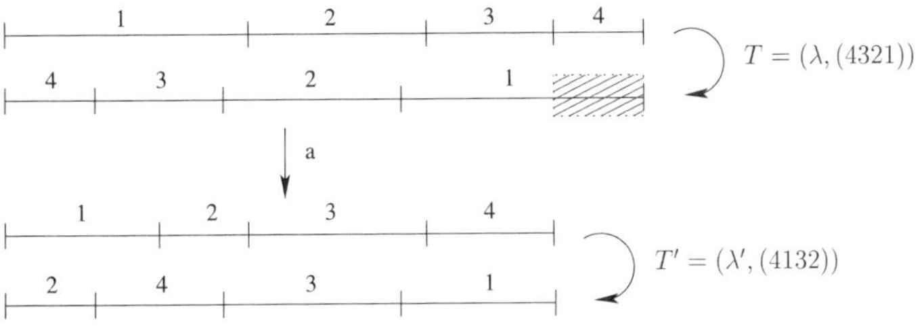

(a) Case $\lambda_\kappa < \lambda_{\pi-1(\kappa)}$

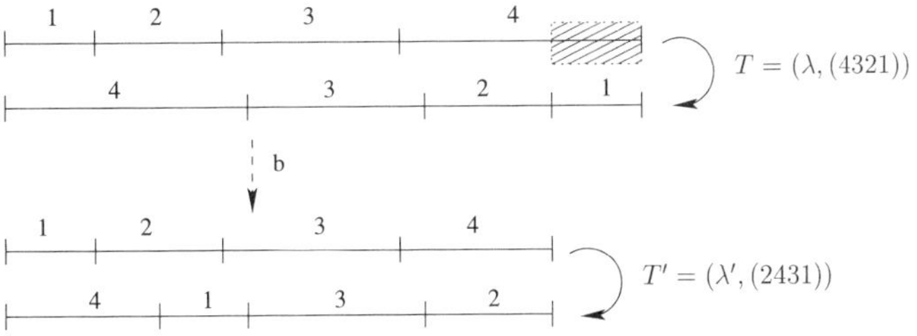

(b) Case $\lambda_\kappa > \lambda_{\pi^{-1}}(\kappa)$

FIGURE 1. One step of the Rauzy–Veech algorithm on $T = (\lambda, (4321))$.

As shown by Keane, the (IDOC) implies minimality. We also assume that π is *irreducible*, i.e., if the subset $\{1, 2, \ldots i\}$ is π-invariant, then $i = \kappa$, since this is a necessary condition for minimality.

2.1. Rauzy–Veech algorithm and the associated cocycle. Starting with $T = T^{(0)}$, the Rauzy–Veech algorithm produces a sequence of IET $T^{(n)}$ which are induced maps of T onto a sequence of nested subintervals $I^{(n)} \subset I$. At the first step, compare the lengths of I_κ and $I_{\pi^{-1}(\kappa)}$, i.e., of the last subintervals before and after the transformation. Assume $\lambda_\kappa \neq \lambda_{\pi^{-1}\kappa}$ and put $\lambda_{\min} \doteqdot \min\{\lambda_\kappa, \lambda_{\pi^{-1}(\kappa)}\}$. Cut an interval of the length $\lambda_{\min}$ from the right end of I and define $T^{(1)}$ to be the first return map of $T^{(0)}$ onto $I^{(1)} = [0, 1 - \lambda_{\min}]$. It is easy to see that in general the induced first return map of T on a subinterval $I' \subset I$ is again an IET, of at most $\kappa + 2$ intervals. One Rauzy-step is defined so that $T^{(1)}$ is an exchange of exactly the same number κ of subintervals. Denote it by $T^{(1)} = (\lambda^{(1)}, \pi^{(1)})$ (see Figure 1(a) and Figure 1(b)).

Define inductively $T^{(n)} = (\lambda^{(n)}, \pi^{(n)})$ to be the induced map of $T^{(n-1)}$ on $I^{(n)}$, as long as $\lambda_\kappa^{(n-1)} \neq \lambda^{(n-1)}_{(\pi^{(n-1)})^{-1}(\kappa)}$. It can be seen that for a generic T, $T^{(n)}$ is defined for any n. The precise (generic) condition to guarantee that the algorithm never ends is exactly the (IDOC).

The new data $(\lambda^{(1)}, \pi^{(1)})$ can be expressed in terms of (λ, π). Distinguish two cases, a or b, according to $\lambda_\kappa < \lambda_{\pi^{-1}\kappa}$ or $\lambda_\kappa > \lambda_{\pi^{-1}\kappa}$. To describe the new permutation $\pi^{(1)}$, introduce the following operators a and b on S_κ:

$$a(\pi)(j) = \begin{cases} \pi(j) & j \le \pi^{-1}(\kappa); \\ \pi(\kappa) & j = \pi^{-1}(\kappa) + 1; \\ \pi(j-1) & \text{otherwise}; \end{cases} \tag{2.1}$$

$$b(\pi)(j) = \begin{cases} \pi(j) & j \le \pi(\kappa); \\ \pi(j)+1 & \pi(\kappa) < \pi(j) < \kappa; \\ \pi(\kappa)+1 & \pi(j) = \kappa. \end{cases} \tag{2.2}$$

We introduce the following matrices to describe the new lengths. Denote by Id the identity $\kappa \times \kappa$ matrix and by $E_{i,j}$ the matrix whose only nonzero entry is $(E_{i,j})_{ij} = 1$. Introduce the auxiliary permutation $\tau_s \in S^\kappa$, $\tau_s = (1\ 2\ \ldots\ s\ s+2\ \ldots\ \kappa\ \kappa{+}1)$ if $1 \le s < \kappa - 1$ and $\tau_{\kappa-1} = \mathrm{id}$, which rotates cyclically all elements after the sth one. Denote by $P(\tau_s)$ the matrix associated to the permutation, i.e., $P(\tau_s)_{ij} = \delta_{i\tau_s(i)}$. The two Rauzy–Veech elementary matrices associated to (λ, π) are defined by

$$\begin{cases} A(\pi, a) & = & (Id + E_{\pi^{-1}(\kappa),\kappa}) \cdot P(\tau_{\pi^{-1}(\kappa)}); \\ A(\pi, b) & = & Id + E_{\kappa,\pi^{-1}(\kappa)}. \end{cases} \tag{2.3}$$

The matrix $A(\pi, b)$ has only one 1 out of diagonal, at entry $(\pi^{-1}(\kappa), \kappa)$, while $A(\pi, a)$ is obtained by an analogous matrix with only one nonzero entry at $(\kappa, \pi^{-1}(\kappa))$ by shifting cyclically all columns after the $\pi^{-1}(\kappa)^{\text{th}}$ one. Notice that both matrices have nonnegative entries and determinant one:

$$\underset{\substack{\uparrow \\ \pi^{-1}(\kappa)}}{\overset{A(\pi,b)}{\begin{pmatrix} 1 & & & & & 0 \\ & 1 & & & & \\ & & \ddots & & & \\ & & & \ddots & & \\ & & & & 1 & \\ 0 & & 1 & & & 1 \end{pmatrix}}} \quad \underset{\substack{\uparrow \\ \pi^{-1}(\kappa)}}{\overset{A(\pi,a)}{\begin{pmatrix} 1 & & & & & 0 \\ & \ddots & & & & \\ & & 1 & 1 & & \\ & & & 0 & \ddots & \\ & & & & \ddots & 1 \\ 0 & & 1 & & & 0 \end{pmatrix}}} \leftarrow \ \pi^{-1}(\kappa).$$

The induced IET $T^{(1)}$ is then given by

$$(\lambda^{(1)}, \pi^{(1)}) \doteqdot \begin{cases} \left(A^{-1}(\pi,\, a) \cdot \lambda,\, a(\pi)\right), & \lambda_\kappa < \lambda_{\pi^{-1}(\kappa)}; \\ \left(A^{-1}(\pi,\, b) \cdot \lambda,\, b(\pi)\right), & \lambda_\kappa > \lambda_{\pi^{-1}(\kappa)}. \end{cases} \tag{2.4}$$

The matrix $A(\pi, a)$ has a more complicated form than $A(\pi, b)$ because subintervals following $I_{\pi^{-1}(\kappa)}$ undergo a relabeling in the new IET, the ith becoming the $(i+1)^{\text{th}}$ (when $i \neq \kappa$). See e.g. Figure 1(a).

When iterating the inductive step, we associate to each $T^{(n)}$ the product of *elementary Rauzy–Veech matrices* used at each step, i.e.,

$$B^{(n)} \doteqdot A^{(1)} \cdot A^{(2)} \cdot \ldots \cdot A^{(n)}, \tag{2.5}$$

where $A^{(i)} = A\left(\pi^{(i)}, a\right)$ or $A^{(i)} = A\left(\pi^{(i)}, b\right)$ according to whether $\lambda^{(i)}_\kappa < \lambda^{(i)}_{\pi^{-1}\kappa}$ or $\lambda^{(i)}_\kappa > \lambda^{(i)}_{\pi^{-1}\kappa}$ is the elementary matrix associated to $T^{(i)}$. We define in this way a cocycle $A^{(n)}$ for the Rauzy–Veech algorithm, with values in $SL(\kappa, \mathbb{Z})$, the *Rauzy–Veech cocycle.*

Iterating the lengths relation in (2.4), we get the formula for the length vector of $T^{(n)}$:

$$\lambda^{(n)} = \left(B^{(n)}\right)^{-1} \lambda. \tag{2.6}$$

An IET is uniquely determined by the associated sequence of elementary matrices $A^{(n)}$, $n \in \mathbb{N}$, exactly when it is uniquely ergodic (see [**V3**]). Therefore, a.e. T is characterized by its Rauzy–Veech matrices.

2.1.1. *Rauzy graphs and renormalized Rauzy–Veech map.* Not all permutations of S_κ can be obtained by iterating the algorithm. The *Rauzy class* of π, denoted by $\mathscr{R}(\pi) \subset S_\kappa$, is the set of all permutations obtained iterating the operators a and b starting from π. Rauzy classes can be visualized in terms of a directed labeled graph, the *Rauzy graph.* Vertices are in one-to-one correspondence with permutations of $\mathscr{R}(\pi)$; arrows connect permutations obtained one from the other by applying a or b and are labeled according to the type, a or b, respectively.

Using the norm $|\lambda| = \sum_{i=1}^{\kappa} \lambda_i$, the initial lengths belong to the simplex $\Delta_{\kappa-1}$ of vectors $\lambda \in \mathbb{R}_+^\kappa$ such that $|\lambda| = 1$. The space of all possible IET obtained from (π, λ) is therefore $\Delta_{\kappa-1} \times \mathscr{R}(\pi)$.

Consider the *Rauzy–Veech map*, i.e., the map on the space of IET $\Delta_{\kappa-1} \times \mathscr{R}(\pi)$ which associates to T the induced IET after one step of the algorithm including the following renormalization:

$$\mathscr{R}\left((\lambda^{(0)}, \pi^{(0)})\right) \doteqdot \left(\frac{\lambda^{(1)}}{|\lambda^{(1)}|}, \pi^{(1)}\right).$$

Veech proved that $\mathscr{R}$ admits an invariant measure, absolutely continuous with respect to the Lebesgue measure, which is nevertheless infinite. Zorich defined an acceleration of the algorithm, obtained by grouping together all the following steps of the Rauzy–Veech algorithm of the same type, a or b.

Zorich acceleration admits a finite invariant measure [**Z**].

2.2. Towers and their structure. In this subsection we define the towers which allows us to retrieve T from $T^{(n)}$ and $B^{(n)}$. Note first that entries of $B^{(n)}$ have a dynamical meaning in terms of return times. Namely, denote by $I_j^{(n)}$, $1 \le j \le \kappa$, the subintervals of $T^{(n)}$. The entry $B_{ij}^{(n)}$ gives the number of visits of the orbit of any point $x \in I_j^{(n)}$ to the interval I_i of the original partition before its first return in $I^{(n)}$.

Therefore, the norm $|Z_j^{(n)}|$ of the jth column of $B^{(n)}$, i.e.,

$$|Z_j^{(n)}| \doteqdot \sum_{i=1}^{\kappa} B_{ij}^{(n)}, \tag{2.7}$$

gives the return time of any $x \in I_j^{(n)}$ to $I^{(n)}$.

Define

$$Z_j^{(n)} \doteqdot \bigcup_{l=0}^{|Z_j^{(n)}|-1} T^l I_j^{(n)}. \tag{2.8}$$

When T is ergodic, $\bigcup_{j=1}^{\kappa} Z_j^{(n)}$ is a nontrivial T-invariant set; therefore the sets $Z_j^{(n)}$, $1 \le j \le \kappa$, give a partition of the whole I. Each $Z_j^{(n)}$ can be visualized as a tower over $I_j^{(n)} \subset I^{(n)}$, of height $|Z_j^{(n)}|$ (see Figure 2). A floor of the tower, denoted by $Z_{j,l}^{(n)}$, is defined by

$$Z_{j,l}^{(n)} \doteqdot T^l I_j^{(n)}, \qquad l = 0, \dots, |Z_j^{(n)}| - 1. \tag{2.9}$$

Notice that when $n = 0$, $Z_j^{(0)} = I_j^{(0)}$, i.e., towers of step zero are simply the subintervals of the IET. The original T is an integral map over $I^{(n)}$; under the action of T every floor $Z_{j,l}^{(n)}$, but the top one ($l \neq |Z_j^{(n)}| - 1$), moves one step up, while $T\left(Z^{(n)}_{j,|Z_j^{(n)}|-1}\right) = T^{(n)}\left(I_j^{(n)}\right)$.

The Rauzy–Veech algorithm can be visualized as acting on the towers, in terms of *stacking* of towers. One step corresponds to cutting the last tower before the permutation, i.e., $Z_\kappa^{(n)}$, and stacking it over $Z^{(n)}_{\pi^{-1}\kappa}$. In the case a, when $\lambda_\kappa^{(n)} < \lambda^{(n)}_{\pi^{-1}\kappa}$, $Z_\kappa^{(n)}$ is completely cut and stacked above $Z^{(n)}_{\pi^{-1}\kappa}$, at its right end (see Figure 2(a)). In the case b, $\lambda_\kappa^{(n)} > \lambda^{(n)}_{\pi^{-1}\kappa}$, only the right portion of $Z_\kappa^{(n)}$ of width $\lambda^{(n)}_{\pi^{-1}\kappa}$ is cut and stacked completely above $Z^{(n)}_{\pi^{-1}\kappa}$ (see Figure 2(b)).

2.2.1. *Partitions of the towers.* Let us investigate the recursive structure of a tower $Z_{j_0}^{(n)}$ for given n in terms of the towers $Z_j^{(m)}$ of previous steps $m < n$. It is clear from the stacking description of the algorithm that each tower $Z_{j_0}^{(n)}$ consists of pieces of towers $Z_j^{(m)}$.

We define the following system of measurable partitions $\xi_m = \xi_m(Z_{j_0}^{(n)})$ of the tower $Z_{j_0}^{(n)}$ in terms of the subtowers $Z_i^{(m)}$, $0 \le m \le n$. The elements of the partition are complete blocks of floors of $Z_{j_0}^{(n)}$ which are all contained inside the same tower $Z_j^{(m)}$: namely, for each floor $Z_{j_0,l}^{(n)}$ which is contained in $I^{(m)}$, construct an element $C_m \subset \xi_m$ in the following way. If $Z_{j_0,l}^{(n)} \subset I_j^{(m)}$,

$$C_m \doteqdot \bigcup_{i=0}^{|Z_j^{(m)}|-1} T^i Z_{j_0,l}^{(n)}. \tag{2.10}$$

The set of all such C_m gives a partition ξ_m of $Z_{j_0}^{(n)}$. Clearly for each $C_m \in \xi_m$ there is a unique j such that $C_m \subset Z_j^{(m)}$.

Partitions $\xi_{m'}$, $m' < m$, are refinements of ξ_m. Indeed, ξ_{m-1} is obtained from ξ_m by splitting some of the elements $C_m \in \xi_m$ into two elements of ξ_{m-1} (see Figure 3) . More precisely, if C_m is contained in a tower $Z_j^{(m)}$ not involved in the stacking at step $m-1$, it remains unchanged, while if $C_m \subset Z^{(m)}_{\pi^{(m)-1}\kappa}$ in case b or $C_m \subset Z^{(m)}_{\pi^{(m)-1}(\kappa)+1}$ in case a, it splits into two. Notice that $\xi_0(Z_{j_0}^{(n)})$ is the partition of the tower into floors $Z_{j_0,l}^{(n)}$.

If we follow the increasing order of m, then we get a decreasing sequence of partitions which is the object which A. M. Vershik likes so much.

As remarked at the beginning of 2.2, $B_{ij}^{(n)}$ is the number of visits of $x \in I_j^{(n)}$ to I_i. Therefore $B_{ij}^{(n)}$ gives also the number of $C_0 \in \xi_0(Z_j^{(n)})$ such that $C_0 \subset Z_i^{(0)} = I_i$.

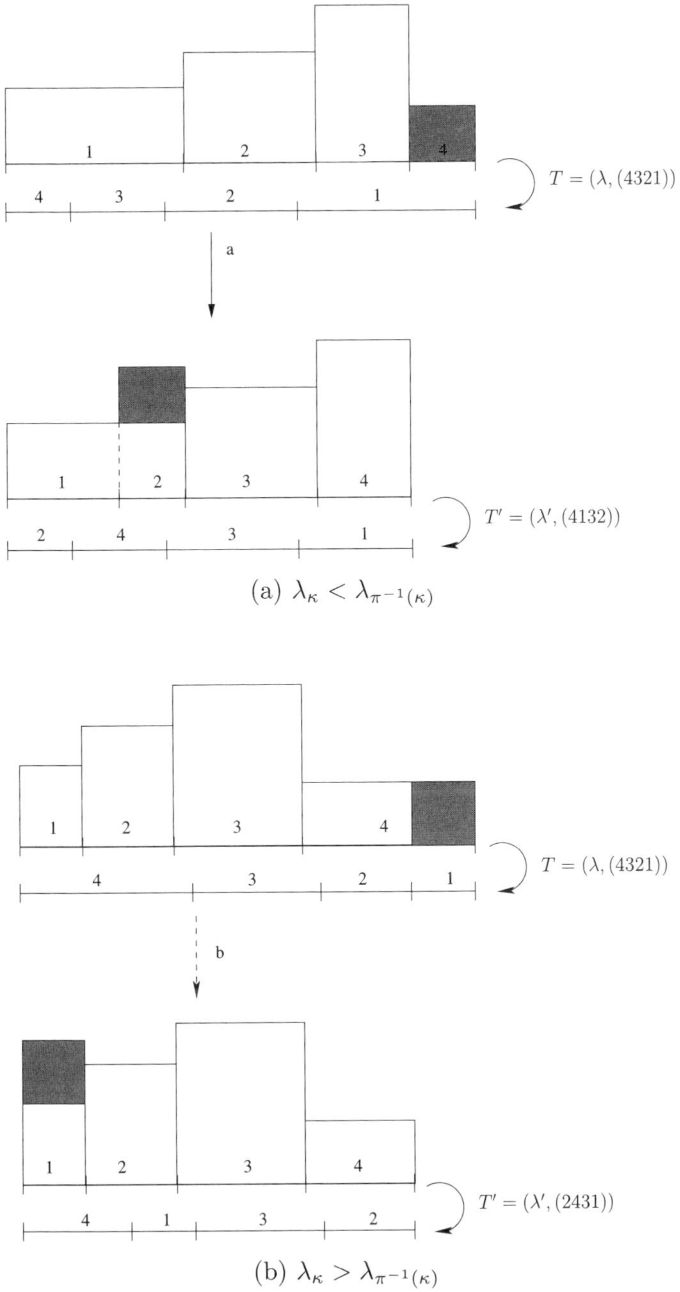

FIGURE 2. Stacking for $T = (\lambda, (4321))$.

More generally, consider products of Rauzy–Veech matrices from m to n, $m < n$, i.e.,

$$B^{(m,n)} \doteqdot A^{(m)} \cdot A^{(m+1)} \cdot \ldots \cdot A^{(n-1)} \cdot A^{(n)}. \tag{2.11}$$

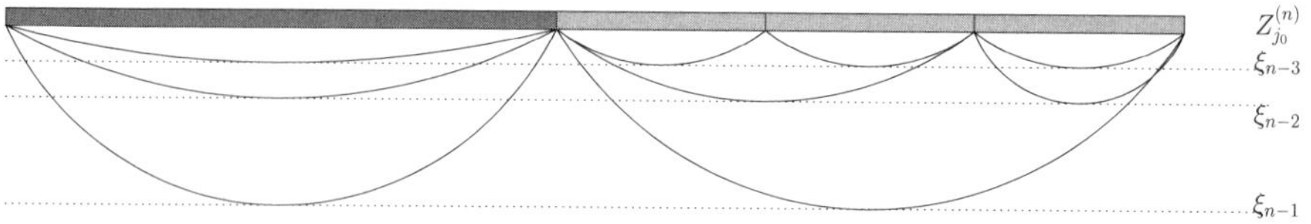

FIGURE 3. An example of partitions $\xi_m(Z^{(n)}_{j_0})$ (arcs show partition elements).

The entries of $B^{(m,n)}$ have the following meaning for the partition $\xi_m(Z^{(n)}_j)$:

$$B^{(m,n)}_{ij} = \text{Card}\left\{C_m \in \xi_m(Z^{(n)}_j)| \quad C_m \subset Z^{(n)}_i\right\}. \tag{2.12}$$

3. The condition for continuous spectrum.

In subsection 3.1 we work out an expression for the spectral measure ρ_A of the characteristic function χ_A, where A is a floor $Z^{(n)}_{j,l}$ of the towers defined in 2.2. This expression is used in subsection 3.2 to prove the condition for the continuity of ρ_A (Proposition 3.3) and consequently the condition for continuous spectrum (Proposition 3.2).

3.1. An expression for discrete components of the spectral measure ρ_A. Throughout this section let $A = Z^{(n_0)}_{j,l}$ be a fixed floor of a tower of step n_0.

The value of the spectral measure ρ_A of χ_A of the point $\omega \in [0, 2\pi[$ is given by the limit

$$\rho_A(\omega) = \lim_{N\to\infty} \frac{1}{N} \sum_{k=0}^{N-1} \mu(A \cap T^k A) e^{-i\omega k}, \tag{3.1}$$

which always exists (see [**H**]).

We can write

$$R_N \doteqdot \frac{1}{N} \sum_{k=0}^{N-1} \mu(A \cap T^k A) e^{-i\omega k} = \sum_{j=0}^{\kappa} \frac{1}{N} \sum_{k=0}^{N-1} \mu(A \cap T^k A \cap Z^{(n)}_j) e^{-i\omega k} \tag{3.2}$$

where n is so large that $|Z^{(n)}_j| \gg N$ (see below).

For each j the measure

$$\mu(A \cap T^k A \cap Z^{(n)}_j) = \frac{\mu^{(n)}_j}{|Z^{(n)}_j|} N_k(Z^{(n)}_j) + \alpha_1, \tag{3.3}$$

where $\mu^{(n)}_j$ is the measure of the tower $Z^{(n)}_j$, $N_k(Z^{(n)}_j)$ is the number of floors $Z^{(n)}_{j,i} \subset Z^{(n)}_j$ such that $Z^{(n)}_{j,i} \in A$, $Z^{(n)}_{j,i+k} \in A$. The reminder α_1 comes from those floors for which $Z^{(n)}_{j,i} \in A$, $i > |Z^{(n)}_j| - N$. In other words, $T^k Z^{(n)}_{j,i}$ may not be in $Z^{(n)}_j$. It is easy to see that

$$0 \le \alpha_1 \le \frac{\mu^{(n)}_j \cdot N}{|Z^{(n)}_j|},$$

which can be made small if n is sufficiently large.

3.1.1. *Admissible pairs.* For given N take the largest m_1 so that $|Z_j^{(m_1)}| \leq N$, $1 \leq j \leq \kappa$. Consider the partitions $\xi_m(Z_j^{(n)})$ of the tower $Z_j^{(n)}$, $m \leq m_1$.

A pair (C_m, C'_m) of distinct elements of $\xi_m(Z_j^{(n)})$ is called *admissible* if

(i) for any two floors of $Z_j^{(n)}$ such that $Z_{j,k}^{(n)} \in C_m \cap A$ and $Z_{j,k'}^{(n)} \in C'_m \cap A$, we have $0 < k' - k \leq N$;

(ii) the pair (C_m, C'_m) cannot be included in a pair $(C_{m'}, C'_{m'})$ with $m' > m$ satisfying (i), where the inclusion among the pairs means that $C_m \subset C_{m'}$ and $C'_m \subset C'_{m'}$.

At step m_1, since (ii) is trivially verified, admissible pairs are precisely the pairs (C_{m_1}, C'_{m_1}) satisfying (i), i.e., such that the distance between A-floors of C_{m_1} and A-floors of C'_{m_1} is at most N. Here *distance* between two floors means the number of floors of the tower $Z_j^{(n)}$ in between the two considered. Notice also that the condition $k' - k > 0$ in (i) guarantees that the pair is ordered.

For all $m < m_1$, by condition (i), admissible pairs are either pairs of elements of ξ_m which are neighboring, i.e., both contained in the same C_{m+1}, or pairs (C_m, C'_m) of elements quite far from each other, i.e., $C_m \subset C_{m+1}$ and $C'_m \subset C'_{m+1}$ where (C_{m+1}, C'_{m+1}) is not admissible and therefore C_{m+1} and C'_{m+1} contain A-floors at distance bigger than N.

The contribution to the sum (3.2) coming from the term $N_k(Z_j^{(n)})$ in the right-hand side of (3.3) can be obtained in the following way. First take a floor $Z_{j,k}^{(n)} \in Z_j^{(n)} \cap A$ and consider the sum over all $k' > k$ such that $Z_{j,k'}^{(n)} \in Z_j^{(n)} \cap A$ and $k' - k < N$. Then sum over k too. Each term enters with weight $e^{-i\omega(k'-k)}$. Each of the considered pairs of floors $(Z_{j,k}^{(n)}, Z_{j,k'}^{(n)})$ is contained in exactly one admissible pair. In order to see this, notice that $(Z_{j,k}^{(n)}, Z_{j,k'}^{(n)})$ is a pair of elements of $\xi_0(Z_j^{(n)})$ satisfying (i) of admissibility. Consider the pair with maximal $m \leq m_1$ such that $(Z_{j,k}^{(n)}, Z_{j,k'}^{(n)}) \subset (C_m, C'_m)$. The pair (C_m, C'_m) is then admissible and contains $(Z_{j,k}^{(n)}, Z_{j,k'}^{(n)})$.

Hence we can rearrange the summation appearing as the left-hand side of the equation below in the following way:

$$\sum_{k=0}^{N-1} N_k(Z_j^{(n)}) e^{-i\omega k} = \sum_{m \leq m_1} \sum_{(C_m, C'_m)}^{(\mathrm{adm})} \sum_{\substack{Z_{j,k}^{(n)} \in C_m \cap A \\ Z_{j,k'}^{(n)} \in C'_m \cap A}} e^{-i\omega(k'-k)} \tag{3.4}$$

where $\sum^{(\mathrm{adm})}$ means the summation over all admissible pairs (C_m, C'_m) of elements of $\xi_m(Z_j^{(n)})$.

3.1.2. *Definition of φ_ω^+ and φ_ω^-.* We define the following quantities associated to a given tower $Z_j^{(m)}$:

$$\varphi_\omega^{\pm}(Z_j^{(m)}) \doteqdot \frac{1}{A(Z_j^{(m)})} \sum_{Z_{j,k}^{(m)} \in A} e^{-i\omega d^{\pm}(k)} \tag{3.5}$$

where $A(Z_j^{(m)})$ is the number of floors $Z_{j,k}^{(m)} \in A$, which coincides with the number of terms in the sum, and $d^+(k)$ (respectively $d^-(k)$) is the number of floors between

$Z^{(m)}_{j,k}$ and the first (last) floor of $Z^{(m)}_j$ (i.e., $d^+(k) = k$ and $d^-(k) = |Z^{(m)}_j| - k$). More generally, for $C_m \in \xi_m(Z^{(n)}_j)$, let

$$\varphi^{\pm}_{\omega}(C_m) \doteqdot \varphi^{\pm}_{\omega}(Z^{(m)}_i), \qquad A(C_m) \doteqdot A(Z^{(m)}_i), \tag{3.6}$$

where i is the unique index such that $C_m \subset Z^{(m)}_i$.

Denoting by $d(C_m, C'_m)$ the distance between C_m and C'_m, i.e., the number of floors of the tower $Z^{(n)}_j$ between C_m and C'_m, we get

$$\sum_{\substack{Z^{(n)}_{j,k} \in C_m \cap A \\ Z^{(n)}_{j,k'} \in C'_m \cap A}} e^{-i\omega(k'-k)} = A(C_m) \cdot A(C'_m) \cdot \varphi^-_{\omega}(C_m) \cdot \varphi^+_{\omega}(C'_m) \cdot e^{-i\omega d(C_m, C'_m)}. \tag{3.7}$$

Putting together (3.2), (3.3), (3.4), and (3.7), we get the following expression for R_N:

$$\begin{aligned} R_N = \sum_{j=0}^{\kappa} \sum_{m \le m_1} \sum_{(C_m, C'_m)}^{(\mathrm{adm})} \frac{\mu^{(n)}_j A(C_m) \cdot A(C'_m)}{N \cdot |Z^{(n)}_j|} \\ \cdot \varphi^+_{\omega}(C_m) \cdot \varphi^-_{\omega}(C_m) \cdot e^{-i\omega d(C_m, C'_m)} + \alpha_2 \end{aligned} \tag{3.8}$$

where α_2 is the reminder coming from α_1 in (3.3) and it is still small since $|\alpha_2| \le \alpha_1$.

3.1.3. *The main contribution to the sum.* The main contribution to (3.8) comes from m's which are close to m_1.

LEMMA 3.1. *Given $\epsilon > 0$, if $m_2 < m_1$ is such that $\frac{|Z^{(m_2)}_j|}{N} < \epsilon/2$ for each $1 \le j \le \kappa$, then*

$$\left| \sum_{j=0}^{\kappa} \sum_{m \le m_2} \sum_{(C_m, C'_m)}^{(\mathrm{adm})} \frac{\mu^{(n)}_j A(C_m) A(C'_m)}{N |Z^{(n)}_j|} \varphi^+_{\omega}(C'_m)\, \varphi^-_{\omega}(C_m)\, e^{-i\omega d(C_m, C'_m)} \right| \le \epsilon.$$

PROOF. To estimate the absolute value of the last sum, it is enough to estimate

$$\sum_{j=0}^{\kappa} \sum_{m \le m_2} \sum_{(C_m, C'_m)}^{(\mathrm{adm})} \frac{\mu^{(n)}_j A(C_m) A(C'_m)}{N |Z^{(n)}_j|}, \tag{3.9}$$

since $|\varphi^{\pm}_{\omega}| \le 1$.

Rewrite the sum over the pairs in (3.9) and over steps $m \le m_2$ again in terms of floors, in a similar way to (3.4). Take a floor $Z^{(n)}_{j,k} \in A$. Consider the floors $Z^{(n)}_{j,k'}$ such that the pair of floors $\left(Z^{(n)}_{j,k}, Z^{(n)}_{j,k'}\right)$ is contained in an admissible (C_m, C'_m) occurring in the sum (3.9). The number of such floors will be at most $2 \max_{j=1}^{\kappa} |Z^{(m_2)}_j|$, because admissible pairs of step at most m_2 are either contained in the same C_{m_2} or are contained in a nonadmissible $\left(C_{m_2}, C'_{m_2}\right)$. Therefore, (3.9) is less than

$$\sum_{j=0}^{\kappa} \frac{\mu^{(n)}_j}{N |Z^{(n)}_j|} \sum_{Z^{(n)}_{j,k} \in A} 2 \max_{j=1,\ldots,\kappa} |Z^{(m_2)}_j| \le \sum_{j=0}^{\kappa} \frac{\mu^{(n)}_j}{|Z^{(n)}_j|} |Z^{(n)}_j| \frac{2 \max_{j=1,\ldots,\kappa} |Z^{(m_2)}_j|}{N} \tag{3.10}$$

which, by using the hypothesis of the lemma and the relation $\sum_{j=0} \mu^{(n)}_j = 1$, is less than ϵ. $\square$

3.1.4. *Recursive formulas for* $\varphi_\omega^\pm$ *and definition of* $\psi_{\omega,A}$. From definition (3.5) it follows that

$$\varphi_\omega^+(Z_j^{(m)}) = \overline{\varphi_\omega^-(Z_j^{(m)})}e^{-i\omega|Z_j^{(m)}|} \tag{3.11}$$

so that $|\varphi_\omega^+(Z_j^{(m)})| = |\varphi_\omega^-(Z_j^{(m)})|$. Therefore, introduce

$$\psi_{\omega,A}^{(n)} \doteqdot \max_{j=1,\dots,\kappa} |\varphi_\omega^+(Z_j^{(m)})| = \max_{j=1,\dots,\kappa} |\varphi_\omega^-(Z_j^{(m)})|, \tag{3.12}$$

where the A in the notation reminds us that $\varphi_\omega^\pm$ and therefore $\psi_{\omega,A}^{(n)}$ depend also on the floor A fixed at the beginning of 3.1.

The following crucial *recursive formulas* are satisfied by $\varphi_\omega^\pm$:

$$\varphi_\omega^\pm(Z_j^{(m+1)}) = \frac{A(Z_{j_1}^{(m)})}{A(Z_j^{(m+1)})}\varphi_\omega^\pm(Z_{j_1}^{(m)}) + \frac{A(Z_{j_2}^{(m)})}{A(Z_j^{(m+1)})}\varphi_\omega^\pm(Z_{j_2}^{(m)})e^{-i\omega Z_{j_1}^{(m)}} \tag{3.13}$$

where

$$\text{for } \varphi_\omega^+ : \quad \begin{cases} j = (\pi^{(m)})^{-1}(\kappa)+1 \\ j_1 = (\pi^{(m)})^{-1}(\kappa) \quad \text{case } a \\ j_2 = \kappa \end{cases} \qquad \begin{cases} j = (\pi^{(m)})^{-1}(\kappa) \\ j_1 = (\pi^{(m)})^{-1}(\kappa) \quad \text{case } b \\ j_2 = \kappa \end{cases}$$

$$\text{for } \varphi_\omega^- : \quad \begin{cases} j = (\pi^{(m)})^{-1}(\kappa)+1 \\ j_1 = \kappa \quad \text{case } a \\ j_2 = (\pi^{(m)})^{-1}(\kappa) \end{cases} \qquad \begin{cases} j = (\pi^{(m)})^{-1}(\kappa) \\ j_1 = \kappa \quad \text{case } b \\ j_2 = (\pi^{(m)})^{-1}(\kappa). \end{cases}$$

For all other j's, $\varphi_\omega^\pm$ remain the same, even if there can be a relabeling of indexes in the case a of the algorithm, i.e.:

$$\varphi_\omega^\pm(Z_j^{(m+1)}) = \begin{cases} \varphi_\omega^\pm(Z_j^{(m)}) & \text{in case } a, \text{ for } j \le \pi^{(m)-1}(\kappa); \\ \varphi_\omega^\pm(Z_{j+1}^{(m)}) & \text{in case } a, \text{ for } j > \pi^{(m)-1}(\kappa); \\ \varphi_\omega^\pm(Z_j^{(m)}) & \text{in case } b. \end{cases}$$

These formulas can be easily proved by analyzing the result of one step of the Rauzy Veech algorithm (Figure 2) and the defining equation (3.5) for $\varphi_\omega^\pm$. For example, for φ_ω^+, when the tower $Z_{j_2}^{(m)}$ is stacked over $Z_{j_1}^{(m)}$ to obtain $Z_{j_2}^{(m)}$, the contribution $e^{-i\omega d^+(k)}$ coming from the floors of $Z_{j_2}^{(m)}$ has to be multiplied by $e^{-i\omega Z_{j_1}^{(m)}}$.

For brevity, we will often use the notation

$$p_{j_1}^{(m)} \doteqdot \frac{A(Z_{j_1}^{(m)})}{A(Z_j^{(m+1)})}, \qquad p_{j_2}^{(m)} \doteqdot \frac{A(Z_{j_2}^{(m)})}{A(Z_j^{(m+1)})}$$

where clearly, from the definition of $A(Z_j^{(m)})$, $p_{j_1}^{(m)} + p_{j_2}^{(m)} = 1$.

The function $\psi_{\omega,A}^{(n)}$ is nondecreasing in n, since from (3.13), for each j

$$|\varphi_\omega^\pm(Z_j^{(m+1)})| \le |p_{j_1}^{(m)} \cdot \varphi_\omega^\pm(Z_{j_1}^{(m)})| + |p_{j_2}^{(m)} \cdot \varphi_\omega^\pm(Z_{j_2}^{(m)})e^{-i\omega Z_{j_1}^{(m)}}| \le \psi_{\omega,A}^{(n-1)}. \tag{3.14}$$

Therefore the following limit exists and we can define

$$\psi_{\omega,A} \doteqdot \lim_{n\to\infty} \psi_{\omega,A}^{(n)}.$$

3.2. Continuity of the spectral measure ρ_A and of the spectrum. In this subsection we state the condition for continuous spectrum using the quantity $\psi_{\omega,A}$ defined in 3.1 and prove its sufficiency reducing it to a condition on the continuity of spectral measures ρ_A. The necessity is discussed in 3.2.1.

PROPOSITION 3.2. *Let T be a minimal and ergodic IET. Consider the towers $Z_j^{(n)}$, $n \in \mathbb{N}$, $1 \leq j \leq \kappa$, obtained iterating the Rauzy–Veech algorithm. If $\psi_{\omega,A} = 0$ for any $0 < \omega < 2\pi$ and any floor $A = Z_{j,l}^{(n)}$, then T has continuous spectrum.*

When T is minimal, it is known that the partition of I into floors of $Z_j^{(n)}$, $1 \leq j \leq \kappa$, converges to the trivial partition as n tends to infinity. If one proves that for any fixed n and any floor A of the towers $Z_j^{(n)}$ the spectrum of the characteristic function χ_A is continuous, then this clearly implies weak mixing of T. Therefore the proof of Proposition 3.2 is reduced to proving the following proposition:

PROPOSITION 3.3. *Let $A = Z_{j,l}^{(n)}$ be a floor of a tower, let ρ_A be the spectral measure of χ_A, and let $\omega \in]0, 2\pi[$.*

If $\psi_{\omega,A} = 0$, then the $\rho_A(\omega) = 0$, i.e., the spectral measure is continuous at ω.

PROOF. Given $\epsilon > 0$, since $\psi_{\omega,A} = 0$, we can choose m_2 s.t. $\psi_{\omega,A}^{(m_2)} \leq \sqrt{\epsilon}$. Now choose N so that the hypothesis of Lemma 3.1 holds, i.e., $|Z_j^{(m_2)}|/N < \epsilon/2$ for each j.

From (3.8), by Lemma 3.1,

$$
\begin{aligned}
|R_N| &\leq \epsilon + \left| \sum_{j=0}^{\kappa} \sum_{m_2 < m \leq m_1} \sum_{(C_m, C'_m)}^{(\mathrm{adm})} \frac{\mu_j^{(n)} A(C_m) A(C'_m)}{N |Z_j^{(n)}|} \varphi_\omega^+(C_m)\, \varphi_\omega^-(C_m) \right| \\
&\leq \epsilon + (\psi_{\omega,A}^{(m_2)})^2 \left| \sum_{j=0}^{\kappa} \sum_{m_2 < m \leq m_1} \sum_{(C_m, C'_m)}^{(\mathrm{adm})} \frac{\mu_j^{(n)} A(C_m) A(C'_m)}{N |Z_j^{(n)}|} \right| \qquad (3.15)
\end{aligned}
$$

where the second inequality follows simply from the definition of $\psi_{\omega,A}^{(m)}$ and the fact that it is decreasing.

Rearranging again the sum over admissible couples as a sum over floors, as in (3.4), we get

$$
\sum_{m_2 \leq m \leq m_1} \sum_{(C_m, C'_m)}^{(\mathrm{adm})} A(C_m) A(C'_m) \leq \sum_{k=0}^{N-1} N_k(Z_j^{(n)}) \leq N \cdot |Z_j^{(n)}|. \qquad (3.16)
$$

Moreover $\sum_{j=0}^{\kappa} \mu_j^{(n)} = 1$. Therefore, the second term in (3.15) is bounded by $|\psi_{\omega,A}^{(m_2)}|^2 < \epsilon$ and $|R_N| \leq 2\epsilon$. □

3.2.1. *A formula for the spectral measure $\rho_A(\omega)$.* For a typical IET a refinement of Proposition 3.3 is true: the measure $\rho_A(\omega)$ of a point can be expressed in terms of $\psi_{\omega,A}$.

PROPOSITION 3.4. *For Lebesgue-a.e. $\lambda \in \Delta_{\kappa-1}$,*

$$
\rho_A(\omega) = \mu^2(A) \psi_{\omega,A}^2 \qquad (3.17)
$$

where μ is the Lebesgue measure on I.

The precise conditions on λ under which the formula (3.17) can be proved are unique ergodicity of (λ, π) and density of its orbit under $\mathcal{R}$. These conditions are generic as follows, respectively, from [**V3, M**] and from the conservativity of $\mathcal{R}$ proved in [**V3**].

The proof of Proposition 3.4 will be discussed in another paper.

REMARK 3.5. Notice that Proposition 3.4 implies that for a typical IET the condition in Proposition 3.2 is not only sufficient, but also necessary for the continuity of the spectrum.

References

[AF] A. Avila, G. Forni, *Weak mixing for interval exchange transformations and translations flows,* Ann. of Math., to appear, preprint, 2004, arXiv:math.DS/0406326.

[H] E. Hopf, *Ergodentheorie,* Springer, Berlin, 1937.

[Ka] A. B. Katok, *Interval exchange transformations and some special flows are not mixing,* Israel J. Math. **35** (1980), 301–310.

[Ke] M. Keane, *Interval exchange trasformations,* Math. Z. **141** (1975), 25–31.

[KS] A. B. Katok, A. M. Stepin, *Approximations in ergodic theory,* Uspehi Mat. Nauk **137**:5 (1967), 81–106.

[M] H. Masur, *Interval exchange transformations and measured foliations,* Ann. of Math. 115 (1982), 169–200.

[NR] A. Nogueira, D. Rudolph, *Topological weak-mixing of interval exchange maps,* Erg. Th. Dynam. Sys. **17** (1997), 1183–1209.

[R] G. Rauzy, *Échanges d'intervalles et trasformations induites,* Acta Arithmetica **34** (1979), 315–328.

[SU] Ya. G. Sinai, C. Ulcigrai, *Weak mixing in interval exchange transformations of periodic type,* Lett. Math. Phys. **74** (2005), 111–133.

[V1] W. A. Veech, *Interval exchange transformations,* J. d'Analyse Mathématique **33** (1978), 222–272.

[V2] W. A. Veech, *Projective Swiss cheeses and uniquely ergodic interval exchange transformations,* Ergodic theory and dynamical systems, I, College Park, MD, 1979-80. Birkhäuser, 1981, pp. 113–193.

[V3] W. A. Veech, *Gauss measures for transformations on the space of interval exchange maps,* Ann. of Math. **115** (1982), 201–242.

[V4] W. A. Veech, *The metric theory of interval exchange transformations. I. Generic spectral properties,* Amer. J. Math. **107** (1984), 1331–1359.

[Z] A. Zorich, *Finite Gauss measure on the space of interval exchange transformation. Lyapunov exponents,* Ann. Inst. Fourier Grenoble **46** (1996), 325–370.

MATHEMATICS DEPARTMENT, PRINCETON UNIVERSITY, PRINCETON, NJ 08544, USA
E-mail address: bufetov@math.princeton.edu

MATHEMATICS DEPARTMENT, PRINCETON UNIVERSITY, PRINCETON, NJ 08544, USA,
and
LANDAU INSTITUTE OF THEORETICAL PHYSICS, RUSSIAN ACADEMY OF SCIENCES
E-mail address: sinai@math.princeton.edu

MATHEMATICS DEPARTMENT, PRINCETON UNIVERSITY
E-mail address: ulcigrai@math.princeton.edu

Amer. Math. Soc. Transl.
(2) Vol. **217**, 2006

Several Nonstandard Remarks

Ivan Fesenko

ABSTRACT. On interaction between nonstandard mathematics as part of model theory and several areas of modern mathematics and mathematical physics.

> Progress does not consist of replacing a theory that is wrong with one that is right. It consists of replacing a theory that is wrong with one that is more subtly wrong.
>
> Hawkin's Theory of Progress

This text aims to present and discuss a number of situations in analysis, geometry, number theory and mathematical physics which can profit from developing their nonstandard description or interpretation and then using it to prove standard results and/or establish standard theories.

Entirely nonstandard proofs are quite rarely of much use in standard mathematics. One of the main obstructions is that it is hard to find a satisfactory standard shape of hyperstructures. The value of nonstandard mathematics is in serving as a "guiding star" and often offering a conceptually simple and elegant interpretation and generalization of classical structures. This provides a broader view of the standard theory and sometimes leads to new concrete standard results.

The text can be divided into five parts. The first short part comprises Sections 1 and 2. Section 1 contains a calculation of the ring of hyperintegers, for which I was unable to find a reference. The calculation implies that the endomorphism ring of hyperintegers is highly noncommutative, unlike the ring of endomorphisms of integers. This supports the well-known expectation that noncommutative methods might prove useful in the study of commutative structures in various parts of mathematics including number theory. Section 2 recalls a hyperfinite (or more generally hyperdiscrete) approximation principle, new applications/examples of which are proposed in later parts.

2000 *Mathematics Subject Classification.* Primary 03H05, 03H15, 03H99, 03C99, 11G99, 11R99, 46L85, 46L87, 46L89, 46L99.

Key words and phrases. Model theory and its applications, nonstandard mathematics and its applications.

The second part consists of Sections 3–7 where some analytical and topological constructions are discussed from the nonstandard mathematics point of view. Compact operators on a Hilbert space are interpreted in Section 3 as infinitesimally small elements in the algebra of linear operators on the hyperspace, which leads to a number of questions on nonstandard interpretation (as standard images of hyperobjects) of some important concepts of operator algebra theory in Section 4. Two nonequivalent topologies on a hypertopological space are recalled in Section 5. In Section 6, to a "noncommutative space" we associate a hypercommutative one. This leads to many open questions about new commutative interpretations of various objects of noncommutative differential geometry; see Section 7.

The third part consists of Sections 8–9. Section 8 lists several known points of interaction between noncommutative differential geometry and number theory, each possessing a suitable nonstandard commutative interpretation. Section 9 contains a research programme of a hyperarithmetical approach towards realization of Manin's Alterstraum of real multiplication. It is based on using hyperlattices, the hyper-$\wp$-function and hyperdivision points of hyperelliptic curves. Quantum tori and quantum theta-functions, known to have relevance for this programme, are expected to be images of hypertori and hyper-theta-functions, respectively.

The fourth part comprises Sections 10–13. Section 10 raises a question of existence of a hypercommutative description of a number of standard noncommutative theories. It also contains links of the current work to the theory of analytic Zariski structures by Zilber and geometric stability theory. Section 11 and 12 present several more examples of nonstandard interpretations of arithmetic algebraic geometry structures. Relations or applications of nonstandard mathematics to the central object of number theory—zeta-functions—are brought up in Sections 12 and 13.

The last part discusses several applications in quantum physics, a very promising direction of further research.

The work concentrates on presenting main points rather than on already available technical details: those should appear elsewhere.

The origin of this work came from the study of translation invariant measures on generalized loop spaces with values in formal power series over reals and associated harmonic analysis with its applications in arithmetic geometry, see [**Fe1**, Remarks in Sections 4 and 13]. At later stages the work aimed at understanding a series of papers at the intersection of noncommutative differential geometry and commutative number theory, and reasons underlying applications of the former to the latter. One of the conclusions proposed in this work is a principle that whenever there is such an application, there is always in the background an entirely commutative description of the situation, and often in such situations it can be seen as the shadow image of a hypercommutative description.

Talks on parts of this work were delivered at seminars in Cambridge, Bonn, St. Petersburg and Oxford, and I thank its participants for their remarks. Some recent works [**B, BS1, T, Cl**] are related to preliminary versions of this text. I am especially grateful to P. Cartier, Yu. I. Manin, M. Marcolli, B. Zilber and A. M. Vershik for their comments, explanations and suggestions.

The classical text on nonstandard mathematics is the Robinson book [**Ro1**]. For a short modern introduction into hyperobjects (mostly hyperreals) see [**Gl**] or the shorter [**AFHL**, Chapters 1–3]. For recent reviews of various applications of nonstandard mathematics in analysis, statistics, differential equations and physics; see [**NAiP, NA, NAiA, NAWM**]; for some results in nonstandard number theory 25 years ago, see e.g. [**RR, Mc**] and references therein. These are just some examples of the vast literature on the subject. For an introduction into hypercategories see [**BS1**]. In Sections 1, 3–4, 6–14 we mainly discuss situations not covered by the previous references.

It is a pleasure to dedicate this text to Anatolij Moiseevich Vershik, whose broad and encyclopedic knowledge of mathematics (rather a Moscow mathematical school quality) is harmonically combined with the best St. Petersburg mathematical traditions.

1. Endomorphisms of hyperintegers

We start with the basic object—the abelian group $\mathbb{Z}$: its endomorphism ring is commutative canonically isomorphic to the ring of integers. The situation is drastically different when one calculates the endomorphism ring of the group ${}^*\mathbb{Z}$ of hyperintegers. There is a natural epimorphism induced by ${}^*\mathbb{Z}/n \to \mathbb{Z}/n$,

$$^*\mathbb{Z} \to \prod \mathbb{Z}_p = \widehat{\mathbb{Z}},$$

whose kernel equals $\bigcap_{n \in \mathbb{Z}\setminus\{0\}} n\,{}^*\mathbb{Z}$ and is an infinitely divisible group of uncountable cardinality. Hence

$$^*\mathbb{Z} \simeq \widehat{\mathbb{Z}} \times \mathbb{Q}^\kappa,$$

with uncountable κ. This follows easily from the well-known relation of ${}^*\mathbb{Z}$ and $\mathbb{Z}_p$; see for example [**Ro5**] or [**Gl**, Chapter 18]. Alternatively use the fact that the abelian group of integers is elementarily equivalent to the saturated abelian group $\widehat{\mathbb{Z}} \times \mathbb{Q}^\kappa$, ${}^*\mathbb{Z}$ is saturated, and elementarily equivalent saturated structures of the same cardinality are isomorphic; see e.g. [**EF**].

Thus, the endomorphism ring of ${}^*\mathbb{Z}$ is noncommutative. In particular, one can find two endomorphisms $A, B\colon {}^*\mathbb{Z} \to {}^*\mathbb{Z}$ such that both A and B induce an identity operator on $\widehat{\mathbb{Z}}$ and $A \circ B - B \circ A$ is the identity operator on
${}^*\mathbb{Z} \setminus \widehat{\mathbb{Z}}$.

Note that by transfer the ring of internal endomorphisms of ${}^*\mathbb{Z}$ is the commutative ring ${}^*\mathbb{Z}$. See also Section 10.

If one wants to get a hyperfinite instead of hyperdiscrete group, then choose $\omega \in {}^*\mathbb{N} \setminus \mathbb{N}$. There are two injective homomorphisms

$$\mathbb{Z} \to {}^*\mathbb{Z}/\omega^*\mathbb{Z} \to {}^*\mathbb{Z};$$

the last is not canonical and follows from a description similar to the previous description ${}^*\mathbb{Z}/\omega^*\mathbb{Z} \simeq \widehat{\mathbb{Z}} \times \mathbb{Q}^\kappa$. The middle object is hyperfinite.

2. Hyperfinite approximation principle

One of the most important methodological tools to use in nonstandard mathematics is the hyperfinite (more generally, hyperdiscrete) approximation: for a usual object X, try to find a hyperfinite (or hyperdiscrete) object Y and morphisms

$$X \longrightarrow Y \longrightarrow {}^*X,$$

with $X \longrightarrow {}^*X$ being the diagonal embedding. See e.g. [**Gl**, Chapter 19]. By transfer, the hyperobject ${}^*X$ has many properties similar to X; inducing those on

Y and using the fact that Y has many properties similar to finite objects, one can then attempt to interpret various classical constructions on X.

For example, the existence of Haar measure on locally compact abelian groups can be established in the shortest way by inducing it from the hypercounting measure on a covering hyperfinite abelian group (e.g. [**NA, NAiA, Gr1**]); the same is also true for the Fourier transform [**Gr2**]. Hyperfinite measure space theory due to works of Loeb and Anderson has found many applications in measure theory. Notice that in hypertheory one often works with approximate morphisms and functors (modulo the halo), e.g. [**Gr2, YO**].

A generalization of Haar measure on the additive group of higher local fields [**Fe1**] can be viewed as induced from the counting measure on hyperhyperfinite abelian groups (or hyper-Haar measure on hyper-locally-compact groups).

For more examples see the following sections.

3. Compact operators as infinitesimally small elements

For a Hilbert space X such a hyperfinite space Y does exist, e.g. [**AFHL**, Chapter 3]. Compact operators on X are restrictions of hyperfinite operators on Y which enjoy the following additional property: they map finite points to nearstandard, e.g. [**AFHL**, 2.3]. We have algebra homomorphisms

$$\mathcal{L}(X) \to \mathcal{L}({}^*X), \quad r\colon \mathcal{L}({}^*X) \to \mathcal{L}(Y);$$

the second is given by $A \mapsto P \circ A\colon Y \to Y$ where $P\colon Y \to {}^*X$ is a projection. In particular, the algebra $\mathcal{K}(X)$ of compact operators on X can be viewed as a summand of the hyperfinite algebra $\mathcal{L}(Y)$ of linear operators on Y. The space $\mathcal{K}(X)$ contains the subspace S generated by the images of $\mathcal{L}(Z)$, Z a finite subspace, and $\mathcal{K}(X)$ is the halo of S with respect to the hypernorm on $\mathcal{L}(Y)$. So the compact operators do play the role of infinitesimal objects. Compare this with the psychologically different point of view of Connes in [**Co4**, VIII] where the two classes are rather contrasted with each other (and note that the Dixmier trace has also a natural nonstandard interpretation; see [**Ro10**]).

4. Nonstandard interpretation of operator algebras

A number of results on stellar and especially operator algebras (which are in a sense algebraic objects) is likely to have a nice (and hopefully useful) interpretation in the language of internal algebras of hyperfinite spaces, with the possibility of having a simple conceptual picture of what is really going on in the existing, often technically complicated, classical proofs.

For example, the Morita equivalence studied by Rieffel [**Rf1, Rf2, Rf3**] for stellar algebras $\mathcal{A}, \mathcal{B}$, which in the simplest case (if A, B have countable approximate identities, cf. [**Rf1**]) is equivalent to the stable isomorphism $\mathcal{A}\otimes\mathcal{K}(H) \simeq \mathcal{B}\otimes\mathcal{K}(H)$ for a separable Hilbert space H, must have a very simple interpretation in $\mathcal{L}(Y)$ (we view $\mathcal{A}, \mathcal{B}$ inside $\mathcal{L}(X)$).

Existence of a nonstandard interpretation would agree with the completeness of such algebras and would correspond to the well-known fact that the shadow image of an internal subset in a Hausdorff space is closed. From this point of view it seems more likely that one really needs completed versions of objects like cross algebras, see the last pages of [**Mn2**], as opposed to incomplete ones, even for applications to algebraic geometry and number theory. See also Section 6.

5. Two topologies of a hyperspace

Recall that there are various complications about topologies of the hyperspace ${}^*X$ for a topological space X with topology τ: this hyperspace does not have a canonical topology, and there are at least two important topologies on it: one called the S-topology with basis of fundamental neighbourhoods generated by ${}^*U$, U runs through open subsets in X, and the other called the Q-topology (i.e., ${}^*\tau$-topology) with basis of neighbourhoods consisting of ${}^*T$ where T is the set of a fundamental neighbourhoods in τ. (As usual, we always assume an appropriate saturation level for the hyperuniverses.) The space ${}^*X$ is compact with respect to the S-topology, and normally the space ${}^*X$ is not Hausdorff with respect to the S-topology. The restriction of the S-topology to X coincides with τ. The space ${}^*X$ is Hausdorff with respect to the Q-topology, and the restriction of the Q-topology to X normally does not coincide with τ.

The shadow map from the nearstandard point of ${}^*X$ to X is defined using the S-topology and hence is not quite compatible with the Q-topology. This makes many applications or constructions of nonstandard mathematics quite cumbersome, since the Q-topology is more natural in many issues.

6. Hypercommutative space associated to a noncommutative space

When the quotient space X/Y of a metric topological space X is not Hausdorff, it is sometimes called a "noncommutative space". In the case of $X/\mathbb{Z}$ some of its invariants are studied by working with the noncommutative algebra of functions which is an appropriate completion of the cross product of continuous or smooth functions on X with the action of $\mathbb{Z}$. A central initial example is the noncommutative algebra of functions associated to the degenerate torus $\mathbb{C}/T_\theta$ where $T_\theta = \mathbb{Z} + \theta\mathbb{Z} \subset \mathbb{R}$, $\theta \in \mathbb{R} \setminus \mathbb{Q}$. The torus splits into the product of $\mathbb{R}$ and $\mathbb{R}/T_\theta$ and the latter can be viewed as the quotient of the compact space $\mathbb{R}/\mathbb{Z}$ by the induced via multiplication by θ action of integers. Via the exponential map the torus is isomorphic to $\mathbb{C}^\times/\langle\exp(2\pi i\theta)\rangle$ and the corresponding stellar or operator algebra is the appropriate completion of the algebra generated over $\mathbb{C}$ by x, y, x^{-1}, y^{-1} satisfying $xy = \exp(2\pi i\theta)\, yx$.

Now let θ_n be a sequence of complex numbers with nonzero imaginary part such that $\theta_n \to \theta$. It is known that noncommutative objects can be viewed as degenerate objects, as certain imprecise limits of commutative objects (see Section 8). In the particular case of the previous paragraph one can easily give more mathematical meaning to this. Let Θ be the equivalence class of (θ_n) in ${}^*\mathbb{C}$. Consider the stellar or operator algebra of functions on the hypercommutative torus $T_\Theta = {}^*\mathbb{C}/({}^*\mathbb{Z} + \Theta{}^*\mathbb{Z})$. The stellar algebra is Morita equivalent to the algebra of internal continuous ${}^*\mathbb{C}$-valued functions on T_Θ. It is generated by X, Y satisfying $XY = \exp(2\pi i\Theta)\, YX$. The shadow image of this algebra is the noncommutative algebra of functions of T_θ. Note that the Morita equivalence is not compatible with the shadow map. So we can view the function algebra of the noncommutative object T_θ as a quotient of the hyperfunction algebra of the hypercommutative T_Θ. For the latter, one gets via the transfer principle analogues of many properties of commutative tori, and then using the shadow map, one can reduce those to properties of the quantum torus, including its invariants supplied by operator algebra theory.

One has a similar description if S is a dense subspace and subgroup of a metric topological space and group T such that, as above, S is a "limit" of S_n and T/S_n are Hausdorff.

Denote $G = {}^*\mathbb{Z}/\omega{}^*\mathbb{Z}$ for $\omega \in {}^*\mathbb{N} \setminus \mathbb{N}$. This is a hyperfinite group and there are two injective homomorphisms $\mathbb{Z} \to G \to {}^*\mathbb{Z}$; the last is not canonical and follows from a description similar to Section 1's description $G \simeq \mathbb{Q}^\kappa \times \widehat{\mathbb{Z}}$. For a metric topological space X with an α-action of integers, the topological space ${}^*X$ has an action of hyperintegers. Endow ${}^*X$ with the Q-topology and consider the operator algebra $\mathcal{A}({}^*X)$ generated by (internally) continuous functions ${}^*X \to {}^*\mathbb{C}$.

Now notice that the space ${}^*X/{}^*\mathbb{Z}$ and hence ${}^*X/G$ are Hausdorff. Thus one can use the "covering" ${}^*X/G \to X/\mathbb{Z}$ (more precisely, the shadow map applied to the algebra of functions on ${}^*X/G$) to study refined invariants of the "noncommutative space" $X/\mathbb{Z}$ (i.e. invariants of the associated noncommutative algebra) by means of hyperclassical invariants of the commutative space ${}^*X/G$. In particular, we have $(\mathcal{A}({}^*X)^G)'' \simeq \mathcal{A}({}^*X) \times_\alpha G$ and the stellar algebra of ${}^*X/G$ is Morita equivalent to the cross product of the stellar algebra of ${}^*X$ and G. Both assertions follow from the corresponding properties for finite groups, by transfer. The hyperobject $\mathcal{A}({}^*X) \times_\alpha G$ behaves "better" than the original standard object, and it can be used to reinterpet well-known results about the operator cross algebra $\mathcal{A}(X) \times_\alpha \mathbb{Z}$.

7. Nonstandard interpretation of elements of noncommutative differential geometry

It would be interesting to investigate how much geometry of noncommutative spaces ([**Co1, Co2**] and references therein) can be interpreted as appropriate standard images of geometry of appropriate hypercommutative spaces. Notice that the periodic cyclic homology and KK-constructions can be defined as analogous to infinitesimal homology in algebraic geometry (see [**Cu**] and references there), which in particular gives a nice simple interpretation of the bivariant Chern character. On the other hand, the infinitesimal homology can be viewed as a standard image of a nonstandard homology; see also Section 11 for nonstandard interpretations of direct and inverse limit constructions. Of course, one of the key problems is to obtain a hyperinterpretation of the spectral triple in noncommutative geometry.

8. Nonstandard interpretations of interactions between noncommutative differential geometry and number theory

Turning to some existing applications of noncommutative differential geometry to number theory, one can ask to what extent nonstandard interpretations are useful there. For example, both deformations and representation theory of noncommutative objects are supposed to have lifts to hypercategories and possibly be closely linked up there.

The attitude of viewing noncommutative spaces as boundaries, degenerate objects as sometimes belonging to the boundary of an appropriate modular space, and the boundary being "a bridge to the world of noncommutative differential geometry" proved to be fruitful [**So, SV, MM1**]. This is also related to the holographic principle whose importance in quantum field theory [**O**] and mathematics [**MM2**] becomes more and more evident.

When one works with boundaries, compactification issues come immediately into play. It is well known that nonstandard constructions often provide elegant ways to compactify spaces; cf. [**Ro8**] and a number of later works. A compactification of a space S can be found (sometimes in a "natural" way) inside the space ${}^*S$. Viewing $PGL_2(\mathbb{Z}) \setminus (\mathbb{P}^1(\mathbb{R}) \setminus \mathbb{P}^1(\mathbb{Q}))$ as an invisible part of the modular curve [**MM1**], averages and limits of modular symbols [**MM1, Mr**], all are compatible with the nonstandard point of view.

9. A nonstandard approach to real multiplication

A programme to develop a theory of real multiplication using new (for number theory) structures was recently raised by Manin [**Mn3**]. His own approach advocates the usefulness of noncommutative (quantum) tori and noncommutative (quantum) theta-functions. Here we describe another approach to real multiplication based on using appropriate hyperobjects. One advantages of this approach is that it includes new integral structures playing a central role; those integral structures seem to be very difficult to obtain using quantum objects of [**Mn3**].

The tori in Section 6 have ring structure if the lattice is generated by a quadratic irrationality. Let K_n be a sequence of quadratic imaginary number fields, and let $\theta_n = a_n + b_n \in K_n$, $a_n \in \mathbb{Q}$, $b_n \in i\mathbb{R} \setminus \{0\}$, be such that a_n tend to a quadratic real irrationality θ, lying in a real quadratic field F, and b_n tend to 0. Form a hypercomplex number Θ corresponding to (θ_n), and consider a hypertorus

$$T_\Theta = {}^*\mathbb{C}/L_\Theta, \quad L_\Theta = {}^*\mathbb{Z} + \Theta{}^*\mathbb{Z}.$$

The analytic parameterization in the basic theory of elliptic curves has its hyperanalogue: define the hyper-Weierstraß-function

$$\wp_\Theta(z) = \frac{1}{z^2} + {}^*\!\!\sum_{l \in L_\Theta \setminus \{0\}} \left(\frac{1}{(z-l)^2} - \frac{1}{l^2} \right),$$

which for z being the equivalence class of (z_n), $z_n \in \mathbb{C}$, is the equivalence class of $(\wp_{\theta_n}(z_n))$. Consider a hyperelliptic (not in the standard sense of this combination of adjectives) curve E_Θ over ${}^*\mathbb{C}$ whose points are $(\wp_\Theta(z), \wp'_\Theta(z))$, $z \in T_\Theta$.

Via the transfer principle, abelian extensions of the hyperfield $\mathcal{K}$ corresponding to the sequence (K_n) are explicitly described by hypercomplex multiplication theory. The Galois group of F^{ab}/F is isomorphic to a quotient of $\mathcal{K}^{\mathrm{ab}}/\mathcal{K}$ and making this explicit would give an explicit description of the maximal abelian extension of F, i.e. what could be called a theory of real multiplication. Various standard images of hyperobjects in the hypercomplex multiplication theory associated to $\mathcal{K}$ and E_Θ, or their modifications, supply new analytic, geometric and arithmetic structures (including, importantly, various integral structures) for the theory of real multiplication. Some of them can also be used for an interpretation of noncommutative structures in [**Mn3**], e.g. quantum theta-functions (e.g. [**Mn2**], [**Mn3**, Section 2], [**Sz**]) as standard images of hyper-theta-functions. For more details see [**T**].

10. Hypercommutative versus noncommutative

In the previous case of nonstandard extension of interaction of noncommutative geometry and number theory the noncommutative description is closely related to a nonstandard commutative one. *How many more situations with commutative hyperobjects in the background can be detected among other applications of noncommutative differential geometry, at least to number theory?* The material of Section 1 might be the starting point of explaning why one gets noncommutative standard objects from commutative hyperobjects.

There are important links between nonstandard mathematics and a part of model theory which support each other. The theory and technique of analytic Zariski structures developed by Zilber [**Z1**] lead to a good understanding (elimination of quantifiers in a geometric language) of limit objects; elimination of quantifiers is equivalent to the understanding of automorphisms of a saturated model. It is conjectured in [**Z2**] that an expanded structure $(\mathbb{C}, +, \times, G)$ on the real curve $G_\theta = \exp((1+i)\mathbb{R})\exp(\theta 2\pi i\mathbb{Z})$ is superstable if and only if θ is a real quadratic irrationality. Assuming Shanuel's conjecture, it is shown that the structure is indeed superstable and the indeterministic real curve G_θ is definable. If θ is rational, the ordinary real curve G_θ is not stable.

11. More instances of hyperdiscretization

(1) Viewing local fields as subquotients of a hyperglobal field, one can try to systematically reformulate in the hyperlanguage and then further develop appropriate classical local theories in number theory and algebraic geometry (e.g. [**Ro7, Ro8, Ro9**]). The list of such theories may include infinite-dimensional algebraic geometry (see, e.g., [**B**]), resolution of singularities, and many other theories in which ind and pro objects play an essential role.

(2) Applying the same general mechanism, one obtains a discretization and algebraization of analytic and topological theories. Of course, nonstandard analysis is the first primary example of this trend in modern mathematics. In particular, in the case where such a theory exists over Archimedean fields only, one can try to extend the lower level hyperdiscrete theory to an arbitrary base, thus hoping to obtain a reasonable analogue over other fields.

For example, objects in complex algebraic geometry can be viewed as coming from hyperdiscrete algebraic geometry over ${}^*\mathbb{Q}(i)$.

(3) By gluing lifts of objccts from local theories over different places of a global object to hyperlevel, one might expect to have a unified theory in which the difference between Archimedean and non-Archimedean places disappears. Would this lead to a conceptually better presentation of Arakelov's theory? Recent results [**CM**] on relations of the latter and noncommutative differential geometry, which are relied upon for Manin's interpretation of hyperbolic geometry as a sort of Archimedean Arakelov geometry [**Mn1**], are well suitable for a nonstandard interpretation as well.

(4) For the study of hyper-étale cohomology theory, see the recent work [**BS2**]. This theory has applications to l-adic cohomology, when one uses ultraproducts of finite fields for the coefficients of the cohomology. The latter object is quite useful, as was demonstrated in the famous work [**AK**] and following papers; see e.g. [**FJ**].

12. Hyperglobal objects and adelic objects

In nonstandard arithmetic one can interpret the adele and idele groups using hyperfractional ideals in hyperalgebraic number fields [**Ro2, Ro3, Ro4, Ro5, Ro6**]; in particular, $\mathbb{A}_\mathbb{Q}$ can be viewed as a subquotient of ${}^*\mathbb{Q}$ and the latter is sometimes easier to work with. This, together with the previous remarks, leads to the natural question: *can one derive a hyperapproach to one-dimensional and higher-dimensional class field theory?*

Interpreting zeros of the Riemann zeta-function (cf. [**Co3**]), it might be useful to work not only with the space $\mathbb{A}/\mathbb{Q}^\times$ but with the much larger and more interesting (and easier to study) space ${}^*\mathbb{Q}/\mathbb{Q}^\times$ as well.

Remarks in the previous section and standard images of hyperglobal objects and of function spaces on them are closely related to "q-world" objects in the approach by Haran [**H**]. "Quantum" q-objects which specialize both to p-adic and Archimedean objects can be viewed as standard images of hyperrational objects. For example [**H**] put

$$\zeta_q(s) = \prod_{n\geq 0}(1-q^{s+n})^{-1}.$$

Then

$$\zeta_q(s) \to \zeta_p(s) = 1/(1-p^{-s}), \quad \text{if } q, s \to 0 \text{ and } q^s \to p^{-s}$$

for prime p, and

$$\zeta_q(s) \to \Gamma(s) = \zeta_\infty(s)\pi^{s/2}, \quad \text{if } q \to 1 \text{ and } (1-q^s)/(1-q) \to s.$$

For similar examples see [**BF**, Section 14]. See also Section 14.

Several constructions in Haran's theory can be viewed as shadow images of hyperdiscrete objects, and it would be very interesting to systematically investigate this.

13. Nonstandard approach to complex analysis and zeta-functions

Not so much is known about nonstandard complex analysis (just [**Ro1**, Chapter VI], [**Fr, Ca**]), unlike nonstandard real analysis, topology and metric spaces theory. Investigation in this area and the study of hyperanalytic number theory might be very useful for a conceptual interpretation of some central objects in analytic number theory including zeta-functions. This study will undoubtfully be very useful for the work sketched in Section 9.

In relation to zeta-functions, a "hyperdream" is *whether various types of quite different zeta-functions in number theory (complex-valued and p-adic-valued) could originate from a single source.* In the case of one-dimensional theory such a source could be the hyper-zeta-functions of hyper-A-fields K (say $\zeta_{{}^*K}({}^*s)$). It would be then this source from which zeta values drop as anticipated by Kato in [**K**]. For a first work on hyper-zeta-functions, see [**Cl**].

14. Relations with physics

Hyperdiscrete constructions very well correspond to the way physicists argue in quantum physics: the combination of both discrete and continuous properties in hyperdiscrete objects is extremely promising for applications in mathematical physics: hyperdiscrete objects are ideally suited to describe the familiar type of wave–particle behaviour in physics through shadow images of nonstandard objects.

14.1. Divergent integrals ubiquitous in field theories can be naturally viewed as hypercomplex unlimited numbers, and various renormalization procedures could have enlightening nonstandard interpretations. In particular, nonrigorous physical constructions could be given mathematically sound justification. It is very surprising that almost nothing has been done in this direction. For first steps in nonstandard interpretations of aspects of quantum physic theories, see e.g. [**AFHL, Y**] and references therein.

14.2. Much of what is known about quantum field theory comes from perturbation theory and applications of Feynman diagrams for calculation of scattering amplitudes. Recall that Feynman "never felt order by order was anything but an approximation to the path integral" [**Sr**, p. 460]. The Feynman path integral is extremely difficult to give a mathematically sound theory, even though there are many attempts to produce such; each of them has some drawbacks (see [**Fe2**, Section 18]). In particular, the value of the integral has a rigorous mathematical meaning as a hypercomplex number, manipulations with which do produce standard complex numbers seen in the recipes of Feynman; see e.g. [**AFHL**]. However, several attempts to construct and apply a nonstandard theory of the path integral (e.g. [**L1, L2, N1, N2**]) have not yet led to a successful and practically useful general theory. Recall that Wiener measure (often used in mathematical approaches to the path integral) can be viewed as the Loeb measure associated to the hyperrandom discrete walk, e.g. [**NA, NAiA, NAiP**]. One of the rigorous approaches to a *shift invariant* measure behind the Feynman integral might show up as a generalization of the measure on two-dimensional local fields [**Fe1**] which itself is induced from a hyper-Haar measure. See [**Fe2**] for the measure and integration on algebraic loop spaces.

14.3. A string with infinitesimally small diameter can be viewed inside the halo of a standard path, and the path is its shadow. It is interesting to see if (parts of) string theory could be viewed as shadow images of hyper-two-dimensional quantum field theory, by passing from hypercomplex numbers to power series over $\mathbb{C}$ in which a variable corresponds to an infinitesimally small positive element and then working with the theory which takes values in formal power series in X and at the last stage replacing X with the constant $\sqrt{\alpha'}$ as in string theory (for a similar approach see [**Fe2**]).

14.4. Space-time structure. As explained in the number of examples in Sections 3, 4, and 6, various parts of operator theory and noncommutative differential geometry can be viewed as shadow images of appropriate hyperconstructions. Noncommutative differential geometry seems to find more and more relations and applications in string theory. So the physicist's opinion, "One of the great advantages of noncommutative geometry is precisely to put on the same ground the discrete and the continuous cases" [**I**, p. 245], is entirely compatible with the previous statement. In relation to Section 10 we mention an important map in string theory, called the Seiberg–Witten map [**SW**]. It relates a noncommutative gauge field with an ordinary gauge field in string theory. Observe that from the physicist point of view [**SW**, 6.4] focusing on the full algebra of operators acting on an open string leads naturally to finitely generated projective modules over a commutative torus, and the noncommutative torus appears when one "attempts to focus attention on just one end of the open string". Little seems to be known about analogues of this map in pure mathematics; those could be closely related to links between noncommutative descriptions and standard images of hypercommutative descriptions of the objects in the previous sections.

14.5. From the point of view of the previous material, one can refine Riemann's words in [**Ri**]: "in a discrete manifold the ground of its metric relations is given in the notion of it, while in a continuous manifold this ground must come from outside. Either therefore the reality which underlies space must form a discrete manifold, or we must seek the ground of its metric relations outside it, in binding forces which act upon it"—if the continuous manifold is actually given as an image of a hyperdiscrete manifold, then it has the natural ground of its metric induced from the natural hyperdiscrete one.

References

[AFHL] S. Albeverio, J. E. Fenstad, R. Hoegh–Krohn, T. Lindstom, *Nonstandard methods in stochastic analysis and mathematical physics,* Academic Press, Orlando, 1986.

[AK] J. Ax, S. Kochen, *Diophantine problems over local fields. I,* Amer. J. Math. **87** (1965), 605–630.

[B] C. Birkar, *Elements of nonstandard algebraic geometry I,* preprint, Nottingam, `www.maths.nott.ac.uk/personal/pmxcb1/nsag/`.

[BF] L. Brekke, P. G. O. Freund, *p-adic numbers in physics,* Physics Rep. **233** (1993), 1–66.

[BS1] L. Brünjes, C. Serpé, *Enlargement of categories,* preprint SFB478 Münster, 286, 2003.

[BS2] L. Brünjes, C. Serpé, *Nonstandard étale cohomology,* preprint SFB478 Münster, 288, 2003.

[Ca] J.-L. Callot, *Calcul infinitésimal complexe,* prepubl. Inst. Rech. Math. Av. 1995/13, Strasbourg, 1995.

[Cl] B. Clare, *Properties of the hyper Riemann zeta-function,* preprint, Nottingham, 2004.

[Co1] A. Connes, *Noncommutative geometry,* Academic Press, 1994.

[Co2] A. Connes, *Geometry from the spectral point of view,* Lett. Math. Physics **34** (1995), 203–238.

[Co3] A. Connes, *Trace formula in noncommutative geometry and the zeros of the Riemann zeta function,* Selecta Math. (N.S.) **5** (1999), 29–106.

[Co4] A. Connes, *Noncommutative geometry – year 2000,* GAFA 2000 (Tel Aviv, 1999), Geom. Funct. Anal. 2000, Special Vol., Part II, 481–559.

[CM] C. Consani, M. Marcolli, *Noncommutative geometry, dynamics, and ∞-adic Arakelov geometry,* preprint, 2002, `arXiv:math.AG/0205306`.

[Cu] J. Cuntz, *Quantum spaces and their noncommutative topology,* Notices AMS **48** (2001), 793–799.

[EF] P. C. Eklof, E. R. Fisher, *The elementary theory of abelian groups,* Ann. Math. Logic *4* (1972), 115–171.

[Fe1] I. Fesenko, *Analysis on arithmetic schemes. I,* Docum. Math., Extra Volume (Kato's 50th birthday) (2003), 261–284.

[Fe2] I. Fesenko, *Measure, integration and elements of harmonic analysis on generalized loop spaces,* preprint, 2003, `www.maths.nott.ac.uk/personal/ibf/aoh.ps`.

[Fr] A. Fruchard, *Complex analysis,* Nonstandard analysis in practice, Universitext, Springer, 1995, pp. 23–50.

[FJ] M. Fried, M. Jarden, *Field arithmetic,* Springer, Berlin, 1986.

[FS] M. Fried, C. Sacerdote, *Solving Diophantine problems over all residue class fields of a number field and all finite fields,* Ann. of Math. (2) **104** (1976), 203–233.

[FV] I.B. Fesenko, S.V. Vostokov, *Local fields and their extensions,* 2nd ed., AMS, 2002.

[Gl] R. Goldblatt, *Lectures on the hyperreals,* Springer, 1998.

[Gr1] E. Gordon, *Nonstandard analysis and locally compact abelian groups,* Acta Appl. Math. **25** (1991), 221–239.

[Gr2] E. Gordon, *Nonstandard methods in commutative harmonic analysis,* AMS. Transl. Math. Monogr. 164, 1997.

[H] Sh. Haran, *The mysteries of the real prime,* Clarendon Press, Oxford, 2001.

[I] B. Iochum, *The impact of NC geometry in particle physics,* Noncommutative geometry and the standard model of elementary particle physics (Hesselberg, 1999), Lecture Notes in Phys. 596, Springer, Berlin, 2002, pp. 244–259.

[JL] G.W. Johnson, M.L. Lapidus, *The Feynman integral and Feynman's operational calculus,* Oxford Univ. Press, 2000.

[K] K. Kato, *Lectures on the approach to Iwasawa theory for Hasse–Weil L-functions via B_{dR},* Lect. Notes Math. **1553** (1993), 50–163.

[L1] K. Loo, *Nonstandard Feynman path integral for the harmonic oscillator,* J. Math. Phys. **40** (1999) 5511–5521.

[L2] K. Loo, *A rigorous real-time Feynman path integral and propagator,* J. Phys. A **33** (2000), 9215–9239.

[Mc] A.J. Macintyre, *Nonstandard number theory,* Proc. ICM 1978, vol. 1, 253–262.

[MM1] Yu.I. Manin, M. Marcolli, *Continued fractions, modular symbols and noncommutative geometry,* Sel. Math. (N.S.) **8** (2002), 475–521.

[MM2] Yu.I. Manin, M. Marcolli, *Holography principle and arithmetic of algebraic curves,* Adv. Theor. Math. Phys. **5** (2002), 617–650.

[Mn1] Yu.I. Manin, *Three-dimensional hyperbolic geometry as ∞-adic Arakelov geometry,* Invent. Math. **104** (1991) 223–243.

[Mn2] Yu. I. Manin, *Mirror symmetry and quantization of abelian varieties,* Moduli of abelian varieties, C. Faber et al. (eds.), Progress in Math. 195, Birkäuser, 2001, pp. 231–254.

[Mn3] Yu.I. Manin, *Real multiplication and noncommutative geometry,* preprint, 2002, `arXiv: math.AG/0202109`.

[Mr] M. Marcolli, *Limiting modular symbols and the Lyapunov spectrum,* J. Number Theory **98** (2003) 348–376.

[N1] T. Nakamura, *Path space measures for Dirac and Schrödinger equations: nonstandard analytical approach,* J. Math. Phys. **38** (1997), 4052–4072.

[N2] T. Nakamura, *Path space measure for the $3+1$-dimensional Dirac equation in momentum space,* J. Math. Phys. **41** (2000), 5209–5222.

[NA] *Nonstandard analysis: theory and applications,* L.O. Arkeryd et al. (eds.), Kluwer, 1997.

[NAiA] *Nonstandard analysis and its applications,* N. Cutland (ed.), London Math. Soc. Student Texts 10, Cambridge Univ. Press, 1988.

[NAiP] *Nonstandard analysis in practice,* F. Diener and M. Diener (eds.), Springer, 1995.

[NAWM] *Nonstandard analysis for the working mathematician,* P.A. Loeb and M. Wolff (eds.), Mathematics and its Applications 510, Kluwer Academic Publishers, Dordrecht, 2000.

[O] R. Oeckl, *A "general boundary" formulation for quantum mechanics and quantum gravity,* preprint, 2003, `arXiv:hep-th/0306025`.

[Ri] B. Riemann, *Über die Hypothesen, welche der Geometrie zu Grunde liegen, Habilitationsschrift, 1854,* Abhandlungen der Königlichen Gesellschaft der Wissenschaften zu Göttingen **13** (1868), `www.emis.de/classics/Riemann/`.

[Rf1] M.A. Rieffel, *Morita equivalence operator algebras,* Proc. Symp. Pure Math. **38** (1982), Part 1, 285–298.

[Rf2] M.A. Rieffel, *Morita equivalence for C^*-algebras and W^*-algebras,* J. Pure Applied Algebra **5** (1974), 51–96.

[Rf3] M. A. Rieffel, *Applications of strong Morita equivalence to transformation group C^*-algebras,* Proc. Symp. Pure Math. **38** (1982), Part 1, 299–310.

[Ro1] A. Robinson, *Non-standard analysis,* reprint of the 2nd (1974) edition, with a foreword by W.A. Luxemburg, Princeton Landmarks in Math., Princeton Univ. Press, 1996.

[Ro2] A. Robinson, *A new approach to the theory of algebraic numbers,* Atti Accad. Naz. Lincei Rend. Cl. Sci. Fis. Mat. Natur. (8), **40** (1966), 222–225; II, Atti Accad. Naz. Lincei Rend. Cl. Sci. Fis. Mat. Natur. (8), **40** (1966), 770–774.

[Ro3] A. Robinson, *On some applications of model theory to algebra and analysis,* Rend. Mat. e Appl. (5) **25** (1966), 562–592.

[Ro4] A. Robinson, *Nonstandard arithmetic,* Bull. Amer. Math. Soc. **73** (1967), 818–843.

[Ro5] A. Robinson, *Non-standard theory of Dedekind rings,* Nederl. akad. Wetensch. Proc. Ser. A **29** (1967), 444–452.

[Ro6] A. Robinson, *Topics in nonstandard algebraic number theory,* Applications of model theory to algebra, analysis, and probability (Intern. Sympos. Pasadena Calif. 1967), Holt, Rinehart and Winston, New York, 1969, pp. 1–17.

[Ro7] A. Robinson, *Germs,* Applications of model theory to algebra, analysis, and probability (Intern. Sympos. Pasadena Calif. 1967), Holt, Rinehart and Winston, New York, 1969, pp. 138–149.

[Ro8] A. Robinson, *Compactification of groups and rings and nonstandard analysis,* J. Symbolic Logic **34** (1969), 576–588.

[Ro9] A. Robinson, *Enlarged sheaves,* Lect. Notes Math. 369 (1974), 249–260.

[Ro10] A. Robinson, *On generalized limits and linear functionals,* Pacific J. Math. **14** (1964), 269–283.

[RR] A. Robinson, P. Roquette, *On the finiteness theorem of Siegel and Mahler concerning Diophantine equations,* J. Number Theory **7** (1975), 121–176.

[So] Y. Soibelman, *Quantum tori, mirror symmetry and deformation theory,* Letters in Math. Physics **56** (2001), 99–125.

[Sr] S. S. Schweber, *QED and the men who made it: Dyson, Feynman, Schwinger, and Tomonaga,* Princeton Univ. Press, 1994.

[SV] Y. Soibelman, V. Vologodsky, *Noncommutative compactifications and elliptic curves,* preprint, 2002, `arXiv:math.AG/0205117`.

[SW] N. Seiberg, E. Witten, *String theory and noncommutative geometry,* J. High Energy Phys. 1999, no. 9, Paper 32, 93 pp. (electronic).

[Sz] A. Schwarz, *Theta-functions on noncommutative tori,* preprint, 2001, `arXiv:math.QA 0107186`.

[T] L. Taylor, *The double sine function and theta functions on the noncommutative torus,* preprint, Nottingham, 2005.

[Y] H. Yamashita, *Nonstandard methods in quantum field theory I: a hyperfinite formalism of scalar fields,* Intern. J. Th. Phys. **41** (2002), 511–527.

[YO] H. Yamashita, M. Ozawa, *Nonstandard representations of the canonical commutation relations,* Rev. Math. Phys. **12** (2000), 1407–1427.

[Z1] B. Zilber, reprints and preprints, `www.maths.oxford.ac.uk/~zilber`.

[Z2] B. Zilber, *Pseudo-analytic structures, quantum tori and non-commutative geometry,* preprint, 2003, `www.maths.oxford.ac.uk/~zilber`.

DEPARTMENT OF MATHEMATICS, UNIVERSITY OF NOTTINGHAM, NOTTINGHAM NG7 2RD, UNITED KINGDOM

E-mail address: `ibf@maths.nott.ac.uk`

Amer. Math. Soc. Transl.
(2) Vol. **217**, 2006

The Interface Between Probability Theory and Additive Number Theory (Local Limit Theorems and Structure Theory of Set Addition)

Gregory A. Freiman and Alexander A. Yudin

ABSTRACT. We develop a new method for studying distributions of sums of independent random variables. This method is based on results obtained from inverse additive number theory. The notion of an "isomorphism" of random variables with supports in spaces of different dimensions is used, paving the way to new types of local limit theorems.

1. A short review of results in the field of local limit theorems

We list here several of the main local limit theorems [LLT] emphasizing those whose joint features spurred on the new approach in this paper.

THEOREM 1.1 (B. Gnedenko [**Gn**]). *Let ξ be a random variable (r.v.) with $\mathbb{P}(\xi \in \mathbb{Z}) = 1$ and $\xi_1, \xi_2, \ldots, \xi_n$ be independent random variables distributed as ξ. Let $S_n = \sum_{i=1}^n \xi_i$, $E\xi = a$, and $E(\xi - E\xi)^2 = \sigma^2$. Let the maximal step of ξ equal 1, i.e., if $q, r \in \mathbb{N}$ and $\mathbb{P}(\xi \in q\mathbb{Z} + r) = 1$, then $q = 1$.*
Uniformly on k, we have

$$(1.1) \qquad \sigma\sqrt{n}\mathbb{P}(S_n = k) - \frac{1}{\sqrt{2\pi}} \exp\left(-\frac{(k - na)^2}{2\sigma^2 n}\right) \to 0, \qquad \text{if } n \to \infty .$$

A general formulation of the limit problem may be found in M. Loeve [**L**]: There is a series of sums of independent random variables

$$S_n = \sum_{i=1}^n X_i .$$

Find conditions for which

$$(1.2) \qquad \mathcal{L}\left(\frac{S_n - ES_n}{\sigma S_n}\right) \to N(0,1) ,$$

where

$$N(0,1) = \frac{1}{\sqrt{2\pi}} e^{-x^2/2}.$$

2000 *Mathematics Subject Classification.* Primary 11P99; Secondary 60F99.
Key words and phrases. Structure theory of set addition, local limit theorems.
This work was supported by RFFI grant 05-01-00617.

We can treat (1.2) in the following way: transform the random variable ξ, centralizing it, $\xi \to \xi - a$, and then use the affine transformation in $\mathbb{R}^2$,

$$x' = \frac{x}{\xi\sqrt{n}} , \quad y' = \sigma\sqrt{n}y ,$$

with corresponding transformation of the random variable.

In a further generalization, the random values

$$\frac{S_n}{b_n} - a_n$$

were studied, and only affine transformations of the random variable S_n were used.

A condition of an arithmetical nature for which Gnedenko's theorem is valid is when the maximal step equals 1.

A natural generalization of Gnedenko's theorem to the case of differently distributed summands is given in [**Pr**].

THEOREM 1.2. (Y. Prokhorov). *Let*

$$\xi_1, \xi_2, \dots, \xi_n, \dots \tag{1.3}$$

be a series of independent random variables such that

$$\begin{aligned} &1. \quad \forall_i \mathbb{P}(\xi_i \in \mathbb{Z}) = 1 , \\ &2. \quad \forall_i |\xi_i| \le C , \end{aligned} \tag{1.4}$$

where C is an absolute constant.

For LLTs for the sequence (1.3) to hold, it is necessary and sufficient for the set

$$A = \left\{ k \in \mathbb{Z} : \sum_{i=1}^{\infty} \mathbb{P}(\xi_i = k) = \infty \right\}$$

to have the maximal step equal to 1.

In Y. Rozanov's paper [**Ro**], condition (1.4) of uniform boundedness is weakened.

In [**MFY**] the following is proved:

THEOREM 1.3. (G.A. Freiman, D.A. Moskvin, A.A. Yudin) *Let*

$$\xi_{1n}, \xi_{2n}, \dots, \xi_{nn}, \quad n = 1, 2, \dots, \tag{1.5}$$

be a triangular array of independent random variables distributed as ξ_{1n} for every given n, $\mathbb{P}(\xi_{1n} \in \mathbb{Z}) = 1$, $E\xi_{1n} = a_n$, $E(\xi_{1n} - a_n)^2 = \sigma_n^2$.

Suppose that

1. *for (1.5) the central limit theorem is valid;*
2. *$\sigma_n^2 = O(n^\rho)$, where*

$$\rho < \frac{\ln(2+c)}{\ln 2} - 1 ,$$

$$c < 1;$$

3. *for $q = 2, 3, \dots$ and ω positive and sufficiently large*

$$\max_{1 \le r \le q} \mathbb{P}(\xi_{1n} \equiv r (\operatorname{mod} q)) < 1 - \omega \max\left(\frac{\rho_n^2}{n\sigma_n^4}, \frac{1}{n^{1-\mu}} \right) \ln n ,$$

where

$$\rho_n = E|\xi_{1n} - a_n|^3 \ ,$$

$$\mu = \frac{(1+\rho)\ln 2}{\ln(2+c)} \ .$$

Then the LLT for (1.5) *holds.*

2. Analysis of results of §1 and directions for future study

The comparison and analysis of conditions formulated in [**Pr, Ro, MFY, MS1, MS2**] (we do not give the formulations of theorems from [**MS1**] and [**MS2**] because they are cumbersome) lead us to the following conclusions.

1. In each theorem there are conditions describing "compactness of a random variable". In [**Pr**] the random variables have to be uniformly bounded. In [**Ro, MFY, MS1, MS2**] there are conditions on the rate of growth of variance, of the third moment, and in [**Gn**], only the condition of existence of variance is needed.

2. Conditions characterizing the arithmetical properties of the support of a random variable are always present. For example, values taken by random variable ξ_{1n} have to be well distributed between classes mod q for every q and the values

$$\max_r \mathbb{P}(\xi_{1n} \in \{n : n \equiv r (\bmod q)\})$$

must be separated from 1 as in [**Gn, Pr**].

Different forms of arithmetic conditions are discussed in more detail in [**MFY, Mu**].

The aim of this work is to show how the additive structure of the support of a random variable ξ in some cases defines the behavior of

$$\mathbb{P}(S_n = a) \ .$$

Let G be a group where the group operation is denoted by $+$. For $A \subset G$ and $B \subset G$, we define

$$A + B = \{x : x = a + b, \ a \in A, b \in B\} \ ,$$

$$sA = (s-1)A + A \ .$$

The condition of "small doubling" of the support A of a random variable ξ will play the main role here. Namely, we suppose that

$$|2A| = |A + A| < C|A| \ , \tag{2.1}$$

where C is some positive constant.

We will show that the condition of small doubling (2.1) gives a new form of the condition of compactness and will provide a way of obtaining a new type of LLT, valid for distributions for which the usual conditions on random variables from [**Gn, L, Pr, Ro, MFY, MS1, MS2, Mu**] may not be true (for example, the variance may be missing), and LLT in its usual form may not take place at all.

It appears that, if $\operatorname{supp}\xi$ is a set with small doubling, we can build a map of ξ on a random variable ξ' with values in $\mathbb{Z}^s$ for suitable s (see §4). This map, which preserves additive properties of $\operatorname{supp}\xi$, may be constructed in such a way that, for $\operatorname{supp}\xi$, conditions of compactness and of arithmetic types will be fulfilled. Applying LLT to S'_n and knowing the values $\mathbb{P}(S'_n = a')$, we can now find the values $\mathbb{P}(S_n = a)$.

For the history, bibliography and results for the problem of structure of sets with a small doubling, the reader is referred to [**Fr4, Fr2**], [**B**], [**N**] and [**Go**]. See also the website `www.maths.cam.ac.uk` and [**Fr1, Fr3**].

In §§3 and 7, we formulate results for random variables with finite support. Let us mention that this condition does not limit generality. If $\operatorname{supp}\xi$ is infinite, we can always approximate ξ by a random variable ξ' with finite support.

We conclude this section with a discussion of the methods of the proofs of LLT, usually the method of characteristic functions. Let

$$f_\xi(t) = \sum_{k\in\mathbb{Z}} P(\xi = k)e^{2\pi ikt}$$

be a characteristic function of the random variable ξ. To obtain the LLT, we have only to estimate the asymptotics of $f_\xi^n(t)$ in some neighborhood of $t = 0$ and estimates $|f_\xi(t)|^n$ from above for points t which are not in this neighborhood. In [**Gn, Pr, Ro, MS1, MS2, Mu**] these estimates were obtained explicitly with the use of the usual characteristics of a random variable ξ (variance, moments). Only in [**MZ, AZ**] were the conditions and methods proposed, which enabled the study of the behavior of characteristic functions $f_\xi(t)$ in a more subtle way; and in [**MZ**], the behavior of the resolvent of a random walk was discussed. Estimates of precision for asymptotics of LLT may be found in [**Mu, Ga, St, Sh**]. In [**Ga**] it was shown that the precision of LLTs depends on the structure of the set of those t for which $|f_\xi(t)|$ is close to 1. In [**DFY1, DFY2**] it was shown how the distribution of values of a positively definite function (the characteristic function is, evidently, such a function) is defined by the additive arithmetic structure of its support, $\operatorname{supp}\xi$.

3. An example

The discussion of §2 will now be illustrated by a simple example.

Let us consider a triangular array of independent random variables

$$\xi_{1n}, \xi_{2n}, \ldots, \xi_{nn}\ , \tag{3.1}$$

where

$$K_n = \operatorname{supp}\xi_{in} = \{0, 1, 2, 2n+2, 2n+3, 4n+4\}\ .$$

and $\forall_i, 1 \le i \le n$ and $\forall a,\ a \in K_n$ we have

$$\mathbb{P}(\xi_{in} = a \in K_n) = \frac{1}{6}\ .$$

Let us show that the LLT in the usual form (1.1) does not hold. Denoting $\xi_{1n} = \xi$, we have,

$$E\xi = \frac{4}{3}n + O(1),$$
$$E(\xi - E\xi)^2 = b^2 \sim cn^2\ .$$

If for $S_n = \sum_{i=1}^{n} \xi_{in}$ LLT is valid, then

$$\sqrt{cn^3}\,\mathbb{P}(S_n = k) = \frac{1}{\sqrt{2\pi}} \exp\Big(- \frac{(k - \frac{4}{3}n^2)^2}{2cn^3}\Big) + o(1)$$

and for all k such that

$$\left|k - \frac{4}{3}n^2\right| < 2n$$

or for

$$\frac{4}{3}n^2 - 2n \le k \le \frac{4}{3}n^2 + 2n \tag{3.2}$$

we will obtain

$$\sqrt{cn^3}\mathbb{P}(S_n = k) = \frac{1}{\sqrt{2\pi}} + o(1) \ . \tag{3.3}$$

Let us study now the numbers x from S_n. They have the form

$$x = x_1 + x_2 + \cdots + x_n \tag{3.4}$$

where $\forall_j x_j \in \{0, 1, 2, 2n+2, 2n+3, 4n+4\}$ and therefore

$$x_j = c_j + (2n+2)d_j \tag{3.5}$$

and

$$x_j \equiv c_j \pmod{2n+2} \ , \tag{3.6}$$

where c_j may be equal to only one of three values $0, 1$ or 2.

We see that in the case

$$x \equiv c \pmod{2n+2}$$

where c is a residue from the system of nonnegative minimal residues, we have

$$0 \le c \le 2n$$

from (3.4) and (3.6), and $2n + 1 + (2n+2)t \notin S_n$ for any integer t. Therefore,

$$\mathbb{P}(S_n = 2n + 1 + (2n+2)t) = 0 \ .$$

Inequality (3.2) defines an interval containing more than $2n+2$ points. So in this interval we can find a number $k^* \equiv 2n+1 (\mathrm{mod}\ 2n+2)$ for which $\mathbb{P}(S_n = k^*) = 0$, and (3.3) gives us

$$0 = \frac{1}{\sqrt{2\pi}} + o(1) \ ,$$

a contradiction.

We will now show that the LLT in the sense of the discussion in §2 is nevertheless valid. For this, we relate the random variable ξ_{1n} with random variable η with values in $\mathbb{Z}^2$.

Define η in the following way:

$$K' = \operatorname{supp} \eta = \{(0,0), (1,0), (2,0), (0,1), (1,1), (0,2)\}$$

and $\forall a' \in K'$

$$\mathbb{P}(\eta = a') = \frac{1}{6} \ .$$

The map $\varphi : K_n \to K'$;

$$\varphi(0) = (0,0),\ \varphi(1) = (1,0),\ \varphi(2) = (2,0),\ \varphi(2n+2) = (0,1) \ ,$$

$$\varphi(2n+3) = (1,1),\ \varphi(4n+4) = (0,2)$$

(see Figure 1) is a bijection which, because of (3.6), may be written as

$$\varphi(c + (2n+2)d) = (c, d) \ .$$

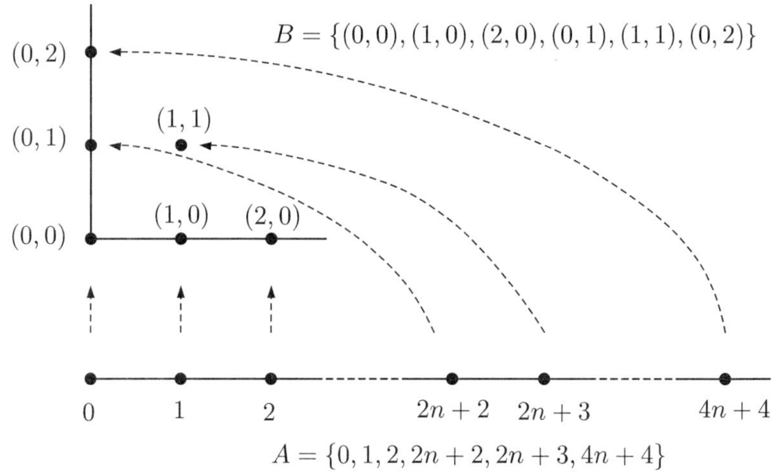

FIGURE 1

For a sequence of independent random variables

$$\eta_1, \eta_2, \ldots, \eta_n, \ldots \tag{3.7}$$

where $\forall_i \eta_i$ is distributed as η, the multidimensional LLT is valid (see [**MPR2, MPR1**]).

These papers showed that the necessary and sufficient condition for validity of LLT for the sequence (3.7) is the following: let a be any fixed element of $\operatorname{supp}\xi$. Then $\operatorname{supp}\xi - a$ generates $\mathbb{Z}^2$. In our example this fact is evident.

Therefore, we will have (see Figure 2)

$$\begin{aligned}&\mathbb{P}(S_n' = \eta_1 + \cdots + \eta_n = h \in \mathbb{Z}^2)\\ &= \frac{1}{2\pi n\sqrt{\Delta}}\left(\exp\left(-\frac{1}{2}Q\left(\frac{h-na}{\sqrt{n}}\right)\right) + o(1)\right) ,\end{aligned} \tag{3.8}$$

where $a = E_\eta$, Q is the quadratic form with the matrix inverse to the covariance matrix of the random variable η, and Δ is the determinant of this covariation matrix.

What kind of progress have we made by introducing the series (3.7)? First, instead of the triangular array (3.1), we obtained a simpler kind of series of identically distributed random variables. Of course, this is not the most general situation, but it may occur often. Second, these random variables appeared to be "good", i.e., the compactness condition and arithmetic conditions were fulfilled automatically, which gave us the possibility of obtaining the value of $\mathbb{P}(S_n')$ in (3.8). Last, we will now show how formula (3.8) will enable us to find the distribution for the array ξ.

Let us build a bijection

$$f : nK_n \to nK' \ .$$

For the element $x \in nK_n$, we have

$$x = x_1 + x_2 + \cdots + x_n, \tag{3.9}$$

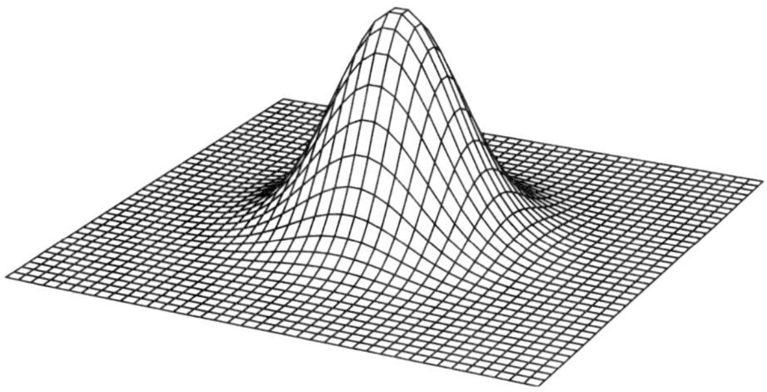

FIGURE 2

where

$$\forall_i x_i \in K_n \ ,$$

and (3.9) is one of the possible representations of the element x.

In view of (3.6), x_i may be presented as

$$x_j = h_{1j} + h_{2j}(2n+2) \ .$$

Define

$$\begin{aligned} f(x) &= \varphi(x_1) + \varphi(x_2) + \cdots + \varphi(x_n) \\ &= \sum_{j=1}^{n} \varphi(h_{1j} + h_{2j}(2n+2)) \\ &= \sum_{j=1}^{n} (h_{1j}, h_{2j}) = (h_1, h_2) \ . \end{aligned}$$

We still have to prove that the map f is defined correctly. Let some other representation of x be

$$x = y_1 + y_2 + \cdots + y_n \ ,$$

where

$$\forall_i \ y_i \in K_n \ .$$

Then

$$y_j = h'_{1j} + h'_{2j}(2n+2) \ ,$$

$$x = \sum_{i=1}^{n} h'_{1j} + (2n+2) \sum_{j=1}^{n} h'_{2j} = h'_1 + (2n+2)h'_2$$

and

$$f(x) = (h'_1, h'_2) \ .$$

We have

$$x = h_1 + (2n+2)h_2 = h'_1 + (2n+2)h'_2.$$

From $h_1 = \sum\limits_{j=1}^{n} h_{1j}$ and $0 \le h_{ij} \le 2$ we get

$$0 \le h_1 \le 2n$$

and, in the same way, we have

$$0 \leq h_1' \leq 2n.$$

It follows that

$$h_1 \equiv h_1' \pmod{2n+2} ,$$
$$h_1 = h_1'$$

and

$$h_2 = h_2' .$$

Let us show that f is a surjection. Take some element from nK':

$$(h_1, h_2) \in K'.$$

Then (3.9) gives

$$(h_1, h_2) = \sum_{j-1}^{n} (h_{1j}, h_{2j}) ,$$

where

$$\forall_j (h_{1j}, h_{2j}) \in K'.$$

We see that for

$$x = x_1 + x_2 + \cdots + x_n ,$$

where

$$x_j = h_{1j} + h_{2j}(2n+2),$$

we have

$$x = \sum_{j=1}^{n} h_{1j} + (2n+2) \sum_{j=1}^{n} h_{2j} = h_1 + (2n+2)h_2$$

and

$$f(x) = (h_1, h_2) .$$

We have shown that $f(x)$ is a surjection.

Let us show that f is an injection. Suppose that

$$f(x) = f(y) = (h_1, h_2) . \tag{3.10}$$

We have

$$x = h_1 + (2n+2)h_2, \quad y = h_1' + (2n+2)h_2$$

and therefore

$$f(x) = (h_1, h_2), \quad f(y) = (h_1', h_2'),$$

whence (3.10) gives

$$h_1 = h_1', \quad h_2 = h_2'$$

and

$$x = y,$$

so that f is an injection.

Now we prove that

$$\mathbb{P}(S_n = x) = \mathbb{P}(S_n' = (h_1, h_2)) ,$$

where

$$f(x) = (h_1, h_2) .$$

Take one of the possible representations of x:

$$x = x_1 + x_2 + \cdots + x_n , \tag{3.11}$$

where

$$x_j \in K_n$$

and

$$x_j = h_{1j} + (2n+2)h_{2j} \ .$$

The event $\xi_j = x_j$ has the probability of $1/6$ and the probability of event x under condition (3.9) is $1/6^n$. If Q is the number of representations (3.11), then

$$\mathbb{P}(S_n = x) = \frac{Q}{6^n}.$$

The value x_j is defined by the value of (h_{1j}, h_{2j}). Then

$$f(x) = (h_1, h_2) = (\Sigma(h_{1j}, h_{2j})) \ ,$$

$$\mathbb{P}\left(\eta_j = (h_{1j}, h_{2j})\right) = \frac{1}{6} \ ,$$

and $\mathbb{P}(S = (h_1, h_2)) = \mathbb{P}\left(S = \sum_{j=1}^{n} (h_{1j}, h_{2j})\right) = 1/6^n$ if (h_{1j}, h_{2j}) is given. The number of all n possible pairs (h_{1j}, h_{2j}) is equal to Q and

$$\mathbb{P}(S = (h_1, h_2)) = \frac{Q}{6^n} \ ,$$

which means

$$P(S_n = x) = \mathbb{P}(S = (h_1, h_2)) \ . \tag{3.12}$$

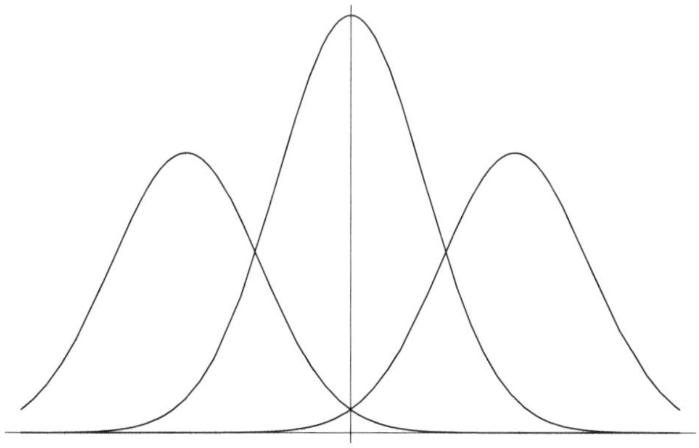

FIGURE 3

As a result (see Figure 3), we obtain formula (3.12) for finding the value of $P(S_n = a)$ from the value of $P(S'_n = h)$. For this we do not need to know the variance of the random variable ξ_{1n} (of the order n^2). It is sufficient to know that the mean and variance of the random variables η_i are uniformly bounded; they have all the moments and values of $\mathbb{P}(S'_u = h)$ that could be found with the required precision.

If the distribution of the random variable S_n were presented graphically on $\mathbb{Z}$, it would be difficult to see any regularity. At the same time, the distribution of

S'_n on $\mathbb{Z}^2$, even for rather small ($n \approx 10$) values of n, is very close to the normal distribution.

We have discussed a rather simple example: the support $K_n = \operatorname{supp} \xi_{1n}$ has a small cardinality; the distribution of ξ_{1n} is uniform.

To obtain results for wider classes of random variables, we need to introduce notions from structure theory of set addition (see the review [**Fr4**] and Nathanson's book [**N**]).

4. Isomorphism of random variables

DEFINITION 4.1. Let $A \subset G_1$ and $B \subset G_2$ and in each of sets G_1 and G_2 an algebraic operation is defined. Sets A and B will be called isomorphic if

1. there exists a bijection $\varphi : A \to B$;
2. bijection φ induces the bijection
$$\chi : 2A \to 2B$$
in the following sense: if $b \in 2A$ and $b = a_1 + a_2$, $a_1, a_2 \in A$, then
$$\chi(b) = \varphi(a_1) + \varphi(a_2) \ .$$

We can formulate the second point in another form:
$$\forall x_1, x_2, x_3, x_4 \in A$$
$$x_1 + x_2 = x_3 + x_4 \underset{\chi^{-1}}{\overset{\chi}{\rightleftarrows}} \varphi(x_1) + \varphi(x_2) = \varphi(x_3) + \varphi(x_4) \ .$$

EXAMPLE 4.2. Let $A = \{0, 1, 3\}$, $B = \{0, 1, 5\}$. Then A is isomorphic to B. We will denote this by $A \sim B$.

EXAMPLE 4.3. $A = \{0, 1, 3\}$ and $B = \{(0, 0), (1, 0), (0, 1)\}$ are isomorphic. $\varphi(0) = (0, 0)$, $\varphi(1) = (1, 0)$, $\varphi(3) = (0, 1)$.

EXAMPLE 4.4. Here we will give an example of two sets which are not isomorphic: $A = \{0, 1, 2\}$, $B = \{(0, 0), (1, 0), (0, 1)\}$. If these two sets were isomorphic, then the sets 2A and 2B would have had the same cardinality, but this is not the case.

In Definition 4.1 we describe the notion of isomorphism for the case when only two sets are added. This is a partial case ($n = 2$) of the following definition.

DEFINITION 4.5. Subsets $A \subset G$ and $B \subset G_2$ will be called isomorphic of the n-th order if

1. there exists a bijection $\varphi : A \to B$,
2. there exists the map $\chi : nA \to nB$ which is induced by φ and which is a bijection.

EXAMPLE 4.6. The sets $A = \{0, 1, n+1\}$ and $B = \{0, 1, n+2\}$ give an example of sets which are isomorphic of the n-th order but are not isomorphic of the $(n+1)$-th order.

Now let us introduce a definition for the isomorphism of n-th order for random variables.

Let ξ_1 and ξ_2 be random variables with values in $\mathbb{Z}^{s_1}$ and $\mathbb{Z}^{s_2}$, respectively,
$$K_j = \operatorname{supp} \xi_j = \{a \in \mathbb{Z}^{S_j},\ \mathbb{P}(\xi_j = a) \neq 0\} \ , \quad j = 1, 2 \ .$$

DEFINITION 4.7. Random variables ξ_1 and ξ_2 are called s-isomorphic if the sets K_1 and K_2 are s-isomorphic, and for the bijection $\varphi : K_1 \to K_2$ we have for $\forall a \in K_1$

$$\mathbb{P}(\xi_1 = a) = \mathbb{P}(\xi_2 = \varphi(a)) .$$

EXAMPLE 4.8. Let ξ_1 be a random variable uniformly distributed on its support $K_1 = \{-h, -1, 0, 1, h\}$, and for a random variable ξ_2, $\operatorname{supp} \xi_2 = K_2 = \{(0,0), (0,1), (1,0), (-1,0), (0,-1)\}$ also with probabilities $1/5$ at each point.

The bijection

$$\varphi : 0 \to (0,1),\ 1 \to (1,0),\ -1 \to (-1,0),\ h \to (0,1),\ -h \to (0,-1)$$

is an $(h-1)$-isomorphism.

At the same time K_1 and K_2 are not h-isomorphic, and therefore ξ_1 and ξ_2 are not isomorphic either.

The random variable ξ_1 can be presented in a "good" way if we study its s-fold convolutions only up to $S \leq h-1$.

Now we shall discuss the notion of dimension of a set K and of a random variable ξ.

DEFINITION 4.9. Let φ be an isomorphism $A \to \mathbb{Z}^m$ such that there exists no proper subgroup $G \subset \mathbb{Z}^m$ or $G + x$ such that $\varphi(A) \subset G + x$. The maximal number m with this property will be called the "dimension" of the set A with respect to the s-isomorphism φ. We will denote it by $\dim_s A$.

DEFINITION 4.10. The dimension of the support ξ will be called the dimension of ξ, denoted by $\dim_s \xi$.

Some properties of $\dim_s \xi$ are listed below:

1. Dimension $\dim_s \xi$ is invariant under the action on ξ of the group of nonsingular linear transformation g, i.e.,

   $$\dim_s \xi = \dim(g\xi + b) ,$$

   where g is a nonsingular linear transformation, where b is in the domain of values of ξ.
2. $s_1 \geq s_2 \Rightarrow \dim_{s_1} A \leq \dim_{s_2} A$.
3. Let $A \subset B$. Is it true that

   $$\dim_s A \leq \dim_s B?$$

   The answer is negative, for example,

   $$A = \{1, 2, 2^2, 2^3, 2^4, 2^5\},$$

   $$B = \{0, 1, 2, 3, \ldots\}$$

   We have $A \subset B$, $\dim_2 B = 1$, $\dim_2 A = 5$.
4. Is it true that, for $A_1 \subset \mathbb{Z}^m$ and $A_2 \subset \mathbb{Z}^m$ which are s-isomorphic, there exists a nonsingular affine map $g : \mathbb{Z}^m \to \mathbb{Z}^m$, such that $g(A_1) = g(A_2)$?

   The answer is negative. The sets $A = \{(0,0), (2,0), (0,1), (1,1)\}$ and $B = \{(0,0), (3,0), (4,0), (0,1)\}$ are 2-isomorphic but not affine equivalent.

PROBLEM. Find an estimate of

$$\dim_s(\{1, 2^k, 3^k, \ldots, n^k\})$$

as a function of s, k and n.

An important step in the study of the s-isomorphism of discrete random variables was made in [**DFM**].

5. A new type of local limit theorem

The final result of Gnedenko's theorem is formula (1.1). Formula (3.8) is an additional example, when for the series of random variables (3.7), $\eta_1, \eta_2, \ldots, \eta_n$, the formula for $P(S'_n)$ was obtained, and it may be used to compute this probability. Other forms of the formula for $P(S_n)$ are possible. They may depend on the structure of the random variables that are being added, on their dimension, and on the precision needed. In all these cases it may be stated that an LLT takes place in an "explicit" form. Now the LLT for series of random variables can be formulated for cases when, in general, the explicit form of the LLT is not applicable.

THEOREM 5.1. *Let us suppose that the triangular array* ξ

$$\begin{array}{llll} \xi_{11} & & & \\ \xi_{21} & \xi_{22} & & \\ & \cdots & & \\ \xi_{n1} & \xi_{n2} & \cdots & \xi_{nn} \end{array}$$

is isomorphic to the triangular array η

$$\begin{array}{llll} \eta_{11} & & & \\ \eta_{21} & \eta_{22} & & \\ & \cdots & & \\ \eta_{n1} & \eta_{n2} & \cdots & \eta_{nn} \end{array}$$

and the LLT for η *takes place in an explicit form. Then for* ξ *we obtain the LLT induced by the LLT for* η. *The conditions ensuring the description of distributions for the scheme* ξ *are the conditions ensuring that the LLT for the scheme* η *is true. These conditions may already have the usual character (existence of variance, conditions on the order of its growth, or the order of growth of the determinant of the covariation matrix of r.v.* η *and so on).*

The problem of finding the series η for which a good LLT exists (if at all) in a family of isomorphic series may not be simple.

In the example from §3, the very structure of $\operatorname{supp}\xi$ helped to find the map of scheme ξ to the plane $\mathbb{Z}^2$.

Look at two more examples.

Let ξ be a scheme of series where ξ_{n1} have uniform distribution on the set

$$\{0, 1, 2, 2n+1, 2n+2, 2n+3, 4n+2, 4n+3\}\ .$$

Let ρ be a scheme of series, where ρ_{n+1} is uniformly distributed on the set

$$\{0, 2n, 3n+1, 4n, 6n+1, 6n+2, 7n+1, 8n+8\}.$$

To find a two-dimensional scheme η isomorphic to ξ is not a difficult task, as $\operatorname{supp}\xi_{1n}$ is very similar to $\operatorname{supp}\xi_{1n}$ in the scheme of §3.

The situation is quite different for the scheme ρ.

The schemes ρ and ξ are isomorphic (an exercise for the reader). However, starting from the scheme ρ, it may not be simple to find an isomorphic scheme on the plane.

We have shown how to use the theorem in only a few cases.

Naturally, we can ask how wide the domain of applications of this theorem is. How large is the number of classes of the isomorphic schemes ξ for which an explicit form of LLT is applicable?

In §7 we will try to develop the first approach to this problem. To insist that two different schemes be isomorphic up to the n-th order is a rather strong condition to impose.

In §6 we develop the notion of isomorphism of subsets, (s, δ, ϵ)-isomorphism, with the help of which we can strongly lessen the order of the isomorphism in question.

6. (s, δ, ϵ)–isomorphism of random variables

DEFINITION 6.1. Let P and Q be two probabilistic measures with finite supports A and B, respectively. Let $s \in \mathbb{Z}_+$, and let δ and ϵ be two nonnegative real numbers. We shall say that the measures P and Q are (s, δ, ϵ)-isomorphic if

1. there exists a bijection
$$\varphi : A \to B ;$$
2. for two subsets
$$A^{(s)} \subseteq sA, \quad B^{(s)} \subseteq sB ,$$
there exists a bijection
$$\varphi^{(s)} : A^{(s)} \to B^{(s)} ;$$
3. (i)
$$P^{*s}(A^{(s)}) \geq 1 - \delta ,$$
$$Q^{*s}(B^{(s)}) \geq 1 - \delta ;$$
(ii) for $x_1, \dots, x_s, y_1, \dots, y_s \in \mathcal{A}$ we have
$$x_1 + x_2 + \cdots + x_s = y_1 + y_2 + \cdots + y_s \iff \varphi(x_1) + \cdots + \varphi(x_s) = \varphi(y_1) + \cdots + \varphi(y_s);$$
(iii) $\forall x \in A^{(s)} |P^{*s}(x) - Q^{*s}(\varphi^{(s)}(x))| \leq \epsilon.$

This notion has already been used in [**DFM**]. We will now return to the Example 4.8 and show how the results of this section can be generalized.

Examine the triangular array
$$\xi_{1s}, \xi_{2s}, \dots, \xi_{ss}, \quad s = 1, 2, \dots$$
where
$$\mathcal{A} = \operatorname{supp} \xi_{1s} = \{-h(s), -1, 0, 1, h(s)\} , \quad \forall h(s) \in \mathbb{N}$$
and p_0, p_1, p_2 are positive numbers, for which
$$p_2 = P(\xi_{1s} = h(s)) = P(\xi_{1s} = -h(s)) ,$$
$$p_1 = P(\xi_{1s} = 1) = P(\xi_{1s} = -1) ,$$
$$p_0 = 1 - p_1 - p_2 .$$

Let η be a two-dimensional random variable and

$$B = \operatorname{supp} \eta = \{(-1,0),(0,-1),(0,0),(1,0),(0,1)\} \ ,$$
$$p_1 = P(\eta = (1,0)) = P(\eta = (-1,0)) \ ,$$
$$p_2 = P(\eta = (0,1)) = P(\eta = (0,-1)) \ .$$

We will study the following cases:

Case 1.

$$s \leq h(s) - 1 \ . \tag{6.1}$$

Define the map $\varphi^{(1)} : B \to A$: $\varphi^{(1)}((0,-1)) = -h(s)$, $\varphi^{(1)}((-1,0)) = -1$, $\varphi^{(1)}((0,0)) = 0$, $\varphi^{(1)}((1,0)) = 1$, $\varphi^{(1)}((0,1)) = h(s)$.

In view of (6.1) the map $\varphi^{(1)}$ induces a one-to-one map

$$\varphi^{(s)} : sB \to sA \ .$$

The map $\varphi^{(s)}$ is induced also on sB by a map φ of $\mathbb{Z}^2$ with the basis $e_1 = (1,0)$ and $e_2 = (0,1)$ on the line $\mathbb{Z} \subset \mathbb{Z}^2$ ($\mathbb{Z} = \{ke_1, k \in \mathbb{Z}$) which is a projection parallel to the vector $(h(s), -1)$.

For the probabilistic distributions we have

$$\forall x \in B^{(s)} \quad P^{*s}(\varphi^s(x)) = Q^{*s} \ .$$

In this first case, distributions P and Q are $(S,0,0)$-isomorphic.

Case 2.

$$h(s) \leq s < \alpha h^2(s) \ ,$$

where α is a sufficiently small positive number.

In this case, the map $\varphi^{(s)} : sB \to sA$ is no longer a bijection, and we can illustrate here the usefulness of the notion of (s,ϵ,δ)-isomorphism to its full extent.

We have to define the sets $B^{(s)}$ and $\mathcal{A}^{(s)}$ from condition 2 of Definition 6.1.

The full preimage G_m of the number $m \in \mathbb{Z}$ under the map φ is

$$G_m = \{(m,0) + \ell(h(s),-1), \ell \in \mathbb{Z}\}. \tag{6.2}$$

For each one of these m which will be in $\mathcal{A}^{(s)}$ we will choose only one point in G_m, in this way building a one-to-one map $\varphi^{(s)}(B^{(s)}) = \mathcal{A}^{(s)}$. This choice will be realized in the following way.

For the random variable η the LLT is valid and

$$\begin{aligned} P(\eta_1 + \cdots + \eta_s = (m_1,m_2)) &= P(H_s = (m_1,m_2)) \\ &= \frac{1}{8\pi p_1 p_2 s} \exp\left(-\frac{1}{2}\left(\frac{m_1^2}{2p_1 s} + \frac{m_2^2}{2p_2 s}\right)\right) + R \ , \end{aligned}$$

where the error R is small and we will disregard it.

In view of

$$P(\xi_{1s} + \cdots + \xi_{ss} = m) \doteq P(M_s = m) = P(H_s \in G_m)$$

we get

$$\begin{aligned} P(M_s = m) &= \sum_{(m_1,m_2)\in G_m} P(H_s = (m_1,m_2)) \\ &= \frac{1}{8\pi p_1 p_2 s} \sum_{(m_1,m_2)\in G_m} P(H_s = (m_1,m_2)) \ . \end{aligned}$$

Now let us find the error if we omit all the summands in the last sum for which

$$\frac{m_1^2}{2p_1 s} + \frac{m_2^2}{2p_2 s} > \lambda \,. \tag{6.3}$$

Easy computation shows that

$$\frac{1}{8\pi p_1 p_2 s} \sum_{\frac{m_2^2}{2p_1 s} + \frac{m_2^2}{2p_2 s} > \lambda} \exp\left(-\frac{1}{2}\left(\frac{m_1^2}{2p_1 s} + \frac{m_2^2}{2p_2 s}\right)\right) \le \frac{1}{\sqrt{p_1 p_2}} e^{-\frac{\lambda}{2}} \,. \tag{6.4}$$

Now we choose λ in such a way that the set G_m and the set of points for which

$$\frac{m_1^2}{2p_1} + \frac{m_2^2}{2p_2} \le \lambda \tag{6.5}$$

intersect at no more than one point.

This will be the case if we choose

$$\lambda = \frac{1}{\alpha p_1 p_2}. \tag{6.6}$$

We can now show that the distributions P and Q are (s, δ, ϵ)-isomorphic for

$$\delta = \frac{1}{\sqrt{p_1 p_2}} e^{-\frac{1}{\alpha p_1 p_2}} \tag{6.7}$$

and

$$\epsilon = \delta \,. \tag{6.8}$$

We define the set $B^{(s)}$ as follows:

$$B^{(s)} = \left\{ (m_1, m_2) : \frac{m_1^2}{2p_1 s} + \frac{m_2^2}{2p_2 s} \le \lambda \right\} . \tag{6.9}$$

The line

$$(m_1, m_2) + \ell(h(s), -1)$$

intersects the line $s(1, 0)$ at the point

$$m = m_1 + m_2 h(s) \,. \tag{6.10}$$

We have $\varphi^{(s)}((m_1, m_2)) = m$ where m is defined by (6.10). For all points on the line G_m except the point in $B^{(s)}$, see (6.9), condition (6.5) is valid and the map $\varphi^{(s)} B^{(s)} = \mathcal{A}^{(s)}$ is now completely defined.

From (6.6) and estimate (6.4) we get the values δ and ϵ in (6.7) and (6.8).

Case 3.

$$\alpha h^2(s) \le s \,.$$

In this case the classical LLT is valid, i.e.,

$$P(M_s = m) = \frac{1}{\sigma\sqrt{2\pi s}} \exp\left(-\frac{m^2}{2\sigma^2 s}\right) (1 + o(1)) \,, \tag{6.11}$$

where

$$\sigma^2 = 2h^2(s) p_2 + p_1 \,.$$

We would like to stress that the LLT for H_s gives a more exact estimate than (6.11). In this connection it is important to mention the following: the distribution ξ_{1s} depends on three parameters p_2, p_2 and h, whereas the distribution of H_s (the main term of it) depends only on

$$E\xi_{11}^2 = 2p_2 h^2(s) + 2p_1 \,.$$

We see that, from the information on the limit distribution of M_s, we cannot obtain estimates for all parameters of ξ_{11}, but it is possible to do it from the distribution H_s .

7. Density characteristics of probability distributions

The most important and most common characteristic of a random variable ξ is the variance

$$E(\xi - E\xi)^2 = \delta^2 .$$

Knowledge of σ^2, with the help of Chebyshev's inequality, enables us to find a segment on which a substantial part of the probability mass is concentrated. We will say that variance is a "density condition".

The main idea of this paper is to propose the use of density conditions which may be applied even in cases where variance does not exist.

Let K be a finite set, $K \subset \mathbb{Z}$, $|K| = \operatorname{card} K$.

We say that the set K has a "small doubling" if

$$|2K| < c|K| ,$$

where $c \in \mathbb{R}^+$.

To have some understanding of the behavior of cardinality of $2K$ defined by the structure of K, we allow $K \subset \mathbb{R}^n$. Instead of cardinality of K we will use its measure.

Let, first, K be an interval I in $\mathbb{R}^1$. Then

$$\mu(2I) = 2\mu(I) .$$

If $A \subset \mathbb{R}^n$, then

$$\mu(2A) \geq 2^n \mu(A)$$

and equality occurs if A is a convex set (Brunn–Minkowski theorem).

We see that in $\mathbb{R}^n$ the sets with small doubling are the convex sets. A remarkable fact is that in the case of small doubling for $K \subset \mathbb{Z}$, the structure of K is such that it may be described as a dense subset of a convex set.

THEOREM 7.1. *Let $K \subset \mathbb{Z}$, $|K| < +\infty$ and*

$$|2K| < 2|K| - 1 + b ,$$

where

$$0 \leq b \leq |K| - 3 .$$

Then K is a part of arithmetic progression of length $|K| + b$, i.e., $K \subset \mathcal{L}$, where

$$\mathcal{L} = \{a, a + d, a + 2d, \ldots, a + (k + b - 1)d\}$$

where $a, d \subset \mathbb{Z}$, $d > 0$.

COROLLARY 7.2. Take $b = |K| - 3$. We obtain that if

$$|2K| \leq 3|K| - 4 ,$$

then K is part of an arithmetic progression of length $2|K| - 3$.

For the proof, see [**Fr1**, page 11]. Further results on the structure of sets with small doubling can be found in [**Fr4, Ma, K, Ru**].

THEOREM 7.3. *Let ξ be a scheme*

$$\begin{matrix} \xi_{11} & & & \\ \xi_{21} & \xi_{22} & & \\ & \dots & & \\ \xi_{n1} & \xi_{n2} & \dots & \xi_{nn} . \end{matrix}$$

Let $K_n = \operatorname{supp} \xi_{n1}$, $p_{a,n} = \mathbb{P}(\xi_{n_1} = a \in K_n)$ and there exist a constant $p > 0$ such that

$$\forall_n \exists K_n \; p_{a,n} \geq p .$$

Let

$$|2K_n| \leq 3|K| - 4 .$$

Then for the scheme ξ the LLT holds, where for the isomorphic scheme η the support η_{n1} is an arithmetic progression with cardinality $2|K| - 3$ and a maximal step equal to one.

Using the conditions of the theorem in [**MFY**], we can verify that the scheme η is isomorphic to the scheme ξ, thus proving our theorem.

Let us stress that the theorem is, in fact, only a partial case of Gnedenko's theorem and is given only to illustrate the method which gives the real results in the Main Theorem to follow.

DEFINITION 7.4. Let the set

$$\mathcal{D} = \{(x_1, x_2, \dots, x_s) \mid x_1 \in \mathbb{Z}, 0 \leq x_i < h_i, h_i \in \mathbb{Z}, h_i \geq 1, 1 \leq i \leq s\}$$

be called a d-dimensional parallelepiped. The number of integer points in $\mathcal{D}$ is $\prod_{i=1}^{s} h_i = |\mathcal{D}|$.

THEOREM 7.5 (G. A. Freiman, 1964). *For every finite subset $K \subset \mathbb{Z}$ for which*

$$|2K| < C|K| , \tag{7.1}$$

where the constant C does not depend on $|K|$, there exist $c = c(C)$, $d \leq [C - 1]$ and a d-dimensional parallelepiped $\mathcal{D}$ with

$$|\mathcal{D}| < c|K|$$

such that

$$K \subset \varphi(\mathcal{D}) ,$$

where φ is an isomorphism of the second order.

The proofs of this theorem can be found in [**B, Fr1, N, Fr2, C**].

Let us discuss what this theorem contributes to the better understanding of the structure of r.v. ξ with finite $\operatorname{supp} \xi$. From the formulation of the theorem it follows that dimension of r.v. is connected with the doubling coefficient of its support. Degree of "compactness" is given by the volume of the parallelepiped $\mathcal{D}$, for which we have an estimate

$$|\mathcal{D}| < c(C)|\operatorname{supp} \xi| .$$

The volume $\mathcal{D}$ shows to what degree we can "compress" $\operatorname{supp} \xi$ under a 2-isomorphism. What is very important here is the connection of the volume of $\mathcal{D}$ with the cardinality of $\operatorname{supp} \xi$ which is defined by the constant $c(C)$. The importance of the study of $c(C)$ was stressed by Gowers in [**Go**]. The best estimate for the moment is

$$\log \frac{|D|}{|\operatorname{supp} \xi|} < \mathcal{A} \cdot C^2 (\log C)^3 ,$$

where A is an absolute positive constant. This result was obtained by M.-C. Chang in her deep work [**C**].

In the book [**Fr1**] the examples of sets K were given which show that

$$c(C) > c^{AC} ,$$

where A is an absolute constant.

Let us point out that the theorem ensures the existence of an isomorphism of second order and the necessity of extending it to higher orders creates additional difficulties. Irregularities of probabilistic distribution, if they exist, must also be taken into account. We now formulate the central result of this paper.

MAIN THEOREM. *Consider the scheme* ξ

$$\begin{array}{llll} \xi_{11} & & & \\ \xi_{21} & \xi_{22} & & \\ \xi_{n1} & \xi_{n2} & \dots & \xi_{nn} \end{array}$$

with the following properties:

1. $\forall_j \mathbb{P}(\xi_{j1} \in \mathbb{Z}) = 1$.
2. $\forall_j \mathbb{P}(\xi_{j1} = a \in \operatorname{supp} \xi_{j1}) \geq p > 0$, *$p$ an absolute constant.*
3.

(7.2) $$\forall_j |2 \operatorname{supp} \xi_{j1}| < C |\operatorname{supp} \xi_{j1}|.$$

4. $\forall_j$ *the 2-isomorphism for* $\operatorname{supp} \xi_{j1}$ *derived from the theorem for the scheme* ξ *is a j-isomorphism.*

Then for the scheme ξ, *the LLT holds.*

PROOF. Condition (7.2) of this theorem is condition (7.1) of Theorem 7.5. We obtain a 2-isomorphism $\varphi : \mathcal{D}_n \subset \mathbb{Z}$ such that

$$\varphi^{-1}(K_n) \subset \mathcal{D}_n .$$

Let η_{n1} be the image of the r.v. ξ_{n1} under the 2-isomorphism map φ^{-1},

$$\varphi^{-1}(\xi_{n1}) = \eta_{n1} .$$

We have obtained a scheme η isomorphic to the scheme ξ. We will now show that for the scheme η the conditions of the multidimensional LLT from [**Fo**] are fulfilled.

Conditions of variance and the third moment are fulfilled in an obvious way in view of the fact that $|K_n|$ are uniformly bounded, which follows from condition 2 of the Main Theorem.

Now, with respect to the arithmetic condition, suppose first that for every sublattice $G \subset \mathbb{Z}^s$ and every $x \in \mathbb{Z}^s$ we have

$$\mathbb{P}(\eta_{n1} \in G + x) \neq 1 .$$

This means that there exists $y \in \operatorname{supp} \eta_{n1}$ and $y \notin G + x$, and then

$$\max_x \mathbb{P}(\eta_{n1} \in G + x) \leq 1 - \mathbb{P}(\eta_{n1} = y) \leq 1 - p .$$

If the arithmetic condition is not fulfilled, i.e., $\mathbb{P}(\eta_{n1} \in G + x) = 1$, then, with the help of a suitable affine transformation, $G + x$ may be transformed to a lattice with the value of the volume of a fundamental parallelepiped equal to 1. Repeating this reasoning, we arrive at an r.v. for which the arithmetic condition is fulfilled. □

The Main Theorem enables us to obtain a new LLT for a wide class of schemes and therefore enables us to compute values of probabilities of n-fold convolutions; the larger the number n the more exact.

In conclusion, we express our gratitude to Professor Jean-Marc Deshouillers for his invaluable help throughout the many years of our collaboration.

Appendix

1. An explicit formula for the example in §3 is as follows:

$$\mathbb{P}(\xi_1 + \cdots + \xi_n = m_1 + (2n+2)m_2) = \frac{3}{\pi \cdot \sqrt{35} \cdot n}$$
$$\cdot \exp\left(-\frac{1}{2n}\left[(m_1 - 2/3n)^2 - \frac{1}{3}(m_1 - 2/3n)(m_2 - 2/3n) + (m_2 - 2/3n)^2\right]\right)$$
$$\cdot (1 + o(1)) \ .$$

2. The formulation of LLT from [MFY]. Let $\xi_{n1}, \xi_{n2}, \ldots, \xi_{nn}$ be a scheme of independent identically distributed random variables, taking integer values and having finite variance. Denote $E(\xi_{n1} - E\xi_{n1})^2 = \sigma_n^2$; $E\xi_{n1} = a_n$.

THEOREM. *Let $E|\xi_{n1}|^3 < \infty$ and $\rho_n = E|\xi_{n1} - a_n|^3$. Suppose that*

(i) *for $n \to \infty$, $\frac{\rho u^2}{u\sigma_n^u} \ln n \to \infty$,*
(ii) *$\sigma_n^2 = O(n^\rho)$, where $\rho < \frac{\ln(2+c)}{\ln 2} - 1$ and $c < 1$,*
(iii) *for $q = 2, 3, \ldots$ and sufficiently large $\omega > 0$*

$$\max_{1 \le r \le q} P(\xi_1 \equiv r(\mathrm{mod} q)) \le 1 - \omega \max\left(\frac{\rho_n^2}{n\sigma_n^4}, \frac{1}{n^{1-\mu}}\right) \ln n \ ,$$

where $\mu = \frac{(1+\rho)\ln 2}{\ln(2+c)}$.

Then for the series $\xi_{n1}, \xi_{n2}, \ldots, \xi_{nn}$ the LLT takes place.

3. Formulation of LLT from [Fo].

THEOREM. *Let $\left\{\vec{\xi}_\ell^{(n)}\right\}_{\ell=1}^n$ be the scheme of independent vectors*

$$\vec{\xi}_e^{(n)} = \left(\xi_{\ell 1}^{(n)}, \ldots, \xi_{\ell d}^{(n)}\right), \quad \ell = 1, 2, \ldots, n, \ n = 1, 2, \ldots \ .$$

We take the vector of mathematical expectations of vectors of series to be zero, and we take variances of components $\sigma_{n_i}^2$, $i = 1, 2, \ldots, d, \ldots$ to be finite, but dependent on the number of series n. Denote by $\|\rho_{ij}^{(n)}\|$ the matrix covariations of the quadratic form connected with $\|\rho_{c_j}^{(n)}\|$ which we suppose to be positively definite:

$$f(\overline{x}) = (2\pi)^{-d/2}(\det \|\rho_{ij}\|)^{-1/2} \cdot \exp\left\{-1/2 \sum_{i,j=1}^{d} \alpha_{ij} x_i x_j\right\},$$

where $\|\alpha_{ij}^{(n)}\| = \|\rho_{ij}^{(n)}\|^{-1}$. *Standard notations are*

$$\mathbb{P}\left\{\overline{\xi}_1^{(n)} = \overline{m}\right\} = p(\overline{m})\ , \quad \overline{m} = (m_1, \dots, m_d) \in \mathbb{Z}^d\ ,$$

$$\overline{s}_u = \sum_{k=1}^{u} \overline{\xi}_k^{(n)} = (S_{n1}, \dots, S_{nd})\ ,$$

$$P\{\overline{s}_n = \overline{z}\} = P_n(\overline{z})\ , \quad \overline{z} = (z_1, \dots, z_d) \in \mathbb{Z}^d\ ,$$

$$B_{ni}^2 = DS_{ni} = n\sigma_{ni}^2\ , \quad i = 1, 2, \dots, d\ ,$$

$$\overline{\xi}_n = (S_{n1}/B_{n1}, \dots, S_{nd}/B_{nd})\ ,$$

$$\overline{z}/B_n = (z_1/B_{n1}, \dots, zd/B_{nd})\ .$$

We suppose that

$$\beta_{ni} = E|\xi_{1i}^{(n)}|^3 < +\infty,\ i = 1, \dots, d\ ,$$

and that $\forall a \in \mathbb{Z}$, $q \geq 2$ $\forall \overline{a} = (a_1, \dots, a_d)$, $a_i \in \mathbb{Z}$, $i = 1, \dots, d$, $(a_1, \dots, a_d, q) = 1$; $\max_{0 \leq r \leq a-1} \mathbb{P}\left\{\left(\overline{a}, \overline{\xi}^{(n)}\right) \equiv r (mod\ q)\right\} \leq 1 - \alpha_n$,

$$\alpha_n = K \max_{1 \leq i \leq d} \max\left\{\frac{\beta_{ni}^2}{\sigma_{ni}^4, \Delta_n^2 \cdot n}\ , \ \frac{\sigma_{ni}}{\sqrt{n}}\right\} \ln n\ ,$$

where K *is some constant,* $\Delta_n = \det \|\rho_{ij}\|$. *Let* $\beta_{n_i} = O(\sigma_{ni}^3 \Delta_n^2 \sqrt{n})$. *Then there exist a constant* $c(d)$ *and* $k \geq c(d)$ *such that the LLT is valid, i.e.,*

$$B_{n1} B_{n2} \cdots B_{nd} P_n(\overline{z}) - f\left(\frac{\overline{z}}{B_n}\right) \to 0\ .$$

4. Inequality of S. N. Bernstein (Probability Theory, 4th Edition, Gostechizdat, 1946, Chapter IV). In the case of uniformly bounded summands

$$S_n = \sum_{j=1}^{n} \xi_j\ ,$$

one has

$$1 - F_n(xB_n) < \exp(-x^2/4) \quad (0 < x < B_n/H)\ ,$$
$$1 - F_n(xB_n) < \exp(-xB_n/4H) \quad (x > B_n/H)\ .$$

References

[AZ] T. V. Arak, A. Yu. Zaĭtsev, *Uniform Limit Theorems for Sums of Independent Random Variables,* Proc. Steklov Inst. Math. **174** (1998).

[B] Y. Bilu, *Structure of sets with small sumset. Structure theory of set addition,* Astérisque **258** (1999), 77–108.

[C] M.-C. Chang, *A polynomial bound in Freiman's theorem,* Duke Math. Journal **113** (2000), 399–419.

[DFM] J.-M. Deshouillers, G. A. Freiman, W. Moran, *On series of discrete random variables, 1: Real trinomial distributions with fixed probabilities, Structure theory of set addition,* Astérisque **258** (1999), 411–423.

[DFY1] J.-M. Deshouillers, G. A. Freiman, A. A. Yudin, *On bounds for the concentration function. I. Structure theory of set addition,* Astérisque **258** (1999), 425–436.

[DFY2] J.-M. Deshouillers, G. A. Freiman, A. A. Yudin, *On bounds for the concentration function. II,* J. Theoret. Probab. **14** (2001), 813–820.

[Fo] A. S. Fomin, *An arithmetical method of proof of a local theorem for series of independent integer random vectors,* Math. Zametki **28** (1980), 791–800 (in Russian).

[Fr1] G. A. Freiman, *Foundations of a Structure Theory of Set Addition,* Translations of Mathematical Monographs **37**, Amer. Math Soc., Providence, R.I., 1973.

[Fr2] G. A. Freiman, *What is the structure of K if $K + K$ is small?*, Springer Lecture Notes in Math. **1240** (1987), 109–134.

[Fr3] G. A. Freiman, *Structure theory of set addition, II. Results and problems,* Paul Erdös and his Mathematics, I, Budapest (Hungary), 1998, Bolyai Society Mathematical Studies X, Budapest, 2001, pp. 1–18.

[Fr4] G. A. Freiman, *Structural theory of set addition,* Astérisque **258** (1999), 1–33.

[FY1] G. A. Freiman, A. A. Yudin, *The interface between probability theory and additive number theory (Local Limit Theorems and Structure Theory of Set Addition),* Technical Report 2-2003, 2003.

[FY2] G. A. Freiman, A. A. Yudin, *The interface between probability theory and additive number theory (Local Limit Theorems and Structure Theory of Set Addition),* Doklady Rus. Acad. Sci., to appear.

[Ga] N. G. Gamkrelidze, *On local limit theorem for lattice random variables,* Theory Probab. Appl. **9** (1964), 733–736.

[Gn] B. V. Gnedenko, *On local limit theorems of probability theory,* Russian Math. Surveys **3:3** (1949), 187–190 (in Russian).

[Go] W. T. Gowers, *Rough structure and classification,* Geom. Funct. Anal., GAFA2000, Special Volume, Part 1 (2000), 79–117.

[K] M. Kneser, *Ein Satz über abelische Gruppen mit Anwendungen auf die Geometric der Zahlen,* Math. Z. **61** (1955), 429–434.

[L] M. Loeve, *Probability Theory,* D. Van Nostrand Company Inc., Princeton, 1960.

[Ma] A. M. Makbeath, *On measure in sum sets, II. The sum theorem for the torus,* Proc. Cambr. Philos. Soc. **49** (1953), 40–43.

[MFY] D. A. Moskvin, G. A. Freiman, A. A. Yudin, *Structural theory of set addition, and local limit theorems for independent lattice random variables,* Theory Probab. Appl. **19** (1974), 52–62 (in Russian).

[MPR1] D. G. Meĭzler, O. S. Parasuk, F. L. Rvacheva, *A multidimensional local limit theorem of the theory of probability,* Doklady Akad. Nauk SSSR (N.S.) **60** (1948), 1127–1128 (in Russian).

[MPR2] D. G. Meĭzler, O. S. Parasuk, F. L. Rvacheva, *On multidimensional local limit theorem of probability theory,* Ukrainian Mathematical Journal **1** (1949), 9–20.

[MS1] A. A. Mitalauskas, V. A. Statulyavichus, *On local limit theorems, I,* Lith. Math. J. **14** (1974), 628–640.

[MS2] A. A. Mitalauskas, V. A. Statulyavichus, *On local limit theorems, II,* Lith. Math. J. **17** (1977), 550–554.

[Mu] A. B. Mukhin, *Local limit theorems for lattice random variables,* Theory Probab. Appl. **36** (1991), 698–713.

[MZ] R. A. Minlos, E. A. Zhizhina, *A local limit theorem for a nonhomogeneous random walk on a lattice,* Theory Probab. Appl. **39:3 (1994)**, 490–503,

[N] M. Nathanson, *Additive Number Theory. Inverse Problems and the Geometry of Sumsets,* Springer, New York, 1996.

[Pe] V. V. Petrov, *Limit Theorems for Sums of Independent Random Variables,* Moscow, Nauka, 1980.

[Pr] Yu. V. Prokhorov, *On a local limit theorem for lattice distributions,* Dokl. Akad. Nauk SSSR **98** (1954), 535–538 (in Russian).

[Ro] Yu. A. Rozanov, *On a local limit theorem for lattice distributions,* Theory Probab. Appl. **2** (1957), 662–664.

[Ru] I. Z. Ruzsa, *Generalized arithmetic progressions and sumsets,* Acta Math. Hungar. **65** (1994), 379–388.

[Sh] T. L. Shervashidze, *On a uniform estimate of the rate of convergence in the multidimensional local limit theorem for densities,* Theory Probab. Appl. **16** (1971), 741–743.

[St] C. J. Stone, *A local limit theorem for nonlattice multi-dimensional distribution functions,* Ann. Math. Statist. **36** (1965), 546–551.

Sackler Faculty of Exact Sciences, School of Mathematical Sciences, Tel Aviv University, Ramat Aviv, Tel Aviv 69978, Israel
E-mail address: grisha@post.tau.ac.il

Vladimir State Pedagogical University, pr. Stroitelej 11, Vladimir 600005, Russia
E-mail address: aayudin@vgpu.vladimir.ru

Amer. Math. Soc. Transl.
(2) Vol. **217**, 2006

Plancherel Measure on Shifted Young Diagrams

Vladimir Ivanov

ABSTRACT. Consider random Young diagrams with a fixed number n of boxes, where the probability distribution on diagrams is determined by the Plancherel measure. As n goes to infinity, the boundary of the (suitably scaled) random diagram concentrates near a curve Ω (Logan and Shepp, 1977, and Vershik and Kerov, 1977). In 1993, Kerov announced a central limit theorem describing Gaussian fluctuations of random diagrams around the limit shape Ω. Here we propose an analogous theorem in the case of projective representations of symmetric groups.

Introduction

Projective characters of symmetric groups were first studied by I. Schur [**Sc**] in 1911. In this paper I. Schur constructed the spin-symmetric group that linearizes projective representations of a finite symmetric group and studied the characters of irreducible complex representations of the spin-symmetric group. Later, irreducible representations of the spin-symmetric group were studied in [**J, HH, Mo, N, P, St**].

In 1984, A. Sergeev [**Se1**] introduced a group such that the characters of its irreducible $\mathbb{Z}_2$-graded representations are connected by simple relations with the irreducible characters of the spin-symmetric group. This group is considerably simpler than the spin-symmetric group. The relation between the representations of the spin-symmetric group and those of the Sergeev group were also studied in [**Y**].

The Plancherel measure M_n on Young diagrams with fixed number n of boxes is defined by the formula $M_n(\lambda) = (\dim \lambda)^2/n!$. In 1977, it was independently obtained by B. Logan and L. Shepp [**LS**] and by A. Vershik and S. Kerov [**VK1**] that, as $n \to \infty$, the boundary of the appropriately rescaled random shape λ concentrates near some remarkable fixed curve Ω. In 1993, S. Kerov [**K1**] announced a central limit theorem describing Gaussian fluctuations around the limit shape Ω. The detailed proof of this theorem was presented in [**IO**]. Another approach to this theorem was obtained by J. Fulman in [**F**].

2000 *Mathematics Subject Classification.* Primary 20C30; Secondary 05E05, 05E10, 20C32, 60B10, 60B15.

Key words and phrases. Shifted Young diagrams, projective representations of symmetric groups, Gaussian fluctuations, limit curve, Plancherel measure.

In this text we suggest a projective analog of this cental limit theorem. Instead of the ordinary Young diagrams, we consider the shifted Young diagrams which correspond to the case of projective representations of the symmetric groups. The curve $\Omega(x)$ remains the same but we restrict $x \geq 0$. The main result of this paper (Theorem 6.1) is very similar to Kerov's central limit theorem. In our proof we follow the methods of [**IO**]. All assertions of this work have their counterparts in [**IO**]. So we provide the proofs only for a few of them. Our purpose is to show the difference between the projective case and the ordinary one.

The paper is organized as follows.

In Section 1, we introduce the algebra $\mathbb{A}$ and a system $p_1, p_3, \dots$ of its generators.

In Section 2, we introduce the necessary geometric setting for visualizing fluctuations of shifted Young diagrams. We introduce the generators $\widetilde{p}_2, \widetilde{p}_4, \dots \in \mathbb{A}$ and the "weight grading" of the algebra $\mathbb{A}$, which is well adapted to the operation of rescaling diagrams.

In Section 3, we examine one more system of generators in $\mathbb{A}$, denoted as $p_1^\#, p_3^\#, \dots$. These are character values on cycles in symmetric groups. We study the transitions between all three systems of generators. Here our tools are a projective analog of Wassermann's formula for the value of a spin-symmetric group character on the $(2k+1)$-cycle and Lagrange's inversion formula.

In Section 4, we start the study of the Plancherel measures M_n.

In Section 5, we examine the random variables $p_3^{\#(n)}, p_5^{\#(n)}, \dots$. We refer to [**I**] that, as $n \to \infty$, the variables $p_k^{\#(n)}$, suitably scaled, are asymptotically independent Gaussians. This result is the first version of the central limit theorem.

In Section 6, we obtain our main result: a description of the Gaussian fluctuations around the limit curve Ω. It is derived from the central limit theorem for the generators $p_{2k+1}^\#$ mentioned above. The proof is based on a formula that gives the highest term of the polynomial expressing $p_{2k+1}^\#$ through the (centered and scaled versions of) the generators $\widetilde{p}_{2j}$. Here "highest term" refers to an appropriate filtration of the algebra $\mathbb{A}$, which we call Kerov's filtration.

The author is very grateful to G. Olshanski for suggesting this problem and a lot of important remarks.

1. The algebra of polynomial functions on the set of shifted Young diagrams

Recall first the basic definitions and notation related to strict partitions and shifted Young diagrams; see [**Ma, St**].

DEFINITION 1.1. A *partition* is any infinite sequence $\lambda = (\lambda_1, \lambda_2, \dots)$ of nonnegative integers such that $\lambda_1 \geq \lambda_2 \geq \dots$ and the number of nonzero λ_i's is finite. The sum $\lambda_1 + \lambda_2 + \dots$ is denoted by $|\lambda|$, and usually we set $|\lambda| = n$.

A partition ρ is called *odd* if all nonzero parts of ρ are odd. Following J. Stembridge [**St**], we denote the set of odd partitions of n by OP_n. Set

$$OP = \bigcup_{n \geq 0} OP_n.$$

A partition λ is called *strict* if all nonzero parts of λ are distinct. Following J. Stembridge [**St**], we denote the set of strict partitions of n by DP_n. Let

$$DP = \bigcup_{n \geq 0} DP_n.$$

It is well known (see, e.g., [**Ma**, Chapter III, §8]) that the number of strict partitions of n equals the number of odd partitions of n. In our notation, $|DP_n| = |OP_n|$.

DEFINITION 1.2. For a strict partition one may define the shifted Young diagram besides the simple Young diagram [**Ma**, Chapter I, §1]. If λ is a strict partition, then the set

$$\{(r,s) \in \mathbb{Z}^2 \mid r \leq s \leq \lambda_r + r - 1, 1 \leq r \leq \ell(\lambda)\}$$

is called *the shifted Young diagram of the partition* λ. It is useful to draw unit squares instead of points of $\mathbb{Z}^2$. We assume that the first coordinate axis is directed downwards, the second coordinate axis is directed to the right and the point $(0,0)$ is in the left upper corner of the figure. Denote by $\partial\lambda$ the doubly infinite polygonal line which first goes upwards along the r-axis, next goes along the boundary line separating λ from its complement in $\mathbb{R}^2_+$, and then goes to the right along the s-axis.

We assign to a strict partition a shifted Young diagram, which is denoted by the same symbol. We identify strict partitions and shifted Young diagrams, and we denote by DP the set of all shifted Young diagrams.

For any $\lambda \in DP$ we set

$$\Phi(z;\lambda) = \prod_{i=1}^{\infty} \frac{z+\lambda_i}{z-\lambda_i}, \qquad z \in \mathbb{C}. \tag{1.1}$$

The product is actually finite, because $\lambda_i = 0$ when i is large enough. Therefore, $\Phi(z;\lambda)$ is a rational function in z. We view it as a generating function of λ.

DEFINITION 1.3. The *algebra of polynomial functions on the set DP*, denoted by $\mathbb{A}$, is generated over $\mathbb{R}$ by the coefficients of the above expansion of $\Phi(z;\lambda)$ or, equivalently, of $\ln \Phi(z;\lambda)$. We also assume that $\mathbb{A}$ contains 1.

PROPOSITION 1.4. *We have*

$$\ln \Phi(z;\lambda) = \sum_{k=1}^{\infty} \frac{2p_{2k-1}(\lambda)}{2k-1} z^{-(2k-1)}, \tag{1.2}$$

where

$$p_k(\lambda) = \sum_{i=1}^{\infty} \lambda_i^k. \tag{1.3}$$

PROPOSITION 1.5. *The generators $p_{2k-1} \in \mathbb{A}$ are algebraically independent, so that $\mathbb{A}$ isisomorphic to $\mathbb{R}[p_1, p_3, p_5, \dots]$.*

2. Continual diagrams and their moments

Denote by $\mathbb{R}_+$ the set $\{x \mid x \geq 0,\, x \in \mathbb{R}\}$.

DEFINITION 2.1. A *continual diagram* is a function $\omega(x)$ on $\mathbb{R}_+$ such that

(i) $|\omega(x_1) - \omega(x_2)| \leq |x_1 - x_2|$ for any $x_1, x_2 \in \mathbb{R}_+$ (the Lipschitz condition);

(ii) there exists a point $x_0 \in \mathbb{R}_+$, called the *center* of ω, such that $\omega(x) = x - x_0$ when x is large enough.

The set of all continual diagrams is denoted by $\mathcal{D}$, and the subset of diagrams with center 0 is denoted by $\mathcal{D}^0$.

This definition is due to S. Kerov [**K2, K3, K4**]. We shall deal mainly with the set $\mathcal{D}^0$.

To any $\omega \in \mathcal{D}$ we assign a function $\sigma(x)$:

$$\sigma(x) = \tfrac{1}{2}(\omega(x) - x). \tag{2.1}$$

It also satisfies the Lipschitz condition (i), which implies the existence of the derivative $\sigma'(x)$; it is defined almost everywhere and satisfies $|\sigma'(x)| \leq 1$. By (ii), the function $\sigma'(x)$ is compactly supported.

If $\omega \in \mathcal{D}^0$, then $\sigma(x)$ is compactly supported, too. Note that $\omega(x)$ is uniquely determined by $\sigma'(x)$. Even more, $\omega(x)$ is uniquely determined by the second derivative $\sigma''(x)$, which is understood in the sense of distribution theory.

DEFINITION 2.2. Define the functions $\widetilde{p}_2, \widetilde{p}_4, \ldots$ on $\mathcal{D}^0$ by setting

$$\widetilde{p}_k[\omega] = \int_0^\infty x^k \sigma''(x)dx = -k\int_0^\infty x^{k-1}\sigma'(x)dx = k(k-1)\int_0^\infty x^{k-2}\sigma(x)dx, \tag{2.2}$$

where $\omega \in \mathcal{D}^0$, $k = 2, 4, \ldots$.

DEFINITION 2.3. Given $\lambda \in DP$, we define a piecewise linear function $\lambda(\,\cdot\,)$ as follows. Let (r, s) and $\partial\lambda$ be as in Definition 1.2. Then the graph $y = \lambda(x)$ describes the intersection $(\mathbb{R}_+ \times \mathbb{R}_+) \cap \partial\lambda$ in the coordinates $x = s - r - \frac{1}{2}$, $y = r + s - \frac{1}{2}$. The correspondence $\lambda \mapsto \lambda(\,\cdot\,)$ yields an embedding $DP \hookrightarrow \mathcal{D}^0$.

Note that

$$\int_0^\infty (\lambda(x) - x)dx = 2|\lambda|. \tag{2.3}$$

DEFINITION 2.4. We have $\lambda'(x) = \pm 1$, except for finitely many points, which are exactly the local extrema of the function $\lambda(x)$. These local extrema form two interlacing sequences of points

$$0 < y_0 < x_1 < y_1 < x_2 < \cdots < x_{m-1} < y_{m-1} < x_m\,, \tag{2.4}$$

where the x_i's are the local minima and the y_j's are the local maxima of the function $\lambda(x)$.

The correspondence $\lambda \mapsto \{x_i\} \cup \{y_j\}$ provides one more useful system of parameters for Young diagrams.

PROPOSITION 2.5. *Let $\lambda \in DP$, let $\lambda(\,\cdot\,) \in \mathcal{D}^0$ be the corresponding continual diagram, and consider the local extrema* (2.4). *We have*

$$\widetilde{p}_k[\lambda(\,\cdot\,)] = \sum_{i=1}^{m} x_i^k - \sum_{j=0}^{m-1} y_j^k\,, \quad k = 2, 4, \ldots. \tag{2.5}$$

PROPOSITION 2.6. *Let* $\lambda \in \mathbb{Y}$ *and let* $\{x_i\} \cup \{y_j\}$ *be the local extrema of* $\lambda(\,\cdot\,)$. *The following identity holds:*

$$\frac{\Phi(z-\frac{1}{2};\lambda)}{\Phi(z+\frac{1}{2};\lambda)} = \prod_{k=1}^{m} \frac{z^2 - y_{k-1}^2}{z^2 - x_k^2}. \tag{2.6}$$

PROOF. It follows from the identity

$$\Phi(z;\lambda) = \prod_{k=1}^{m} \frac{\Gamma(z+x_k+\frac{1}{2})\Gamma(z-x_k+\frac{1}{2})}{\Gamma(z+y_{k-1}+\frac{1}{2})\Gamma(z-y_{k-1}+\frac{1}{2})}. \tag{2.7}$$

□

Set

$$\widetilde{p}_k(\lambda) = \widetilde{p}_k[\lambda(\,\cdot\,)], \qquad k = 2, 4, \ldots, \qquad \lambda \in \mathbb{Y},$$

where the right-hand side is given by (2.2).

PROPOSITION 2.7. *The functions* $\widetilde{p}_2(\lambda), \widetilde{p}_4(\lambda), \ldots$ *belong to the algebra* $\mathbb{A}$ *and are related to the functions* $p_1(\lambda), p_3(\lambda)$ *by the relations*

$$\widetilde{p}_{2k} = \sum_{j=0}^{k-1} \binom{2k}{2j+1} 2^{-2j} p_{2k-1-2j}, \qquad k = 1, 2, 3, \ldots. \tag{2.8}$$

COROLLARY 2.8. *For any* $k = 1, 2, \ldots$

$$\frac{\widetilde{p}_{2k}}{2k} = p_{2k-1} + \text{a linear combination of } p_{2k-3}, \ldots, p_1.$$

Conversely, for any $k = 1, 2, \ldots$

$$p_{2k-1} = \frac{\widetilde{p}_{2k}}{2k} + \text{a linear combination of } \widetilde{p}_{2k-2}, \ldots, \widetilde{p}_2.$$

By Corollary 2.8, the elements $\widetilde{p}_2, \widetilde{p}_4, \ldots$ are algebraically independent generators of the algebra $\mathbb{A}$:

$$\mathbb{A} = \mathbb{R}[\widetilde{p}_2, \widetilde{p}_4, \ldots].$$

DEFINITION 2.9 (Cf. [**EO**]). The *weight grading* of the algebra $\mathbb{A}$ is defined by setting

$$\operatorname{wt}(\widetilde{p}_k) = k, \qquad k = 2, 4, \ldots.$$

The weight grading induces a filtration in $\mathbb{A}$, which we call the *weight filtration* and denote by the same symbol $\operatorname{wt}(\,\cdot\,)$. Note that

$$\operatorname{wt}(p_k) = k+1, \qquad k = 1, 3, \ldots,$$

because the top weight homogeneous component of p_k is $\widetilde{p}_{k+1}/(k+1)$; see Corollary 2.8.

DEFINITION 2.10. a) We define an action of the multiplicative group of positive real numbers on the set $\mathcal{D}^0$ by setting

$$\omega^s(x) = s^{-1}\omega(sx), \qquad \omega \in \mathcal{D}^0, \quad s > 0, \quad x \in \mathbb{R}.$$

In other words, the graph of $y = \omega^s(x)$ is obtained from that of $y = \omega(x)$ by the transformation $(x, y) \mapsto (s^{-1}x, s^{-1}y)$.

b) Since $\mathbb{A} = \mathbb{R}[\widetilde{p}_2, \widetilde{p}_4, \ldots]$, we may define the symbol $f[\omega]$ (where ω ranges over $\mathcal{D}^0$) for any $f \in \mathbb{A}$. Specifically, write f as a polynomial in $\widetilde{p}_2, \widetilde{p}_4, \ldots$ and then specialize each $\widetilde{p}_k$ to $\widetilde{p}_k[\omega]$. In this way, we realize $\mathbb{A}$ as an algebra of functions on $\mathcal{D}^0$.

PROPOSITION 2.11. *Let $f \in \mathbb{A}$ be homogeneous with respect to the weight grading, Definition 2.9. Then for any $\omega \in \mathcal{D}^0$ and $s > 0$,*

$$f[\omega^s] = s^{-\operatorname{wt}(f)} f[\omega].$$

PROOF. By the definition of the weight grading, it suffices to check that

$$\widetilde{p}_k[\omega^s] = s^{-k}\widetilde{p}_k[\omega], \qquad k = 2, 4, \dots, \qquad \omega \in \mathcal{D}^0, \quad s > 0. \tag{2.9}$$

Note that the function $\sigma(x) = \frac{1}{2}(\omega(x) - |x|)$ transforms in the same way as $\omega(x)$. Then (2.9) is clear from Definition 2.2. □

3. The elements $p_k^{\#}$

DEFINITION 3.1. For each n we introduce a collection of numbers X_ρ^λ, $\lambda \in DP_n$, $\rho \in OP_n$, by the formula

$$p_\rho = \sum_{\lambda \in DP_n} X_\rho^\lambda P_\lambda. \tag{3.1}$$

Note that our X_ρ^λ corresponds to $X_\rho^\lambda(-1)$ in the notation of [**Ma**, Chapter III]. The inversion of (3.1) takes the form

$$2^{\ell(\lambda)} P_\lambda = \sum_{\rho \in OP_{|\lambda|}} \frac{2^{\ell(\rho)}}{z_\rho} X_\rho^\lambda p_\rho.$$

Set

$$g_\lambda = X_{1^{|\lambda|}}^\lambda. \tag{3.2}$$

The irreducible characters of the spin-symmetric group [**Sc, St**] and the characters of the irreducible $\mathbb{Z}_2$-graded representation of the Sergeev group [**Se1, Se2**] can be expressed in terms of X_ρ^λ; see e.g. [**I**]. Thus, the table X_ρ^λ may be viewed as a projective analog of the character table of the symmetric group.

DEFINITION 3.2. For $k = 1, 3, \dots$, let $p_k^{\#}$ be following function on DP_n:

$$p_k^{\#}(\lambda) = \begin{cases} n^{\downarrow k} \cdot \dfrac{X_{(k,1^{n-k})}^\lambda}{g_\lambda}, & n := |\lambda| \geq k, \\ 0, & n < k, \end{cases} \tag{3.3}$$

where

$$n^{\downarrow k} = n(n-1)\dots(n-k+1)$$

and

$$(k, 1^{n-k}) = (k, 1, \dots, 1) \in OP_n\,.$$

PROPOSITION 3.3. *For any $k = 1, 3, \dots$ and any $\lambda \in DP$, $p_k^{\#}(\lambda)$ equals the coefficient of z^{-1} in the expansion of the function*

$$-\frac{1}{4k}(2z-k)(z-1)^{\downarrow(k-1)}\frac{\Phi(z;\lambda)}{\Phi(z-k;\lambda)} \tag{3.4}$$

in descending powers of z about the point $z = \infty$.

PROOF. This claim may be proved in the same way as in the case of the symmetric group (see [**IO**, §3]). The main difference is the projective analog of the Murnaghan–Nakayama rule [**Ma**, Chapter III, §8, Example 11]. □

We shall employ the following notation. Given a formal series $A(t)$, let

$$[t^k]\{A(t)\} = \text{ the coefficient of } t^k \text{ in } A(t).$$

The next result is a projective analog of the Wassermann formula [**W**].

PROPOSITION 3.4. *For any $k = 1, 3, \dots$, the function $p_k^\#(\lambda)$ introduced in Definition 3.2 belongs to the algebra $\mathbb{A}$. It can be expressed in terms of the generators $p_1, p_3, \dots$ of $\mathbb{A}$ as*

$$(3.5)\quad p_k^\# = [t^{k+1}]\left\{ -\frac{1}{2k}\,(1-\frac{k}{2}t)\prod_{j=1}^{k-1}(1-jt) \cdot \exp\left(\sum_{j=1}^{\infty}\frac{2p_{2j-1}t^{2j-1}}{2j-1}\,(1-(1-kt)^{-(2j-1)})\right)\right\}.$$

The expression (3.5) can be written in the form

$$(3.6)\quad \begin{aligned} p_k^\# &= -\frac{1}{2k}\,[t^{k+1}]\left\{(1+\varepsilon_0(t))\exp\left(-\sum_{j=1}^{\infty}2kp_{2j-1}t^{2j}(1+\varepsilon_j(t))\right)\right\} \\ &= -\frac{1}{2k}\,[t^{k+1}]\left\{(1+\varepsilon_0(t))\sum_{m=0}^{\infty}\frac{(-1)^m}{m!}\Big(\sum_{j=1}^{\infty}2kp_{2j-1}t^{2j}(1+\varepsilon_j(t))\Big)^m\right\}. \end{aligned}$$

Here each $\varepsilon_r(t)$ is a power series of the form $c_1t + c_2t^2 + \dots$, where the coefficients $c_1, c_2, \dots$ do not involve the generators $p_1, p_3, \dots$.

PROPOSITION 3.5. *Let $k = 1, 3, \dots$. In the weight grading, the top homogeneous component of $p_k^\#$ has weight $k+1$ and can be written as*

$$(3.7)\qquad -\frac{1}{k}\,[t^{k+1}]\left\{\exp\Big(-k\sum_{j=1}^{\infty}\frac{\widetilde{p}_{2j}}{2j}\,t^{2j}\Big)\right\}.$$

PROPOSITION 3.6. *Let $a_2, a_3, \dots$ and $b_2, b_3, \dots$ be two families of elements in a commutative algebra. Let*

$$A(t) = 1 + \sum_{j=2}^{\infty} a_j t^j, \qquad B(u) = 1 + \sum_{j=2}^{\infty} b_j u^j$$

be their generating series, and set

$$\widetilde{A}(t) = \ln A(t) = \sum_{j=2}^{\infty}\frac{\widetilde{a}_j}{j}\,t^j\,.$$

Then the following conditions are equivalent:

(i) $b_k = -\frac{1}{k-1}\,[t^k]\{A^{-(k-1)}(t)\}, \quad k = 2, 3, \dots.$

(ii) $\widetilde{a}_k = [u^k]\{B^k(u)\}, \quad k = 2, 3, \dots.$

Using Proposition 3.5 and Proposition 3.6, we may invert formula (3.7).

PROPOSITION 3.7. *For* $k = 2, 4, \dots$

$$
\begin{aligned}
\widetilde{p}_k &= \frac{1}{2}[u^k]\left\{\Big(1+\sum_{j=1}^{\infty} 2p^{\#}_{2j-1}u^{2j}\Big)^k\right\} + \dots \\
&= \frac{1}{2} \sum_{\substack{m_2, m_4, \dots \\ 2m_2+4m_4+\dots=k}} \frac{k^{\downarrow \sum m_{2i}}}{\prod m_{2i}!} \prod_{i\geq 1} (2p^{\#}_{2i-1})^{m_{2i}} + \dots,
\end{aligned}
\tag{3.8}
$$

where the dots mean a polynomial in $p^{\#}_1, p^{\#}_3, \dots, p^{\#}_{k-3}$ *of total weight* $\leq k-1$, *where* $\mathrm{wt}(p^{\#}_i) = i+1$.

4. The Plancherel measure and the law of large numbers

DEFINITION 4.1. Consider the set DP_n of Young diagrams with n boxes, $n = 1, 2, \dots$, and equip it with the measure PM_n, defined by

$$PM_n(\lambda) = 2^{n-\ell(\lambda)} \frac{g_\lambda^2}{n!}, \quad \lambda \in DP_n .$$

This is a probability measure; see [**B**], [**I**, §8]. It is called the *Plancherel measure*.

In the ordinary, nonprojective case see [**VK1, VK2, VK3**] for more details.

DEFINITION 4.2. We define the function $\Omega(x)$ on $\mathbb{R}_+$ by

$$\Omega(x) = \begin{cases} \frac{2}{\pi}(x \arcsin \frac{x}{2} + \sqrt{4-x^2}), & 0 \leq x \leq 2, \\ x, & x \geq 2. \end{cases}$$

Note that both expressions agree at $x = 2$, so that $\Omega(x)$ is continuous. Moreover, the first derivative $\Omega'(x)$ is continuous on the whole $\mathbb{R}$, while $\Omega''(x)$ is not. This is clear from the explicit expressions

$$\Omega'(x) = \tfrac{2}{\pi} \arcsin \tfrac{x}{2}, \quad \Omega''(x) = \tfrac{2}{\pi} \frac{1}{\sqrt{4-x^2}}, \qquad x < 2.$$

Next, we have $|\Omega'(x)| < 1$ for $x < 2$, which implies that $\Omega(x)$ belongs to $\mathcal{D}^0$. Applying (2.2), for $\omega = \Omega$ we get

$$\widetilde{p}_k[\Omega] = -k \int_0^\infty x^{k-1} \left(\frac{\Omega(x)-x}{2}\right)' dx = -\tfrac{k}{2} \int_0^2 x^{k-1} (\tfrac{2}{\pi} \arcsin \tfrac{x}{2} - x) dx.$$

PROPOSITION 4.3. *We have*

$$\widetilde{p}_{2m}[\Omega] = \frac{1}{2}\binom{2m}{m}, \qquad m = 1, 2, \dots .$$

5. The central limit theorem for characters

For any $f \in \mathbb{A}$, we denote by $f^{(n)}$ the random variable defined on the probability space (DP_n, PM_n) and obtained by restricting f to DP_n.

By the symbol $\xrightarrow{d}$ we will denote convergence of random variables in distribution; see e.g. [**Sh**].

The main result of [**I**] is the following.

THEOREM 5.1 (Central limit theorem for characters). *Choose a sequence $\{\xi_k\}_{k=1,2,3,\dots}$ of independent standard Gaussian random variables. As $n \to \infty$, we have*

$$\left\{ \frac{p^{\#\,(n)}_{2k+1}}{n^{\frac{2k+1}{2}}} \right\}_{k=1,2,3,\dots} \xrightarrow{d} \{2^k\sqrt{2k+1}\,\xi_k\}_{k=1,2,3,\dots}. \tag{5.1}$$

In more detail, for any fixed $N = 2, 3, \dots$, the joint distribution of N random variables

$$\eta_k = \frac{p^{\#\,(n)}_{2k+1}}{2^k\sqrt{2k+1}\,n^{\frac{2k+1}{2}}}, \quad 1 \le k \le N, \tag{5.2}$$

weakly tends, as $n \to \infty$, to the standard Gaussian measure on $\mathbb{R}^N$. Note that we could take equally well in (5.1) the random variables

$$\frac{n^{\frac{2k+1}{2}}}{2^k\sqrt{2k+1}} \, \frac{X^\lambda_{(2k+1,1^{n-2k-1})}}{g_\lambda}, \qquad 1 \le k \le N,$$

where $\lambda \in DP_n$ is the random Plancherel diagram.

It will be convenient to extend the algebra $\mathbb{A}$: we adjoin to it the square root of the element $p^\#_1 = p_1$ and then localize over the multiplicative family generated by $\sqrt{p^\#_1}$. Let $\mathbb{A}^{\text{ext}}$ denote the resulting algebra. As a basis in $\mathbb{A}^{\text{ext}}$ one can take the elements of the form

$$\prod_{j\ge1}(p^\#_{2j+1})^{m_{2j+1}} \cdot (p^\#_1)^{m/2}, \qquad m_1(\rho) = 0, \quad m \in \mathbb{Z}. \tag{5.3}$$

We equip $\mathbb{A}^{\text{ext}}$ with a filtration by the formula

$$\deg_1\Big(\prod_{j\ge1}(p^\#_{2j+1})^{m_{2j+1}} \cdot (p^\#_1)^{m/2}\Big) = \sum_{j\ge1}(2j+1)m_{2j+1} + m.$$

This filtration (see [**I**] for the details) is a projective analog of the Kerov degree.

Since $p^{\#\,(n)}_1 \equiv n$, the symbol $f^{(n)}$ makes sense for any $f \in \mathbb{A}^{\text{ext}}$. Specifically, if $f = g(p^\#_1)^{m/2}$ with $g \in \mathbb{A}$ and $m \in \mathbb{Z}$, then $f^{(n)} = g^{(n)} \cdot n^{m/2}$.

From [**I**, Proposition 8.8] we obtain the following result.

PROPOSITION 5.2. *For any $f \in \mathbb{A}^{\text{ext}}$, $\langle f\rangle_n$ is a Laurent polynomial in $n^{1/2}$ whose degree with respect to n is bounded from above by $\frac{1}{2}\deg_1(f)$.*

6. The central limit theorem for shifted Young diagrams

Given $\lambda \in DP_n$, we set

$$\Delta_\lambda(x) = \tfrac{1}{2}(\lambda(\sqrt{2|\lambda|}\,x) - \sqrt{2|\lambda|}\Omega(x)), \qquad x \in \mathbb{R}_+. \tag{6.1}$$

This is a continuous function on $\mathbb{R}_+$ with compact support. Dropping λ, which we consider as a random element from the probability space (DP_n, PM_n), we interpret (6.1) as a random function $\Delta^{(n)}(x)$.

For any polynomial $v \in \mathbb{R}[x^2]$, the integral

$$v^{(n)} = \int_0^\infty v(x)\Delta^{(n)}(x)dx \tag{6.2}$$

makes sense (because $\Delta^{(n)}$ is compactly supported) and is a random variable. We aim to show that the random variables (6.2), where v ranges over $\mathbb{R}[x^2]$, are asymptotically Gaussian.

The result will be stated in terms of the Chebyshev polynomials of the second kind. Instead of the conventional polynomials $U_k(x)$ (see [**Sz, E**]) we prefer to deal with a slightly modified version of them,

$$u_k(x) = U_k(x/2) = \sum_{j=0}^{[k/2]} (-1)^j \binom{k-j}{j} x^{k-2j}, \qquad k = 0, 1, 2, \dots. \tag{6.3}$$

We have

$$u_k(2\cos\theta) = \frac{\sin((k+1)\theta)}{\sin\theta} \tag{6.4}$$

and

$$\int_{-2}^{2} u_k(x) u_l(x) \frac{\sqrt{4-x^2}}{2\pi} = \delta_{k,l}, \qquad k, l = 0, 1, 2, \dots. \tag{6.5}$$

THEOREM 6.1 (Central limit theorem for Young diagrams). *According to equation* (6.2), *let*

$$u_{2k}^{(n)} = \int_0^\infty u_{2k}(x) \Delta^{(n)}(dx), \qquad k = 1, 2, \dots,$$

and let, as before, $\xi_1, \xi_2, \dots$ stand for a system of independent standard Gaussians. We have

$$\left\{ u_{2k}^{(n)} \right\}_{k \ge 1} \xrightarrow{d} \left\{ \frac{\xi_k}{\sqrt{2(2k+1)}} \right\}_{k \ge 1}, \qquad n \to \infty.$$

Recall that "$\xrightarrow{d}$" means convergence in distribution.

Note that $u_0^{(n)} \equiv 0$, which explains why we start with $k = 1$, not $k = 0$. The theorem is proved at the end of the section. The scheme of the proof is as follows. We remark that the moments of $\Delta^{(n)}$ (i.e., the random variables $v^{(n)}$, where the v's are monomials) are expressed in terms of the elements $\widetilde{p}_k$, appropriately centered and scaled. To evaluate the asymptotics of the corresponding random variables, we employ Theorem 5.1. The main work reduces to expressing the (centered and scaled) elements $\widetilde{p}_k$ through the elements η_k and vice versa, up to lower degree terms.

Also we will use the following well-known general result. It justifies the moment method, which is a convenient tool for checking convergence in distribution.

PROPOSITION 6.2. *Let $a^{(n)}$ be a sequence of real random variables. Assume that $a^{(n)}$ have finite moments of any order, and the moments converge, as $n \to \infty$, to the respective moments of a random variable a. Finally, assume that a is uniquely determined by its moments, which holds, e.g., if the characteristic function of a is analytic.*

Then $a^{(n)} \xrightarrow{d} a$. Moreover, this claim also holds when the variables in question take vector values, i.e., when each $a^{(n)}$, as well as a, is a system of random variables.

As in §5, it is convenient to deal with the extended algebra $\mathbb{A}^{\mathrm{ext}} \supset \mathbb{A}$. We extend the definition of $\deg_1(\,\cdot\,)$ to $\mathbb{A}^{\mathrm{ext}}$ as explained in §5.

We introduce the elements $q_1, q_3, \ldots \in \mathbb{A}^{\text{ext}}$, which are centered and scaled versions of the elements $\widetilde{p}_2, \widetilde{p}_4, \ldots$:

$$q_{2l-1} = \frac{\widetilde{p}_{2l} - 2^{l-1}\binom{2l}{l}(p_1^{\#})^l}{2l(2p_1^{\#})^{l-1/2}}, \qquad l = 1, 2, \ldots. \tag{6.6}$$

Since $\frac{1}{2}\widetilde{p}_2 = p_1$ and $p_1 = p_1^{\#}$, we have $q_1 = 0$.

PROPOSITION 6.3. *For any* $\lambda \in \mathbb{Y}$,

$$\int_0^\infty x^{2k} \Delta_\lambda(x) dx = \frac{q_{2k+1}(\lambda)}{2k+1}, \qquad k = 0, 1, \ldots. \tag{6.7}$$

PROOF. Set $n = |\lambda|$. By the definition of $\Delta_\lambda(x)$, see (6.1),

$$\int_0^\infty x^{2k} \Delta_\lambda(x) dx = \int_0^\infty x^{2k} \frac{\lambda(\sqrt{2n}\,x) - x}{2} dx - \sqrt{2n} \int_0^\infty x^{2k} \frac{\Omega(x) - x}{2} dx.$$

By Definition 2.2, for any $k = 0, 1, \ldots$,

$$\int_0^\infty x^{2k} \frac{\lambda(\sqrt{2n}\,x) - x}{2} dx = (2n)^{-(2k+1)/2} \int_0^\infty x^{2k} \frac{\lambda(x) - x}{2} dx$$
$$= \frac{1}{(2k+1)(2k+2)} \frac{\widetilde{p}_{2k+2}(\lambda)}{(2n)^{(2k+1)/2}} = \frac{1}{(2k+1)(2k+2)} \frac{\widetilde{p}_{2k+2}}{(2p_1^{\#})^{(2k+1)/2}}(\lambda).$$

By Definition 2.2 and Proposition 4.3, for any $k = 0, 1, \ldots$,

$$\sqrt{2n} \int_0^\infty x^{2k} \frac{\Omega(x) - x}{2} dx = \frac{\sqrt{2n}\, \widetilde{p}_{2k+2}[\Omega]}{(2k+1)(2k+2)}$$
$$= \frac{\sqrt{2n}}{2(2k+1)(2k+2)} \cdot \binom{2k+2}{k+1}.$$

Combining this with the definition of $q_1, q_3, \ldots$, we get (6.7). □

In order to apply Theorem 5.1, we need the expression of $p_{2k+1}^{\#}$ in terms of $q_3, q_5, \ldots$ within lower degree terms. We obtain this in two steps. First, using a trick, we deduce from Proposition 3.7 a formula expressing any q_{2k+1} through $p_3^{\#}, p_5^{\#}, \ldots$, up to lower degree terms. See Proposition 6.4. Next, we invert this formula; see Proposition 6.5. One could derive the result directly from Proposition 3.4 but this way turns out to be more difficult.

PROPOSITION 6.4. *For any* $k = 1, 2, 3, \ldots$,

$$q_{2k+1} = \sum_{j=0}^{k-1} \binom{2k+1}{j} \frac{p_{2k+1-2j}^{\#}}{(2p_1^{\#})^{k-j+\frac{1}{2}}} + \ldots, \tag{6.8}$$

where the dots mean a remainder term with $\deg_1(\cdot)$.

Note that the elements occurring in the numerator of the right-hand side are $p_3^{\#}, p_5^{\#}, \ldots$ but not $p_1^{\#}$.

PROOF. The claim of the proposition is equivalent to the following: for any $k = 2, 3, 4, \dots$,

$$(6.9)\quad \widetilde{p}_{2k} = \sum_{j=0}^{k-2} \frac{(2k)^{\downarrow(j+1)}}{j!}\, p^{\#}_{2k-1-2j}\,(2p_1^{\#})^j + \frac{(2k)!}{k!k!}\,2^{k-1}\,(p_1^{\#})^k$$
$$+ \text{ terms with } \deg_1(\,\cdot\,) < 2k-1.$$

We shall deduce this from Proposition 3.7, which expresses $\widetilde{p}_{2k}$ as a polynomial in $p_1^{\#}, p_3^{\#}, \dots$, up to terms of lower weight. A nontrivial point is how to switch from the weight filtration to the Kerov filtration.

Let us write the exact expansion of $\widetilde{p}_{2k}$ in $p_1^{\#}, p_3^{\#}, \dots$,

$$\widetilde{p}_k = \sum_{\nu} a_\nu\,(p_1^{\#})^{\nu_1}(p_3^{\#})^{\nu_3}\dots,$$

where a_ν are certain coefficients. Let us set

$$\|\nu\| = 2\nu_1 + 4\nu_3 + 6\nu_5 + \dots, \qquad \|\nu\|' = 2\nu_1 + 3\nu_3 + 5\nu_5 + \dots,$$

so that

$$(6.10)\qquad \|\nu\|' = \|\nu\| - (\nu_3 + \nu_5 + \dots).$$

We have

$$\mathrm{wt}((p_1^{\#})^{\nu_1}(p_3^{\#})^{\nu_3}\dots) = \|\nu\|, \quad \deg_1((p_1^{\#})^{\nu_1}(p_3^{\#})^{\nu_3}\dots) \le \|\nu\|'.$$

By Proposition 3.7, we know all coefficients a_ν with the maximal value of $\|\nu\|$ (it equals $2k$), while we need all coefficients with $\|\nu\|' \ge 2k-1$. By (6.10), $\|\nu\|'$ does not exceed $2k$, and there are two possible cases:

- $\|\nu\|' = \|\nu\| = 2k$. Then $\nu_3 = \nu_5 = \dots = 0$, $k = \nu_1$. The corresponding monomial is $(p_1^{\#})^k$.
- $\|\nu\|' = 2k-1$, $\|\nu\| = 2k$. Then $\nu_3 + \nu_5 + \dots = 1$, i.e., exactly one of the numbers $\nu_3, \nu_5, \dots$ equals 1. The corresponding monomial is of the form

$$(6.11)\qquad p^{\#}_{2k-1-2j}(p_1^{\#})^j, \qquad j = \nu_1 = 0, 1, \dots, k-2.$$

In the first and second cases, $\|\nu\|$ takes the maximal value k, and then the coefficients a_ν are known from Proposition 3.7.

Thus, we have proved that the top degree component of $\widetilde{p}_{2k}$ is obtained from the top weight terms as given in Proposition 3.7; we simply keep all terms proportional either to $(p_1^{\#})^k$ or to a monomial of the form (6.11), and we remove all the remaining terms. This procedure leads to (6.9). □

In the next proposition we invert (6.8).

PROPOSITION 6.5. *For any $k \ge 1$,*

$$(6.12)\qquad \frac{p^{\#}_{2k+1}}{2^{k+\frac12}(p_1^{\#})^{k+\frac12}} = \sum_{j=0}^{k-1}(-1)^j \frac{2k+1}{2k+1-j}\binom{2k+1-j}{j} q_{2k+1-2j} + \dots,$$

where the dots mean a remainder term with $\deg_1(\,\cdot\,) < 0$.

PROOF. We employ the following combinatorial inversion formula (see [**R**, §2.4, (10)]):

Let $a_1, a_3, \dots, b_1, b_3, \dots$ be formal variables. Then

$$\left\{ a_k = \sum_{j=0}^{[k/2]} \binom{k}{j} b_{k-2j} \right\}_{k=1,3,\dots} \tag{6.13}$$

$$\iff \left\{ b_k = \sum_{j=0}^{[k/2]} (-1)^j \frac{k}{k-j} \binom{k-j}{j} a_{k-2j} \right\}_{k=1,3,\dots}.$$

Set

$$a_1 = 0, \quad a_{2k+1} = q_{2k+1} \quad (k \geq 1); \qquad b_1 = 0, \quad b_{2k+1} = \frac{p^\#_{2k+1}}{2^{k+\frac{1}{2}} (p^\#_1)^{k+\frac{1}{2}}} \quad (k \geq 1).$$

The relations (6.8) coincide with the first system in (6.13), up to remainder terms of negative degree. These terms can be neglected, because they affect only similar remainder terms in the inverse relations. These inverse relations are given then by the second system in (6.13). This leads to (6.12). □

PROOF OF THEOREM 6.1. We rewrite (6.12) as follows. For any $k \geq 2$,

$$\sum_{j=0}^{k-1} (-1)^j \binom{2k-j}{j} \frac{q_{2k+1-2j}}{2k+1-2j} = \frac{1}{\sqrt{2(2k+1)}} \frac{p^\#_{2k+1}}{2^k (p^\#_1)^{k+\frac{1}{2}} \sqrt{2k+1}} + R_k\,,$$

where $R_k \in \mathbb{A}^{\text{ext}}$ is a certain element such that $\deg_1(R_k) < 0$.

In the left-hand side, we may extend the summation up to k, because $q_1 = 0$. Comparing this with formula (6.3) for $u_{2k}(x)$ and formula (6.7) for the moments of Δ_λ, we conclude that

$$u^{(n)}_{2k} = \frac{1}{\sqrt{2(2k+1)}} \eta^{(n)}_k + R^{(n)}_k, \qquad k \geq 2.$$

As $n \to \infty$, the asymptotics of the (mixed) moments of the random variables $u^{(n)}_2, u^{(n)}_4, \dots$ is the same as that for the random variables $\frac{1}{\sqrt{2(2k+1)}} \eta^{(n)}_k$, $k = 1, 2, \dots$. Indeed, the remainder terms of negative degree do not affect the asymptotics; see Proposition 5.2. As for the moments of the random variables $\eta^{(n)}_k$, their asymptotics has been evaluated in the proof of Theorem 5.1. This concludes the proof. □

References

[B] A. M. Borodin, *Multiplicative central measures on the Schur graph,* J. Math. Sci. **96** (1999), 3472–3477, `http://www.pdmi.ras.ru/znsl/1997/v240.html`.

[E] A. Erdelyi (ed.), *Higher transcendental functions,* Vol. 2, McGraw–Hill, 1953.

[EO] A. Eskin, A. Okounkov, *Asymptotics of numbers of branched coverings of a torus and volumes of moduli spaces of holomorphic differentials,* Invent. Math. **145** (2001), 59–103, `arXiv:math.AG/0006171`.

[F] J. Fulman, *Stein's Method and Plancherel Measure of the Symmetric Group,* Trans. Amer. Math. Soc. **357** (2005), 555–570.

[HH] P. N. Hoffman, J. F. Humphreys, *Projective Representations of the Symmetric Groups,* Oxford University Press, 1992.

[I] V. Ivanov, *Gaussian limit for projective characters of large symmetric groups,* J. Math. Sci. **121** (2004), 2330–2344, `http://www.pdmi.ras.ru/znsl/2001/v283.html`.

[IO] V. Ivanov, G. Olshanski, *Kerov's central limit theorem for the Plancherel measure on Young diagrams,* Symmetric functions 2001, S. Fomin, ed., Kluwer Academic Publishers, 2002, pp. 93–151, arXiv:math.CO/0304010.

[J] T. Józefiak, *Characters of projective representations of symmetric groups,* Expositiones Math. **7** (1989), 193–247.

[K1] S. Kerov, *Gaussian limit for the Plancherel measure of the symmetric group,* Comptes Rendus Acad. Sci. Paris, Série I **316** (1993), 303–308.

[K2] S. V. Kerov, *Transition Probabilities of Continual Young Diagrams and Markov Moment Problem,* Funktsion. Anal. i Prilozhen. **27**:2 (1993), 32–49; English translation: Funct. Anal. Appl. **27** (1993), 104–117.

[K3] S. V. Kerov, *The differential model of growth of Young diagrams,* Proc. St. Petersburg Math. Soc. **4** (1996), 167–194.

[K4] S. Kerov, *Interlacing measures* , Kirillov's seminar on representation theory, G. Olshanski, ed., Amer. Math. Soc., Providence, RI, 1998, pp. 35–83.

[LS] B. F. Logan, L. A. Shepp, *A variational problem for random Young tableaux,* Advances in Math. **26** (1977), 206–222.

[Ma] I. G. Macdonald, *Symmetric functions and Hall polynomials,* 2nd edition, Oxford University Press, 1995.

[Mo] A. O. Morris, *A survey on Hall-Littlewood functions and their application to representation theory,* Lect. Notes Math. **579** (1977), 136–154.

[N] M. L. Nazarov, *Projective representations of the infinite symmetric group* , Representation Theory and Dynamical Systems, Advances in Soviet Mathematics, Amer. Math. Soc., A. M. Vershik, ed., vol. 9, 1992, pp. 115–130.

[P] P. Pragacz, *Algebro-geometric applications of Schur S- and Q-polynomials,* Lect. Notes Math. **1478** (1991), 130–191.

[R] J. Riordan, *Combinatorial identities,* Wiley, N.Y., 1968.

[Sc] I. Schur, *Uber die Darstellung der symmetrischen und der alternierenden Gruppe durch gebrocheme lineare Substitionen,* J. Reine Angew. Math. **139** (1911), 155–250.

[Se1] A. N. Sergeev, *Tensor algebra of the identity representation as a module over Lie superalgebras $Gl(n,m)$ and $Q(n)$,* Mat. Sb. **123** (1984), 422–430.

[Se2] A. N. Sergeev, *Laplace Operators and Representations of Lie Superalgebras,* Ph.D Thesis, Moscow State University, Moscow, 1985 (in Russian).

[Sh] A. Shiryaev, *Probability,* Springer-Verlag, New York, 1996.

[St] J. R. Stembridge, *Shifted tableaux and the projective representations of symmetric groups,* Adv. Math. **74** (1989), 87–134.

[Sz] G. Szegö, *Orthogonal polynomials,* Amer. Math. Soc. Colloquium Publ., Vol. XXIII, N.Y., 1959.

[VK1] A. M. Vershik, S. V. Kerov, *Asymptotics of the Plancherel measure of the symmetric group and the limiting form of Young tableaux,* Doklady Akad. Nauk SSSR **233** (1977), 1024–1027; English translation: Soviet Math. Dokl. **18** (1977), 527–531.

[VK2] A. M. Vershik, S. V. Kerov, *Asymptotic theory of characters of the symmetric group,* Function. Anal. i Prilozhen. **15**:4 (1981), 15–27; English translation: Funct. Anal. Appl. **15** (1985), 246–255.

[VK3] A. M. Vershik, S. V. Kerov, *Asymptotics of the largest and the typical dimensions of irreducible representations of a symmetric group,* Funktsional. Anal. i Prilozhen. **19**:1 (1985), 25–36; English translation: Funct. Anal. Appl. **19** (1985), 21–31.

[W] A. J. Wassermann, *Automorphic actions of compact groups on operator algebras,* Thesis, University of Pennsylvania, 1981.

[Y] M. Yamaguchi, *A duality of the twisted group algebra of the symmetric group and a Lie superalgebra,* J. Algebra **222** (1999), 301–327.

Moscow Independent University, Bolshoy Vlasyevskiy Pereulok 11, Moscow 119002, Russia

E-mail address: vivanov@mccme.ru

Amer. Math. Soc. Transl.
(2) Vol. **217**, 2006

Results and Problems in Enveloping Algebras Arising from Quantum Groups

Anthony Joseph

ABSTRACT. Some results and problems on enveloping algebras motivated by basic structure theorems for their quantum analogues are presented. An ultimate goal is the calculation of KPRV determinants pertaining to the quantization of Richardson orbits.

1. Introduction and notation

1.1. The theory of quantum groups introduced by Drinfeld and by Jimbo nearly 20 years ago appeared at first to do little more than decorate known results in enveloping algebras with powers of q. However this perception soon changed dramatically particularly with the introduction of the canonical or global basis but also on a more humble level through Drinfeld duality, which implied a much tighter relationship of the quantized enveloping algebra to its Hopf dual. In this paper we describe some results and open problems in enveloping algebra theory arising from these aspects of quantum groups.

1.2. In more detail let $\mathfrak{g}$ be a complex semisimple Lie algebra and $U_q(\mathfrak{g})$ the "simply-connected" Drinfeld–Jimbo quantization of the enveloping algebra $U(\mathfrak{g})$ of $\mathfrak{g}$. With respect to a triangular decomposition $\mathfrak{g} = \mathfrak{n} \oplus \mathfrak{h} \oplus \mathfrak{n}^-$, let π denote the set of simple roots and P (resp. P^+) the set of weights (resp. dominant weights). Define an order relation $\prec$ on P^+ through $\mu \prec \nu$ given $\nu - \mu \in P^+$. For each $\mu \in P^+$, let $V(\mu)$ be the simple $U(\mathfrak{g})$ with highest weight μ.

1.3. As an analogue of the basic structure theorem [**Jo1**, 7.1.6] for the locally finite part of $U_q(\mathfrak{g})$ we construct a version of Kostant's harmonic space $\mathbb{H}$ that can be expressed as a direct limit

$$\mathbb{H} = \varinjlim E(\mu)$$

where $E(\mu)$ maps isometrically to $End\, V(\mu)$ under the action of $U(\mathfrak{g})$ on $V(\mu)$. Here the main techniques are the Gelfand–Kirillov construction [**GK**] and the graded injectivity of the Conze embedding [**Jo2**].

2000 *Mathematics Subject Classification.* Primary 17B35; Secondary 17B37.
Key words and phrases. Gelfand–Kirillov twisting, Conze embedding.
This work was supported in part by the Minerva Foundation, Germany, Grant 8466.
The author would like to thank I. Heckenberger and V. Hinich for useful discussions.

1.4. Three interrelated problems arise from this circumstance. First, can one calculate the degrees of the simple $U(\mathfrak{g})$ submodules of a given $E(\mu)$? Can one give an appropriate analogue $\mathbb{H}_{\pi'}$ of $\mathbb{H}$ for any $\pi' \subset \pi$ as is possible in the quantum case [**JT**, 3.6]? Does one have $gr Ann\ V(\mu) \subset gr Ann\ V(\nu)$ given $\mu \prec \nu$? Here the goal was to define and calculate classical analogues of the KPRV determinants described in [**JT**, 3.6], which in turn relates to the problem of quantizing Richardson orbits. This becomes delicate if the appropriate moment map is not birational with normal image (see [**Jo2**, 5.4] and [**JLT**, Section 8]).

1.5. The construction and computation of the KPRV determinants described in [**JLT**] required the graded injectivity of the generalized Conze embedding relative to π'. As noted in [**Jo2**, 5.2, 5.3], quasi-injectivity suffices and this is obtained from a result of B. Broer [**B**, Theorem 6.5]. In that paper graded injectivity was established for $\pi' = \emptyset$ by adapting an earlier construction of Broer; but the further deduction of graded injectivity for arbitrary π' (given in [**Jo2**, 6.2]) is erroneous and it remains a seemingly hard open problem. Here we give a second easier proof of graded injectivity for the case $\pi' = \emptyset$ which avoids spectral sequences and also gives a projective resolution of $Ind\ \Lambda^*\mathfrak{n}^-$.

1.6. Let $\mathfrak{p}_{\pi'}$ (or simply $\mathfrak{p}$) be the parabolic subalgebra defined by π', setting $\mathfrak{p}_\emptyset = \mathfrak{b}$, and let $\mathfrak{p}^-$ (resp. $\mathfrak{b}^-$) denote its opposed algebra. Let $\mathfrak{m}$ (resp. $\mathfrak{r}$) denote the nilradical (resp. Levi factor) of $\mathfrak{p}$. The Chevalley–Kostant construction involving the Clifford algebra [**K**] can be viewed as a quantum deformation introduced before quantum group were invented. It gives a $\mathfrak{g}$ module structure on $\Lambda^*\mathfrak{b}^-$ (as well as on $\Lambda^*\mathfrak{n}^-$) which is a deformation of its $\mathfrak{n}$ module structure, the latter obtained by identifying $\mathfrak{b}^-$ with $\mathfrak{b}^*$ through the Killing form. As a consequence $Ind\ \Lambda^*\mathfrak{b}$ (or $Ind\ \Lambda^*\mathfrak{n}^-$) is projective for this deformed action. Assuming the graded injectivity of the Conze embedding gives a projective resolution of $Ind\ \Lambda^*\mathfrak{p}^-$; but we do not know if the latter admits a deformation which is already projective as in the case $\pi' = \emptyset$.

2. A structural lemma for an analogue of the harmonic space

2.1. Let $\mathfrak{g}$ be a split semisimple Lie algebra over a field k of characteristic zero and M a $\mathfrak{g}$ module. For any subalgebra $\mathfrak{a}$ of $\mathfrak{g}$ let $U(\mathfrak{a})$ (resp. $S(\mathfrak{a})$) denote the enveloping (resp. symmetric) algebra over $\mathfrak{a}$ and set $F_{\mathfrak{a}}(M) = \{m \in M \mid dim\ U(\mathfrak{a})m < \infty\}$. Fix a triangular decomposition $\mathfrak{g} = \mathfrak{n} \oplus \mathfrak{h} \oplus \mathfrak{n}^-$ and set $\mathfrak{b} = \mathfrak{n} \oplus \mathfrak{h}$, $\mathfrak{b}^- = \mathfrak{h} \oplus \mathfrak{n}^-$, which are Borel subalgebras. For each $\lambda \in \mathfrak{h}^*$, let k_λ denote the one-dimensional $\mathfrak{b}$ module of weight λ and $M(\lambda) := U(\mathfrak{g}) \otimes_{U(\mathfrak{b})} k_\lambda$, the Verma module of highest weight λ.

2.2. Let us recall a construction of Gelfand and Kirillov [**GK**] as developed by Conze [**Co**] and modified in [**JLT, Jo2**]. In this we consider $End\ M(\lambda)$ as a $\mathfrak{g}$ module for the diagonal action.

By a result of Kostant the embedding

$$U^\lambda := U(\mathfrak{g})/Ann\ M(\lambda) \hookrightarrow F_{\mathfrak{g}}(End\ M(\lambda))$$

is an isomorphism. Obviously $U^\lambda \hookrightarrow A^\lambda := F_{\mathfrak{b}^-}(End\ M(\lambda))$. The latter also has a ring structure and is indeed a Weyl algebra of index $dim\ \mathfrak{n}^-$. This can be made explicit as follows.

2.3. Let Δ (resp. Δ^+) denote the set of non-zero (resp. positive) roots. As an A^λ module, $M(\lambda)$ identifies with $\mathbf{Q}:= k[q_{-\alpha} : \alpha \in \Delta^+]$. Its canonical generator $\hat{v}_\lambda$ identifies with $1 \in Q$ and is annihilated by the augmentation P_+ of $P := k[p_\alpha : \alpha \in \Delta^+]$ where $p_\alpha = \partial/\partial q_{-\alpha}$. The subalgebra A of $End\ M(\lambda)$ generated by Q and P coincides with A^λ as an algebra.

2.4. For each $\alpha \in \Delta$, let $s_\alpha \in Aut\ \mathfrak{h}^*$ denote the reflection it defines and let $W_{\pi'} := \langle s_\alpha : \alpha \in \pi' \rangle$. Then W_π, or simply W, is the Weyl group. Let ρ denote the half sum of positive roots and set $w.\lambda = w(\lambda + \rho) - \rho$, for all $\lambda \in \mathfrak{h}^*, w \in W$. Let $w_{\pi'}$ denote the unique longest element of $W_{\pi'}$. Set $\alpha^\vee = 2\alpha/(\alpha, \alpha)$, for all $\alpha \in \Delta$. Call $\lambda + \rho \in \mathfrak{h}^*$ regular if $\alpha^\vee(\lambda + \rho) \neq 0$, $\forall \alpha \in \Delta$, and dominant if $\alpha^\vee(\lambda+\rho) \not< 0$, $\forall \alpha \in \Delta^+$. Call $\lambda + \rho$ antidominant if $w_\pi(\lambda + \rho)$ is dominant.

Let $\mathcal{O}$ denote the category of left $U(\mathfrak{g})$ modules having a direct sum decomposition into their weight subspaces and on which $U(\mathfrak{b})$ acts locally finitely. Given $M \in Ob\mathcal{O}$, set $\delta M = F_{\mathfrak{h}}(M^*) \in Ob\mathcal{O}$ where M^* is considered as a left $U(\mathfrak{g})$ module through a fixed Chevalley antiautomorphism κ. The resulting functor δ on $\mathcal{O}$ is exact and contravariant.

Recall a construction of [**JLT**, 2.1]. Set $I = Ann_A \hat{v}_\lambda$. Then A/I is an $A - U(\mathfrak{b})$ module, hence a $U(\mathfrak{b})$ module under diagonal action and moreover isomorphic to $M(\lambda) \mid_{U(\mathfrak{b})} \otimes k_{-\lambda}$.

Now assume $\lambda + \rho$ antidominant. Then $M(\lambda)$ is simple and $M(\lambda) \cong \delta M(\lambda)$. Hence

$$M(\lambda) \mid_{U(\mathfrak{b})} \otimes k_{-\lambda} \cong \delta M(\lambda) \mid_{U(\mathfrak{b})} \otimes k_{-\lambda} \cong \delta M(0).$$

As noted in [**JLT**, 2.4], this leads to an isomorphism

$$A^\lambda \xrightarrow{\sim} F_{\mathfrak{b}^-}(Hom_{U(\mathfrak{b})}(U(\mathfrak{g}), \delta M(0)))$$

of $U(\mathfrak{g})$ modules with respect to the diagonal action on A^λ. Moreover by, say, [**Jo1**, 8.1.6] we further have

$$\begin{aligned} F_{\mathfrak{b}^-}(Hom_{U(\mathfrak{b})}(U(\mathfrak{g}), \delta M(0))) &\xrightarrow{\sim} \delta M(0) \otimes F_{\mathfrak{b}^-}(Hom_{U(\mathfrak{b})}(U(\mathfrak{g}), k_0)) \\ &\xrightarrow{\sim} \delta M(0) \otimes \delta M(0)^\iota \end{aligned}$$

where ι is the product of the Chevalley κ and principal σ antiautomorphisms. In particular, (for $\lambda + \rho$ antidominant) the isomorphism class A of A^λ is independent of λ. Moreover A is injective in the ι transport $\mathcal{O}^-$ of the $\mathcal{O}$ category [**Jo2**, 2.6].

LEMMA. *For each $\mu \in P^+$ one has a unique up to scalars embedding $V(\mu) \otimes k_{-\mu} \hookrightarrow \delta M(0)$.*

PROOF. Drinfeld duality gives, in an appropriate $q \to 1$ limit [**Jo1**, 3.4.6], a non-degenerate $U(\mathfrak{n})$ invariant bilinear form $F : U(\mathfrak{n}) \times S(\mathfrak{n}^-) \to k$. This may be also obtained directly by identifying $\delta M(0) = F_{\mathfrak{h}}(M(0)^*) = F_{\mathfrak{h}}((U(\mathfrak{g})/U(\mathfrak{g})\mathfrak{b})^*)$ with $S(\mathfrak{n}^-)$ and setting $F(a, b) = \varepsilon(\sigma(a)b)$, for all $a \in U(\mathfrak{n}), b \in S(\mathfrak{n}^-)$, where ε is the augmentation.

Let $V(-\mu)$ denote the simple $U(\mathfrak{g})$ module with lowest weight vector $v_{-\mu}$ of weight $-\mu$. Set $\widetilde{V(\mu)} = \{b \in S(\mathfrak{n}^-) \mid F(a, b) = 0,\ \forall a \in Ann_{U(\mathfrak{n})} v_{-\mu}\}$. Then $\widetilde{V(\mu)}$ is a $U(\mathfrak{n})$ submodule of $S(\mathfrak{n}^-)$ isomorphic to

$$(U(\mathfrak{n})/Ann_{U(\mathfrak{n})} v_{-\mu})^* \xrightarrow{\sim} V(-\mu)^* \xrightarrow{\sim} V(\mu).$$

This becomes a $U(\mathfrak{b})$ module isomorphism if we take $1 \in S(\mathfrak{n}^-)$ to have weight μ. This establishes the existence of the required embedding.

Conversely suppose $\widetilde{V(\mu)}$ is a $U(\mathfrak{b})$ submodule of $S(\mathfrak{n}^-)$ isomorphic to $V(\mu)$. Its orthogonal $\widetilde{V(\mu)}^\perp$ with respect to F is a $U(\mathfrak{b})$ submodule of $U(\mathfrak{n})$ and the canonical projection to the quotient factors to a $U(\mathfrak{b})$ module isomorphism

$$\varphi : U(\mathfrak{n})/\widetilde{V(\mu)}^\perp \xrightarrow{\sim} V(\mu)^* \xrightarrow{\sim} V(-\mu).$$

Then $\varphi(1)$ is the unique up to scalars weight vector of $V(-\mu)$ which generates $V(-\mu)$ as a $U(\mathfrak{n})$ module and so must be proportional to $v_{-\mu}$. This forces $\widetilde{V(\mu)}^\perp = Ann_{U(\mathfrak{n})}v_{-\mu}$, as required. □

2.5. Let M be a finite-dimensional $U(\mathfrak{b})$ module. Then

$$M \mapsto Coind\ M := F_{\mathfrak{h}}(Hom_{U(\mathfrak{b})}(U(\mathfrak{g}), M)) = F_{\mathfrak{b}^-}(Hom_{U(\mathfrak{b})}(U(\mathfrak{g}), M))$$

is an exact covariant functor [**Jo2**, 4.1]. An exact covariant functor F on an abelian category of modules commutes with sums of submodules of a given module. Indeed let M_1, M_2 be submodules of M and consider the map $\varphi(m_1, m_2) = m_1 + m_2$ of $M_1 \oplus M_2$ onto $M_1 + M_2$ with kernel $M_1 \cap M_2$. Certainly F commutes with direct sums, so by covariance and exactness we obtain an exact sequence

$$0 \to F(M_1 \cap M_2) \to F(M_1) \oplus F(M_2) \xrightarrow{F(\varphi)} F(M_1 + M_2) \to 0.$$

Yet $ker\ F(\varphi) = F(M_1) \cap F(M_2)$ and $Im\ F(\varphi) = F(M_1) + F(M_2)$ and so the assertion is obtained.

2.6. Take $\mu, \nu \in P^+$ with $\mu \succ \nu$. Then $Ann_{U(\mathfrak{n})}v_{-\mu} \subset Ann_{U(\mathfrak{n})}v_{-\nu}$, which gives a surjection $V(-\mu) \twoheadrightarrow V(-\nu)$ of $U(\mathfrak{n})$ modules. Dualizing using σ gives an injection $V(\nu) \hookrightarrow V(\mu)$ of $U(\mathfrak{n})$ modules and then an injection $V(\nu) \otimes k_{-\nu} \hookrightarrow V(\mu) \otimes k_{-\mu}$ of $U(\mathfrak{b})$ modules. On the other hand, for any $\xi \in \mathbb{N}\pi$ the embedding $V(\mu)_{\mu-\xi} \hookrightarrow \delta M(\mu)_{\mu-\xi}$ becomes an isomorphism for μ sufficiently large. In view of the above and **2.4** we obtain the presentation

$$\delta M(0) = \varinjlim\ V(\mu) \otimes k_{-\mu}$$

where the direct limit is taken with respect to the above maps. Observe that

$$F_{\mathfrak{h}}(Hom_{U(\mathfrak{b})}(U(\mathfrak{g}), k_{-\mu})) \cong (\delta M(\mu))^\iota$$

and is injective in $\mathcal{O}^-$. In view of **2.5** and the above we obtain

$$(*) \qquad A \xrightarrow{\sim} F_{\mathfrak{b}^-}(Hom_{U(\mathfrak{b})}(U(\mathfrak{g}), \delta M(0))) \xrightarrow{\sim} \varinjlim\ V(\mu) \otimes (\delta M(\mu))^\iota,$$

which is as a direct limit of modules injective in $\mathcal{O}^-$. We remark as noted in [**Jo2**, 6.4] that this gives a filtration on A whose associated graded module is again injective in $\mathcal{O}^-$. The proof given in [**Jo2**, 6.4] was not quite complete (because [**Jo2**, 4.7(*)] is erroneous) and needs the following lemma, which is a nice application of canonical bases. See also [**Jo2**, 6.5].

LEMMA. *For any finite subset S of P^+ the submodule*

$$\sum_{\mu \in S} V(\mu) \otimes (\delta M(\mu))^\iota$$

is injective in $\mathcal{O}^-$.

PROOF. Set $J(\mu) = V(\mu)\otimes(\delta M(\mu))^\iota$. Since $\delta M(\mu)^\iota$ is injective in $\mathcal{O}^-$ and $V(\mu)$ is finite dimensional, $J(\mu)$ is injective in $\mathcal{O}^-$. On the other hand, by the construction of **2.4** and the common basis theorem [**Jo1**, 6.2.19] for the dual canonical bases (obtained from the work of Kashiwara or of Lusztig) we conclude that the $V(\mu)\otimes k_{-\mu} : \mu \in P^+$ form a distributive lattice of subspaces of $\delta M(0)$. By **2.5**, the $J(\mu) : \mu \in P^+$ form a distributive lattice of subspaces of A. In particular, for any $\mu \in S$ one has

$$\left(\sum_{\nu\in S\setminus\{\mu\}} J(\nu)\right)\cap J(\mu) = \sum_{\nu\in S\setminus\{\mu\}} J(\nu)\cap J(\mu).$$

Finally one checks (as in [**Jo1**, 7.3.1]) that

$$(V(\mu)\otimes k_{-\mu})\cap(V(\nu)\otimes k_{-\nu}) = V(\mu\cap\nu)\otimes k_{-(\mu\cap\nu)},$$

and so $J(\nu)\cap J(\mu) = J(\mu\cap\nu)$, where $\mu\cap\nu$ is the unique maximal element of P^+ less than both μ and ν. Then the required assertion follows by induction on *card* S. □

2.7. After Duflo, the annihilator of any Verma module $M(\lambda)$ is generated by its intersection with the centre $Z(\mathfrak{g})$ of $U(\mathfrak{g})$. After Kostant, $U(\mathfrak{g})$ is a free $Z(\mathfrak{g})$ module and indeed is freely generated over an $ad\mathfrak{g}$ invariant subspace $\mathbb{H}$ which can be taken to be the span of powers of *ad*-nilpotent elements. We call this the harmonic subspace of $U(\mathfrak{g})$. It identifies with $U(\mathfrak{g})/Ann\ M(\lambda)$ and then with $F_\mathfrak{g}(End\ M(\lambda))$ which is just $F_\mathfrak{g}(A^\lambda)$. In virtue of **2.6**$(*)$ we obtain the following presentation of $\mathbb{H}$ as a $U(\mathfrak{g})$ module.

LEMMA.

$$\mathbb{H} = \varinjlim V(\mu)\otimes V(-\mu).$$

PROOF. Recall that for all $\mu \in P^+$ one has $F_\mathfrak{g}(\delta M(\mu))^\iota \cong V(-\mu)$ and in particular that $K := (\delta M(\mu))^\iota/V(-\mu)$ has no finite-dimensional submodules. Suppose $V(\mu)\otimes K$ admits a finite-dimensional submodule V. Then $0 \neq Hom(V, V(\mu)\otimes K) \xrightarrow{\sim} Hom_{U(\mathfrak{g})}(V\otimes V(\mu)^*, K)$, which gives a contradiction to the first part. Hence $F_\mathfrak{g}(V(\mu)\otimes(\delta M(\mu))^\iota) = V(\mu)\otimes V(-\mu)$. Finally any finite-dimensional submodule V of A is contained in some $V(\mu)\otimes\delta M(\mu)^\iota$ for μ sufficiently large. Hence the assertion. □

2.8. We regard **2.7** as the most reasonable possible analogue of the basic structure theorem [**Jo1**, 7.1.6] for the locally finite part of $U_q(\mathfrak{g})$, though Caldero [**Ca**] has forced a rather artificial version using a $q \to 1$ limit. Setting $E(\mu) = V(\mu)\otimes V(-\mu)$, one may ask if the analogue of [**Jo1**, 7.2.5] also holds. That is, does the natural action of $U(\mathfrak{g})$ on $V(\mu)$ induce an isomorphism of $E(\mu)$ onto $End\ V(\mu)$? We shall settle this positively using Gelfand–Kirillov twisting. This method is quite different from that used in the quantum case, which required a positive definite form coming from canonical bases [**Jo1**, Section 7].

3. Gelfand–Kirillov twisting

3.1. For all $\lambda, \lambda' \in \mathfrak{h}^*$ consider $Hom(M(\lambda), M(\lambda'))$ as a $\mathfrak{g}$ module for diagonal action and set

$$A^{\lambda,\lambda'} = F_{\mathfrak{n}^-}(Hom(M(\lambda), M(\lambda'))) = F_{\mathfrak{b}^-}(Hom(M(\lambda), M(\lambda'))), F^{\lambda,\lambda'} = F_\mathfrak{g}(A^{\lambda,\lambda'}).$$

Of course, $A^{\lambda,\lambda} = A^\lambda$ and $F^{\lambda,\lambda} = U^\lambda$.

Let $\theta^{\lambda,\lambda'}$ be the $U(\mathfrak{n}^-)$ module map sending $\hat{v}_\lambda$ to $\hat{v}_{\lambda'}$. Obviously $\theta^{\lambda,\lambda'} \in A^{\lambda,\lambda'}$. Composition of homomorphisms gives bilinear maps

$$A^{\lambda'',\lambda'} \times A^{\lambda,\lambda''} \longrightarrow A^{\lambda,\lambda'}, \; F^{\lambda'',\lambda'} \times F^{\lambda,\lambda''} \longrightarrow F^{\lambda,\lambda'}$$

for all $\lambda, \lambda', \lambda'' \in \mathfrak{h}^*$. In particular, $A^{\lambda,\lambda'} = \theta^{\lambda,\lambda'} A^\lambda$. Thus ignoring $U(\mathfrak{g})$ module structure, these maps are just multiplication in the common algebra A.

3.2. As noted by Conze [**Co**], the image of the composed map $\theta_\lambda : \mathfrak{g} \hookrightarrow U^\lambda \hookrightarrow A$ lies in

$$(*) \qquad \sum_{\alpha \in \Delta^+} q_{-\alpha} P + P.$$

As a consequence P is a submodule of A under the diagonal action of $\mathfrak{g}$.

Observe that by $(*)$ A^λ acquires a filtration $\mathcal{F}$ of $U(\mathfrak{g})$ modules under diagonal action induced from the degree filtration on Q. Taking λ antidominant, so that $M(\lambda) \mid_{U(\mathfrak{b})} \otimes k_{-\lambda} \cong \delta M(0)$, this filtration on Q induces a filtration $\mathcal{F}$ of $U(\mathfrak{b})$ modules on $\delta M(0)$ and moreover the original filtration on A is recovered by $\mathcal{F}^n A = Coind\ \mathcal{F}^n(\delta M(0))$. In particular, $P = \mathcal{F}^0 A = Coind\ k_0 = \delta M(0)^\iota$. We remark that the induced filtration on $V(\mu) \otimes k_{-\mu} \hookrightarrow \delta M(0)$ is the right Brylinski–Kostant filtration described in [**HJ**].

3.3. We may twist the (diagonal) action of $U(\mathfrak{g})$ on P by identifying the latter with $P\theta^{\lambda,\lambda'} \subset A^{\lambda,\lambda'}$. For all $p \in P$ this action is defined by

$$(x, p\theta_{\lambda,\lambda'}) \mapsto \theta_{\lambda'}(x)(p\theta^{\lambda,\lambda'}) - (p\theta^{\lambda,\lambda'})\theta_\lambda(x).$$

As in **3.2** one obtains $P\theta^{\lambda,\lambda'} = Coind\ k_{\lambda-\lambda'} = (\delta M(\lambda - \lambda'))^\iota$. In particular, $\theta^{\lambda,\lambda-\mu} : \mu \in P^+$ is a lowest weight vector of weight $-\mu$. It generates the submodule $V(-\mu)$ of $P\theta^{\lambda,\lambda-\mu} \cong (\delta M(\mu))^\iota$.

3.4. We can also view P as a left A module in which P acts by left multiplication and $q_{-\alpha}$ by $-\partial/\partial p_\alpha : \alpha \in \Delta^+$. For our present purposes this is not useful. Instead we must view P as a right A module in which P acts by right multiplication and $q_{-\alpha}$ by $\partial/\partial p_\alpha : \alpha \in \Delta^+$. Through θ_λ we obtain a right U^λ module structure on P and we denote the resulting module by P_{U^λ}. Since $Ann\ M(\lambda)$ is a primitive ideal, it is κ-stable after Duflo [**D**, II, Corollary 2] and so P_{U^λ} may be viewed as a left U^λ module, denoted ${}_{U^\lambda}P$, through κ.

LEMMA. ${}_{U^\lambda}P \cong \delta M(\lambda)$.

PROOF. Recall [**JLT**, 1.4(iii)] that $P\theta_\lambda(U(\mathfrak{n}^-)) = A$ and so $(P_{U^\lambda})^{\mathfrak{n}^-} = k$. Consequently $({}_{U^\lambda}P)^n = k$. Now for $h \in \mathfrak{h}$, it is clear that $\theta_\lambda(h)$ is a linear combination of the $q_{-\alpha}p_\alpha : \alpha \in \Delta^+$ and hence necessarily given by

$$\theta_\lambda(h) = h(\lambda) - \sum_{\alpha \in \Delta^+} h(\alpha) q_{-\alpha} p_\alpha.$$

In particular, $1 \in P_{U^\lambda}$ (and hence $1 \in {}_{U^\lambda}P$) has weight λ. Moreover $ch_{U^\lambda}P = ch\ M(\lambda) = ch\ \delta M(\lambda)$. Then as in, say, [**Jo1**, 5.3.6], the assertion follows. □

REMARK. Had we taken the left action of A on P described in the first part of this paragraph, then P as a left U^λ module is isomorphic to $(\delta M(-\lambda - 2\rho))^\iota$. Here $-\lambda - 2\rho \in P^+$ means that $\lambda + \rho$ is regular, antidominant.

3.5. Consider $F^{\lambda,\lambda-\mu} : \lambda \in \mathfrak{h}^*, \mu \in P^+$. Through Frobenius reciprocity one checks that $V(-\mu)$ occurs with multiplicity one in $F^{\lambda,\lambda-\mu}$. (Indeed $V(-\mu)$ is the unique minimal "$\mathfrak{k}$-type" of $F^{\lambda,\lambda-\mu}$ with respect to the norm function $|\mu| := (\mu + \rho, \mu + \rho)^{\frac{1}{2}}$.) Since $\theta^{\lambda,\lambda-\mu} \in F^{\lambda,\lambda-\mu}$, this unique copy of $V(-\mu)$ identifies with a subspace $\widetilde{V(-\mu)}$ of P viewed as a submodule of $A^{\lambda,\lambda-\mu}$. More precisely $V(-\mu) = \theta^{\lambda,\lambda-\mu}\widetilde{V(-\mu)}$ for some $U(\mathfrak{b}^-)$ invariant subspace of P with $\widetilde{V(-\mu)}^{\mathfrak{b}^-} = k$. We remark that $\widetilde{V(-\mu)}$ depends only on μ and not on λ.

Similarly $F^{\lambda-\mu,\lambda} : \mu \in P^+$ admits exactly one copy of $V(\mu)^\lambda$ (as its unique minimal $\mathfrak{k}$-type). However in this case the description of $V(\mu)^\lambda$ is more subtle and in particular it is not generated by $\theta^{\lambda-\mu,\lambda}$ which besides does not belong to $F^{\lambda-\mu,\lambda}$. Again $V(\mu)^\lambda$ (or more properly $V(\mu)^\lambda\theta^{\lambda,\lambda-\mu} \subset P$) depends on λ as well as on μ. Let v_μ^λ denote the unique up to scalars highest weight vector of $V(\mu)^\lambda$.

Set $\widetilde{E(\mu)}^\lambda = V(\mu)^\lambda V(-\mu)$ defined by composition of homomorphisms. It is a $U(\mathfrak{g})$ submodule of U^λ under diagonal action and is generated by $\tau^\lambda(-2\mu) := v_\mu^\lambda v_{-\mu}$. (This notation is suggested by the quantum case [**Jo1**, 7.1.6].) Recalling that $v_{-\mu} = \theta^{\lambda,\lambda-\mu}$, we may compute $\tau^\lambda(-2\mu)$ by observing that $v_\mu^\lambda \in F^{\lambda-\mu,\lambda}$ is the unique up to scalars element of A of highest weight μ. Hence it is the unique up to scalars element of A commuting with the $\sum_{\alpha\in\Delta^+} \alpha(h)q_{-\alpha}p_\alpha$ and satisfying

$$(*) \qquad \theta_\lambda(x)v_\mu^\lambda - v_\mu^\lambda\theta_{\lambda-\mu}(x) = 0, \ \forall \ x \in \mathfrak{n}.$$

EXAMPLE. Take $\mathfrak{g} = \mathfrak{sl}_2$ with standard generators $x, h = \alpha^\vee, y$. Then

$$\theta_\lambda(y) = q, \ \theta_\lambda(h) = \lambda(h) - 2qp, \ \theta_\lambda(x) = \lambda(h)p - qp^2.$$

Set $n = \alpha^\vee(\mu)$ and $c = \lambda(h)$, $z = qp$, $v = v_\mu^\lambda$. Then v is a polynomial $v(z)$ in z and $(*)$ becomes

$$(c - (z-1))v(z) = (c - n - (z-1))v(z).$$

Consequently

$$v_\mu^\lambda = \prod_{i=0}^{\alpha^\vee(\mu)-1} (\lambda(h) - qp - i),$$

up to a non-zero scalar. This shows that v_μ^λ depends on λ. Notice that in the right action of U^λ on P, the element $v_\mu^\lambda v_{-\mu}$ does not annihilate 1 if $\lambda(h) \notin \{0, 1, 2, \cdots, \alpha^\vee(\mu) - 1\}$, in particular if $\lambda = \mu$. We show below that the latter holds in general.

3.6. We say that $\lambda, \lambda-\mu : \lambda \in \mathfrak{h}^*, \mu \in P$ are compatible if they lie in the same facet in the sense of Jantzen [**Ja**, 2.6]. If $\lambda+\rho, \lambda-\mu+\rho$ are regular, this just means that $\lambda+\rho, \lambda-\mu+\rho$ lie in the same Weyl chamber. It is the only case we shall need. The following is well known, but we indicate a proof for completeness. For a little more detail, see [**JLZ**, 6.4].

LEMMA. *Take $\lambda \in \mathfrak{h}^*, \mu \in P^+$ so that $\lambda, \lambda-\mu$ are compatible. Then composition of homomorphisms gives*

(i) $V(-\mu)U^\lambda = F^{\lambda,\lambda-\mu}$;
(ii) $V(\mu)^\lambda F^{\lambda,\lambda-\mu} = U^\lambda$.

PROOF. After Jantzen [**Ja**, 2.6], the hypothesis implies that $M(\lambda-\mu)$ is the direct summand of $V(-\mu)\otimes M(\lambda)$ corresponding to the $Z(\mathfrak{g})$ primary component defined by $W.(\lambda-\mu)$. Consequently $F^{\lambda,\lambda-\mu}$ is the direct summand of $V(-\mu)\otimes F^{\lambda,\lambda}$ corresponding to this primary component with respect to the left action of $Z(\mathfrak{g})$. A comparison (see the claim in [**JLZ**, 6.4]) with composition of homomorphisms gives (i). Similarly $M(\lambda)$ is the appropriate direct summand of $V(\mu)\otimes M(\lambda-\mu)$ and this gives (ii). □

3.7. Recall the notation of **3.5**.

COROLLARY. *Assume* $\lambda, \lambda-\mu$ *compatible. Then* $U^\lambda \widetilde{E(\mu)}^\lambda = \widetilde{E(\mu)}^\lambda U^\lambda = U^\lambda$, *with respect to the left (resp. right) action of* U^λ.

PROOF. The first equality is just a consequence of $\widetilde{E(\mu)}^\lambda$ being a $U(\mathfrak{g})$ module under diagonal action. Finally

$$\begin{aligned}\widetilde{E(\mu)}^\lambda U^\lambda = V(\mu)^\lambda V(-\mu) U^\lambda = V(\mu)^\lambda F^{\lambda,\lambda-\mu}, &\quad \text{by } \mathbf{3.6}, \text{ Lemma, (i)}\\ = U^\lambda, &\quad \text{by } \mathbf{3.6}, \text{ Lemma, (ii)}.\end{aligned}$$

□

3.8. Now consider the right action of $\widetilde{E(\mu)}^\lambda \subset A$ on the identity 1_P of P. Let $\overline{V(\mu)}^\lambda$ denote the unique $\mathfrak{b}^-$-stable complement to kv_μ^λ in $V(\mu)^\lambda$.

LEMMA. (i) $1_P\overline{V(\mu)}^\lambda V(-\mu)=0$;

(ii) $1_P\widetilde{E(\mu)}^\lambda = \widetilde{V(-\mu)}$, *if* $\lambda, \lambda-\mu$ *are compatible.*

PROOF. With respect to the right action of $\theta_\lambda(h)$ the weights of P_{U^λ} lie in $\lambda-\mathbb{N}\pi$. On the other hand, for the right diagonal action (obtained from the left diagonal action through σ) of $U(\mathfrak{g})$ the weights of $\overline{V(\mu)}^\lambda v_{-\mu}$ are strictly positive. Hence $1_P\overline{V(\mu)}^\lambda v_{-\mu}=0$. Right multiplication by $\widetilde{V(-\mu)} \subset P$ gives (i).

Since $v_\mu^\lambda v_{-\mu}$ has weight zero, we can write $1_P v_\mu^\lambda v_{-\mu} = c1_P$, for some scalar c. Right multiplication by $\overline{V(-\mu)}$ gives $1_P v_\mu^\lambda V(-\mu) = c\widetilde{V(-\mu)}$. Moreover the left-hand side equals $1_P\widetilde{E(\mu)}^\lambda$ by (i). If $\lambda, \lambda-\mu$ are compatible and $c=0$, then $1_P \in 1_P U^\lambda = 1_P\widetilde{E(\mu)}^\lambda U^\lambda = 0$, which is absurd. Hence (ii) follows. □

3.9. The conclusion of **3.8**, Lemma, (ii) identifies $1_P\widetilde{E(\mu)}^\lambda = \widetilde{V(-\mu)}$ as a right $U(\mathfrak{b}^-)$ submodule of P_{U^λ} isomorphic to the right $U(\mathfrak{b}^-)$ module $k_{\lambda-\mu}\otimes V(-\mu)^\sigma$, where $k_{\lambda-\mu}$ is of highest weight $\lambda-\mu$ and $V(-\mu)^\sigma$ is the right $U(\mathfrak{b}^-)$ module obtained from the left $U(\mathfrak{b}^-)$ module $V(-\mu)$ using the principal antiautomorphism σ.

Now take $\lambda=\mu$. We conclude that $1_P\widetilde{E(\mu)}^\mu$ identifies as a left $U(\mathfrak{b})$ submodule of ${}_{U^\mu}P$ with $V(-\mu)^\iota = V(\mu)$. Yet ${}_{U^\mu}P \cong \delta M(\mu)$ and so admits a $U(\mathfrak{g})$ submodule isomorphic to $V(\mu)$ which by **2.4** must coincide with $1_P\widetilde{E(\mu)}^\mu$. (Actually **2.4** is not really needed here since we have an explicit expression, namely $\widetilde{V(-\mu)}$, for this subspace.) This gives the following:

PROPOSITION. *Take $\mu \in P^+$. The action of U^μ on $\delta M(\mu)$ restricts to an isomorphism of $\widetilde{E(\mu)}^\mu$ onto End $V(\mu)$.*

PROOF. The above action of $\widetilde{E(\mu)}^\mu$ on the submodule $V(\mu)$ of $\delta M(\mu)$ composed with the surjection $\psi : V(\mu) \otimes V(-\mu) \twoheadrightarrow \widetilde{E(\mu)}^\mu$ defines a $U(\mathfrak{g})$ module map φ of $V(\mu) \otimes V(-\mu)$ into *End* $V(\mu)$. By **3.8** the kernel of φ is a $U(\mathfrak{g})$ submodule contained in $\overline{V(\mu) \otimes V(-\mu)}$, hence zero. Indeed a highest weight vector in a non-zero submodule must start with a term in $v_\mu \otimes V(-\mu)$. A fortiori *ker* $\psi = 0$, and so the map $\widetilde{E(\mu)}^\mu \to$ *End* $V(\mu)$ is bijective. □

3.10. We have still to show that $\widetilde{E(\mu)}^\mu$ (or simply $\widetilde{E(\mu)}$) coincides with $E(\mu)$ defined in **2.8**. A priori this makes no sense since the embedding $U^\lambda \hookrightarrow A$ is only defined for $\lambda + \rho$ antidominant. However *Ann* $M(\lambda)$ is independent of λ in its W. orbit. Thus we may take $\lambda = w_\pi.\mu$ and then $\lambda + \rho = w_\pi(\mu + \rho)$ is antidominant. Then $U^\lambda = U^\mu$ and so it makes sense to consider the action of $\widetilde{E(\mu)}$ on $M(\lambda)$. Observe that $\widetilde{E(\mu)}\hat{v}_\lambda \cong \widetilde{E(\mu)}/\widetilde{E(\mu)} \cap Ann_A \hat{v}_\lambda$ is (for the diagonal action) a $U(\mathfrak{b})$ submodule of $M(\lambda) \mid_{U(\mathfrak{b})} \otimes k_{-\lambda} \cong \delta M(0)$.

LEMMA. *With the above hypotheses and conventions, $\widetilde{E(\mu)}\hat{v}_\lambda$ is isomorphic to $V(\mu) \otimes k_{-\mu}$ as a $U(\mathfrak{b})$ submodule of $\delta M(0)$.*

PROOF. Let $\overline{V(-\mu)}$ denote the unique $U(\mathfrak{b})$-stable complement to $kv_{-\mu}$ in $V(-\mu)$. Then $V(\mu)^\lambda \overline{V(-\mu)}\hat{v}_\lambda = V(\mu)^\lambda v_{-\mu}(\overline{V(-\mu)} \cap P_+)\hat{v}_\lambda = 0$. Thus $\widetilde{E(\mu)}\hat{v}_\lambda = V(\mu)^\lambda v_{-\mu}\hat{v}_\lambda$. Moreover the left-hand side is non-zero by **3.7** and is a $U(\mathfrak{b})$ submodule of $M(\lambda)$. Then $V := \{v \in V(\mu)^\lambda \mid vv_{-\mu}\hat{v}_\lambda = 0\}$ is a proper $U(\mathfrak{b})$ submodule of $V(\mu)^\lambda$ and so contains v_μ^λ if it is non-zero. On the other hand, suppose v is a weight vector of maximal weight such that $v \notin V$. Then $x_\alpha v \in V$ for all $\alpha \in \pi$ and so $vv_{-\mu}\hat{v}_\lambda$ is a highest weight vector of $M(\lambda)$, which by the simplicity of the latter must be proportional to $\hat{v}_\lambda$, forcing $v \in k^* v_\mu^\lambda$. This contradiction forces $V = 0$, proving the lemma. □

3.11. By **2.4** and **3.10**, $\widetilde{E(\mu)}\hat{v}_\lambda$ is the unique $U(\mathfrak{b})$ submodule $V(\mu) \otimes k_{-\mu}$ of $\delta M(0)$. Functoriality then gives a unique $U(\mathfrak{g})$ module map ψ making the diagram

$$\begin{array}{ccc} \widetilde{E(\mu)} & \longrightarrow & \widetilde{E(\mu)}\hat{v}_\lambda \\ {\scriptstyle\psi}\big\downarrow & & \big\| \\ Hom_{U(\mathfrak{b})}(U(\mathfrak{g}), V(\mu) \otimes k_{-\mu}) & \xrightarrow{\theta \mapsto \theta(1)} & V(\mu) \otimes k_{-\mu} \end{array}$$

commutative. Moreover,

$$\begin{aligned} Im\ \psi \subset F_{\mathfrak{g}}(Hom_{U(\mathfrak{b})}(U(\mathfrak{g}), V(\mu) \otimes k_{-\mu})) &= F_{\mathfrak{g}}(V(\mu) \otimes \delta M(\mu)^\iota) \\ &= V(\mu) \otimes V(-\mu) = E(\mu). \end{aligned}$$

Then $\psi(\widetilde{E(\mu)}) = E(\mu)$ by **3.9** and dimensionality. Yet ψ is just the restriction of the $U(\mathfrak{g})$ map taking A^λ to $A = F_{\mathfrak{b}^-}(Hom_{U(\mathfrak{b})}(U(\mathfrak{g}), \delta M(0)))$ defined [**JLT**, 2.1] by the corresponding commutative diagram. In this sense $\widetilde{E(\mu)} = E(\mu)$.

We have proved the following:

THEOREM. *Take $\mu \in P^+, \lambda = w_\pi.\mu$. Then the action of U^λ on $V(\mu)$ restricts to an isomorphism of the subspace $E(\mu) \hookrightarrow U^\lambda \hookrightarrow A$ onto End $V(\mu)$.*

3.12. A basic problem is to compute the v_μ^λ. It is clear from the uniqueness property in their definition that

$$(*) \qquad v_\mu^\lambda v_\nu^{\lambda-\mu} = v_{\mu+\nu}^\lambda, \ \forall \ \lambda \in \mathfrak{h}^*; \ \mu, \nu \in P^+,$$

up to non-zero scalars. Again let $p_{\mu'} : \mu' = -w_\pi \mu$ be the unique up to scalars highest weight vector in $V(-\mu)$. One has $p_{\mu'} \in \theta^{\lambda,\lambda-\mu}\widetilde{V(-\mu)}$. Since $\widetilde{V(-\mu)}\widetilde{V(-\nu)} = \widetilde{V(-(\mu+\nu))}$, these elements form a polynomial algebra with generators defined by the fundamental weights and so are relatively easy to compute. Observe that $v_\mu^\lambda p_{\mu'}$ (resp. $p_{\mu'} v_\mu^{\lambda+\mu}$) is non-zero and the unique up to scalars highest weight vector in the unique component of $V(\mu)^\lambda V(-\mu)$ (resp. $V(-\mu)V(\mu)^{\lambda+\mu}$) is isomorphic to $V(\mu+\mu')$. One might expect that these copies of $V(\mu+\mu')$ coincide, that is $p_{\mu'} v_\mu^{\lambda+\mu} = v_\mu^\lambda p_{\mu'}$ up to scalars, though the full spaces may not; yet even this is not quite correct. Finally analogy with the quantum case suggests that the v_μ^λ are polynomials in some commutative subalgebra of $A^{\mathfrak{h}}$ and to be determined by their zeros arising when $\lambda, \lambda-\mu$ are not compatible.

3.13. Fix a parabolic subalgebra $\mathfrak{p} = \mathfrak{p}_{\pi'}$ and set $\mathfrak{p}' = [\mathfrak{p}, \mathfrak{p}]$. Analogy with the quantum case [**FJ1**] indicates how we might find elements in the image of the semicentre of $U(\mathfrak{p})$ in U^λ. Observe that $\mathfrak{m}$ annihilates $V_{\pi'}(\mu) := U(\mathfrak{r})v_\mu$ and $V_{\pi'}(-w_{\pi'}w_\pi\mu) := U(\mathfrak{r})v_{-w_{\pi'}w_\pi\mu}$. Thus $V_{\pi'}(\mu) \otimes V_{\pi'}(-w_{\pi'}w_\pi\mu)$ admits a $\mathfrak{p}'$ invariant element $\hat{a}_\mu$ exactly when $\mu \in D_{\pi'} := \{\lambda + P^+ \mid (\lambda - w_{\pi'}w_\pi\lambda, \pi') = 0\}$. Specifically we may take the unique up to scalars $\mathfrak{r}' := [\mathfrak{r}, \mathfrak{r}]$ invariant element in $U(\mathfrak{r})a_\mu^\lambda$: $a_\mu^\lambda = v_\mu^\lambda v_{-w_{\pi'}w_\pi\mu}$. This approach might be better than that used in [**FJ2**]. For example by **3.8**, **3.9**, a_μ^μ sends v_μ to a non-zero multiple of $v_{w_{\pi'}w_\pi\mu}$ in $V(\mu)$. Yet is $\hat{a}_\mu \in Im\ U(\mathfrak{p})$?

3.14. In general it is not obvious what are the inclusion relations between the $\widetilde{E(\mu)}^\lambda$. By **3.11** it follows that $\widetilde{E(\nu)}^\nu \subset \widetilde{E(\mu)}^\mu$ whenever $\nu \prec \mu$. In particular, $1 \in \widetilde{E(\mu)}^\mu$. (This is also obvious from **3.9** since the only trivial $\mathfrak{g}$ submodule of U^λ is k.) Our example **3.5** shows that in general $1 \notin \widetilde{E(\mu)}^\lambda$ if $\lambda, \lambda-\mu$ are not compatible.

Let $\mathcal{F}$ be the canonical filtration on $U(\mathfrak{g})$. The above inclusions suggest we may have $gr_\mathcal{F} Ann\ V(\mu) \subset gr_\mathcal{F} Ann\ V(\nu)$, when $\mu \succ \nu$. Let $U(\mathfrak{g})^\star$ denote the Hopf dual of $U(\mathfrak{g})$ given the (decreasing) filtration induced from $\mathcal{F}$. Let $C^{V(\mu)} \subset U(\mathfrak{g})^\star$ denote the space of matrix coefficients of $V(\mu)$. Let $H \subset S(\mathfrak{g})$ be the span of powers of *ad*-nilpotent elements of $\mathfrak{g}$. One has $gr_\mathcal{F} C^{V(\mu)} \subset H$ and their sum [**FJ2**, 8.9(ii)] is H. The above inclusion on graded annihilators is equivalent to $gr_\mathcal{F} C^{V(\mu)} \supset gr_\mathcal{F} C^{V(\nu)}$, whenever $\mu \succ \nu$. Such a result would give an analogue $H_{\pi'}$ of H for any $\pi' \subset \pi$, namely

$$H_{\pi'} = \sum_{\mu \in P^+ \mid (\mu, \pi') = 0} gr_\mathcal{F} C^{V(\mu)},$$

satisfying

$$(*) \qquad [H_{\pi'} : V(\mu)] = dim\ V(\mu)^{\mathfrak{r}_{\pi'}}, \quad for\ all \quad \mu \in P^+,$$

and by [**FJ2**, 8.9(ii)] would contain $(ad\ U(\mathfrak{g}))S(\mathfrak{m})$ (in general strictly). A further hard problem is to calculate the multiplicity of $V(\mu)$ in each graded component of $H_{\pi'}$. This is needed to define and compute the "true" KPRV determinants discussed in [**JLT**, Section 8]. Let G be the algebraic adjoint group of $\mathfrak{g}$ and set $\mathfrak{h}^*_{\pi'} := \{\lambda \in \mathfrak{h}^* \mid (\lambda, \pi') = 0\}$. Recall [**JLT**, 8.2, 8.5] that the space of regular functions on the closure $G\mathfrak{m}_{\pi'}$ of the Richardson orbit defined by π' is just $(ad\ U(\mathfrak{g})S(\mathfrak{m}_{\pi'}))$. The ideal of definition of $G\mathfrak{m}_{\pi'}$ contains $gr_{\mathcal{F}}M_{\pi'}(\lambda) : \lambda \in \mathfrak{h}^*_{\pi'}$ (and coincides with the radical of the latter). If the corresponding moment map is birational with normal image [**Jo2**, 5.4], equality holds for all $\lambda \in \mathfrak{h}^*_{\pi'}$. Otherwise this can at most hold for very special λ which should appear as certain zeros of the "true" KPRV determinants.

Recalling **1.3**, one may define

$$\mathbb{H}_{\pi'} := \sum_{\mu \in P^+ \mid (\mu, \pi') = 0} E(\mu)$$

and this also satisfies $(*)$ above. However this is not an appropriate choice for the KPRV determinants as the degrees in which the $V(\mu) : \mu \in P^+$ occur will generally be too large.

4. Graded injectivity revisited

4.1. We would clearly like $\mathbb{H}$ of **2.8** not to depend on the parameter $\lambda \in \mathfrak{h}^*$. To remove this dependence, recall the filtration on $\mathcal{F}$ on A^λ defined in **3.2**. One easily checks that the isomorphism class of $gr_{\mathcal{F}}A^\lambda$ is independent of the choice of $\lambda \in \mathfrak{h}^*$. As in [**Jo2**, 5.1] we view a basis $\{\lambda_i\}$ of $\mathfrak{h}^*$ as independent variables $\{z_i\}$ generating a polynomial algebra Z isomorphic to $S(\mathfrak{h})$. The resulting algebra A^z is a tensor product $A \otimes_k Z$, where A is our original Weyl algebra (generated by P, Q). However A is not a $U(\mathfrak{g})$ module (for diagonal action), only a $U(\mathfrak{b}^-)$ module. Yet through the injectivity [**Jo2**, 4.10] of $gr_{\mathcal{F}}A^\lambda$ one may find [**Jo2**, 5.1] an isomorphic copy G of $gr_{\mathcal{F}}A^\lambda$ in A^z so that $A^z = G \otimes_k Z$ as a $U(\mathfrak{g})$ module. Then one takes $\mathbb{H} = F_{\mathfrak{g}}(G)$. Notice that this constructs $\mathbb{H}$ in the Gelfand–Kirillov extension $U(\mathfrak{g}) \otimes_{Z(\mathfrak{g})} S(\mathfrak{h})$ of $U(\mathfrak{g})$ rather than in $U(\mathfrak{g})$.

4.2. The proof of the injectivity of $gr_{\mathcal{F}}A^\lambda$ given in [**Jo2**, Section 4] was rather complicated, using the Koszul resolution and spectral sequences. Here we give a much simpler proof. As noted in [**Jo2**, 3.7], we must show exactly that $Coind\ S_n(\mathfrak{g}/\mathfrak{b})$ is injective in $\mathcal{O}^-$ for all $n \in \mathbb{N}$. Set $Ind\ M = U(\mathfrak{g}) \otimes_{U(\mathfrak{b})} M$, for any $U(\mathfrak{b})$ module M. The above is equivalent to showing that $Ind\ S_n(\mathfrak{n})$ is projective in $\mathcal{O}$ for all $n \in \mathbb{N}$. More generally one can ask if $U(\mathfrak{g}) \otimes_{U(\mathfrak{p})} S_n(\mathfrak{m})$ is projective in $\mathcal{O}$ where $\mathfrak{m}$ is the nilradical of a parabolic $\mathfrak{p}$.

4.3. Let V be a vector space and W a subspace of V. Fix $k \in \mathbb{N}$ and for each $0 < j \leq k$ define a map $d_j : \Lambda^{k-j}V \otimes S_j(W) \to \Lambda^{k-j+1}V \otimes S_{j-1}(W)$ by

$$d_j(v_1 \wedge v_2 \wedge \cdots \wedge v_{k-j} \otimes w_1 w_2 \cdots w_j) = \sum_{i=1}^{j} v_1 \wedge v_2 \wedge \cdots \wedge v_{k-j} \wedge w_i \otimes w_1 w_2 \cdots \hat{w}_i \cdots w_j$$

where $\wedge$ denotes omission of the argument. One checks that $ker\ d_j = Im\ d_{j+1}$ for $k > j > 0$. Further suppose that V is a $\mathfrak{p}$ module and W a $\mathfrak{p}$ submodule for some Lie algebra $\mathfrak{p}$. Then $d_j : 0 < j \leq k$ is a $\mathfrak{p}$ module map and we obtain a long exact sequence $0 \to S_k(W) \to \cdots \to S_{k-j}(W) \otimes \Lambda^j V \to \cdots \to \Lambda^k V \to \Lambda^k(V/W) \to 0$ of $\mathfrak{p}$ modules.

4.4. Now take $V = \mathfrak{g}$ with $W = \mathfrak{m}$ the nilradical of a parabolic subalgebra $\mathfrak{p} = \mathfrak{p}_{\pi'}$ of $\mathfrak{g}$. Identify $\mathfrak{g}/\mathfrak{m}$ with $\mathfrak{p}^-$. Set $Ind\ M = U(\mathfrak{g}) \otimes_{U(\mathfrak{p})} M$, for any $\mathfrak{p}$ module M and $M_{\pi'}(0) = Ind\ k_0$. Through, say, [**Jo1**, 8.1.6], one has $Ind(\Lambda^j \mathfrak{g} \otimes S_{k-j}(\mathfrak{m})) \cong \Lambda^j \mathfrak{g} \otimes Ind\ S_{k-j}(\mathfrak{m})$ and so induction gives a long exact sequence

$$(*) \qquad \begin{aligned} 0 \to Ind\ S_k(\mathfrak{m}) \to \cdots \to \Lambda^j \mathfrak{g} \otimes Ind\ S_{k-j}(\mathfrak{m}) \\ \to \cdots \to \Lambda^k \mathfrak{g} \otimes M_{\pi'}(0) \to Ind\ \Lambda^k \mathfrak{p}^- . \end{aligned}$$

Recall that $M_{\pi'}(0)$ is projective in the subcategory $\mathcal{O}_{\pi'}$ of $\mathcal{O}$ of $\mathfrak{g}$ modules which are semisimple under the Levi factor $\mathfrak{r}$ of $\mathfrak{p}$.

Fix $N \in Ob\mathcal{O}_{\pi'}$ and let $r \in \mathbb{N}$ be minimal such that $Ext^i_{\mathcal{O}_{\pi'}}(Ind\ S_r(m), N) \neq 0$ for some $i > 0$. Dimension shifting based on the above exact sequence gives an isomorphism

$$(**) \qquad Ext^{r+t}_{\mathcal{O}_{\pi'}}(Ind\ \Lambda^r \mathfrak{p}^-, N) \xrightarrow{\sim} Ext^t_{\mathcal{O}_{\pi'}}(Ind\ S_r(\mathfrak{m}), N),$$

for all $t > 0$. (This is also obtained from [**Jo2**, 4.4] by refining slightly the argument there.)

4.5. For the moment take $\pi' = \emptyset$, so then $\mathfrak{p} = \mathfrak{b}$, $\mathfrak{m} = \mathfrak{n}$ and $\mathfrak{p}^- = \mathfrak{b}^-$. Then $Ind\ \Lambda^r \mathfrak{b}^-$ admits a filtration with quotients isomorphic to the $M(\sigma)$, where σ is a weight of $\Lambda^r \mathfrak{b}^-$, hence a sum of $\leq r$ distinct elements of Δ^-, while $Ind\ S_r(\mathfrak{n})$ has quotients isomorphic to the $M(\sigma')$, where σ' is a weight of $S_r(\mathfrak{n})$, hence a sum of r elements of Δ^+. Now take N simple, say $N = V(\nu) : \nu \in \mathfrak{h}^*$, being the simple module of highest weight ν. The non-vanishing of $Ext^i_{\mathcal{O}}(M(\mu), V(\nu))$ implies $\mu \in W.\nu$. Hence the non-vanishing of the left-hand side of 4.4$(**)$ implies $\sigma' \in W.\sigma$, for some σ, σ' as above. Yet $\sigma + \rho$ is a half sum of distinct elements of Δ and hence so is any $w(\sigma + \rho)$. Consequently $W.\sigma \in -\mathbb{N}\pi$, forcing $\sigma' = 0$. Then $r = 0$, contradicting the projectivity of $M(0)$. This proves the following

PROPOSITION. *For all $k \in \mathbb{N}$, the induced module $Ind\ S_k(\mathfrak{n})$ is projective in $\mathcal{O}$. Furthermore 4.4$(*)$ is a projective resolution of $Ind\ \Lambda^k \mathfrak{b}^-$.*

4.6. The above resolution is not minimal even for $\mathfrak{sl}(2)$. Yet a similar argument shows that $Ind\ \Lambda^k \mathfrak{n}^-$ admits a projective resolution with j^{th} term $\Lambda^j \mathfrak{g} \otimes Ind\ S_{k-j}(\mathfrak{b})$ of length k. This resolution is at least of minimal length. Indeed $\Omega(\Lambda^* \mathfrak{n}^-) = \Omega(V(\rho)) - \rho$, where $\Omega(M)$ denotes the multiset of weights of a semisimple $\mathfrak{h}$ module M. It is elementary that $Ext^1_{\mathcal{O}}(M(w.0), M(w'.0)) = 0$ if $\ell(w) = \ell(w')$. Thus the $Z(\mathfrak{g})$ primary component of $Ind\ \Lambda^k \mathfrak{n}^-$ corresponding to the maximal ideal defined by $W.0$ is just $\bigoplus_{w \mid \ell(w)=k} M(w.0)$.

Yet $Ext^i_{\mathcal{O}}(M(w.0), V(0)) \cong H^i(\mathfrak{n}, V(0))_{w.0}$, which after the Bott–Borel–Kostant–Weil theorem is one-dimensional if $\ell(w) = i$ and zero otherwise. Hence we have that $Ext^j_{\mathcal{O}}(Ind\ \Lambda^k \mathfrak{n}^-, V(0))$ is zero unless $j = k$, in which case it has dimension $card\{w \in W \mid \ell(w) = k\}$. In another language this fact is known as the diagonality of Hodge cohomology.

One may remark that the Chevalley–Kostant construction [**K**] gives a $\mathfrak{g}$ module structure on $\Lambda^* \mathfrak{n}^-$. One may check that in this the natural filtration $\mathcal{F}$ on $\Lambda^* \mathfrak{n}^-$ gives $\Lambda^* \mathfrak{n}^-$ the structure of a filtered $\mathfrak{b}$ module isomorphic to $V(\rho) \mid_{U(\mathfrak{b})} \otimes k_{-\rho}$ and moreover $gr_{\mathcal{F}}\Lambda^*\mathfrak{n}^-$ has just the previous $\mathfrak{b}$ module structure given on $\Lambda^*\mathfrak{n}^-$. Now $Ind(V(\rho)\mid_{U(\mathfrak{b})}\otimes k_{-\rho}) \cong V(\rho)\otimes M(-\rho)$ is projective in $\mathcal{O}$ and admits a filtration with factors $Ind\ \Lambda^k\mathfrak{n}^-$. We do not know if such a construction is possible in the parabolic case. Probably not.

4.7. Given $\mu \in \mathfrak{h}^*$, dominant and integral with respect to π', let $V_{\pi'}(\mu)$ be the simple $\mathfrak{p}_{\pi'}$ module with highest weight μ and set $M_{\pi'}(\mu) = Ind\ V_{\pi'}(\mu)$. The above combinatorics fails for arbitrary $\mathfrak{p}$ even when one takes account of the known vanishings of the $Ext^i_{\mathcal{O}_{\pi'}}(M_{\pi'}(\mu), V(\nu))$, though **4.4**$(*)$ is still a projective resolution of $\Lambda^*\mathfrak{p}^-$ if the $Ind\ S_{k-j}(m)$ are injective for $0 \leq j \leq k$. We note below an example showing how the combinatorics can fail.

4.8. Take π of type A_{14} and $\pi' = \{\alpha_i \mid i = 1,2,6,7,8,9,13,14\}$. Set $\pi'' = \{\alpha_j \mid j = 3,4,5,10,11,12\}$ and take $S = \{\alpha_1+\alpha_2, \alpha_2, -(\alpha_3+\alpha_4), -(\alpha_4+\alpha_5), -(\alpha_3+\alpha_4+\alpha_5), \alpha_6+\alpha_7+\alpha_8+\alpha_9, \alpha_6+\alpha_7+\alpha_8, \alpha_7+\alpha_8+\alpha_9, -(\alpha_{10}+\alpha_{11}), -(\alpha_{11}+\alpha_{12}), -(\alpha_{10}+\alpha_{11}+\alpha_{12}), \alpha_{13}, \alpha_{13}+\alpha_{14}\}$. Then $S \subset \Delta_{\pi'} \cup \Delta^-$, where $\Delta_{\pi'} := \Delta \cap \mathbb{Z}\pi'$. Moreover $card\ S = 13$ and $\sigma := \sum_{\alpha\in S}\alpha$ is dominant with respect to π'. Hence $M_{\pi'}(\sigma)$ occurs in a Verma flag of $Ind\ \Lambda^{13}\mathfrak{p}^-$. Set $s_i = s_{\alpha_i}$ and $w = s_6 s_9 w_{\pi''}$. Then $\ell(w) = 14$ and $\mu := w.\sigma = \bar{\omega}_3 + \bar{\omega}_{12}$ is dominant. Set $S' = \{\alpha_1+\alpha_2+\alpha_3, \alpha_2+\alpha_3, \alpha_3+\alpha_4, \alpha_4+\alpha_5, \alpha_5+\alpha_6+\alpha_7, \alpha_5+\alpha_6+\alpha_7+\alpha_8+\alpha_9, \alpha_8+\alpha_9+\alpha_{10}, \alpha_6+\alpha_7+\alpha_8+\alpha_9+\alpha_{10}, \alpha_{10}+\alpha_{11}, \alpha_{11}, \alpha_{11}+\alpha_{12}, \alpha_{12}+\alpha_{13}, \alpha_{12}+\alpha_{13}+\alpha_{14}\}$. Then $|S'| = 13$ and lies in $\Delta^+ \setminus (\Delta_{\pi'} \cap \Delta^+)$. Moreover $\sigma' = \sum_{\alpha\in S'}\alpha$ is dominant with respect to π' and satisfies $s_4.\sigma' = \mu$. Hence $M_{\pi'}(\sigma')$ occurs in a Verma flag of $Ind\ S_{13}(\mathfrak{m})$. Assume λ dominant and integral with respect to π' and view $V_{\pi'}(\lambda)^*$ as a left $\mathfrak{p}$ module through σ. Recall that $\mathfrak{m}$ acts by zero on $V_{\pi'}(\lambda)^*$. Then by [**Jo2**, 4.4(3)] we obtain $Ext^i_{\mathcal{O}_{\pi'}}(M_{\pi'}(\lambda), V(\mu)) \cong H^i(\mathfrak{m}, V(\mu)\otimes V_{\pi'}(\lambda)^*)^{\mathfrak{r}} \cong (H^i(\mathfrak{m}, V(\mu))\otimes V_{\pi'}(\lambda)^*)^{\mathfrak{r}}$. Then by a result of Kostant (see [**Jo2**, 4.12 centre of p. 368] for example) we conclude that $Ext^{14}_{\mathcal{O}_{\pi'}}(M_{\pi'}(\sigma), V(\mu))$ and $Ext^1_{\mathcal{O}_{\pi'}}(M_{\pi'}(\sigma'), V(\mu))$ are both one-dimensional. Thus our analysis cannot exclude both sides of 4.4$(**)$ being non-zero for $r = 13, t = 1, N = V(\mu)$. Yet this special case is excluded by the result of Broer [**B**, Theorem 6.5], since $\mu \in P^+$ and so $V(\mu)$ is finite dimensional.

4.9. Recall **2.7** and **3.3**. The subspace $V(\mu)\otimes k_{-\mu} : \mu \in P^+$ of $\delta M(0) \cong S(\mathfrak{n}^-)$ obtains a filtration $\mathcal{F}$ induced from the canonical filtration on $S(\mathfrak{n}^-)$, from which the q-character

$$ch_q V(\mu) := \sum q^m ch(\mathcal{F}^m V(\mu)/\mathcal{F}^{m-1}V(\mu))$$

is defined. Then the q-character of $V(\mu)\otimes(\delta M(\mu))^\iota$ of A is given by

$$ch_q V(\mu)\otimes(\delta M(\mu))^\iota = e^{-\mu} ch_q V(\mu) \prod_{\alpha\in\Delta^+}(1-e^\alpha)^{-1}.$$

It is not known if $gr_{\mathcal{F}} V(\mu)\otimes(\delta M(\mu))^\iota$ is injective in $\mathcal{O}^-$ (despite the fact that both $V(\mu)\otimes(\delta M(\mu))^\iota$ and $gr_{\mathcal{F}} A$ are injective). If this were so, we would obtain

$$ch_q E(\mu) = S(ch_q V(\mu)\otimes(\delta M(\mu)))^\iota : S = \sum_{w\in W} w$$

from [**JLT**, 4.2]. In any case the calculation of the left-hand side remains an open problem.

4.10. Take $\pi' \subset \pi$. Then $\delta M_{\pi'}(0) \cong S(\mathfrak{m}^-_{\pi'})$ obtains a filtration $\mathcal{F}$ induced from the canonical filtration on the latter. However the embedding $\delta M_{\pi'}(0) \hookrightarrow \delta M(0)$ is (surprisingly) not compatible with this filtration and that of **4.9**. For example if $\pi = \{\alpha, \beta\}$ is of type A_2 and $\pi' = \{\beta\}$, then $\delta M(0)$ identifies with $k[q_{-\alpha}, q_{-\beta}, q_{-(\alpha+\beta)}]$; but $\delta M_{\pi'}(0)$ *cannot* identify with $k[q_{-\alpha}, q_{-(\alpha+\beta)}]$ because the action of the root vector x_α of weight α on $q_{-(\alpha+\beta)}$ gives a non-zero multiple of $q_{-\beta}$. Yet $x_\alpha q_{-\alpha}$ is also a non-zero scalar, so we can choose $c \in k^*$ so that $q'_{-(\alpha+\beta)} := q_{-(\alpha+\beta)} - cq_{-\alpha}q_{-\beta}$ satisfies $x_\alpha q_{-(\alpha+\beta)} = 0$. Then $\delta M_{\pi'}(0)$ identifies with $k[q_{-\alpha}, q'_{-(\alpha+\beta)}]$. This required "change of variables" upsets the compatibility of filtrations. It is a main difficulty with the erroneous argument in [**Jo2**, 6.2]. Thus it remains an open problem to show that $gr_{\mathcal{F}} A_{\pi'} : A_{\pi'} = F_{\mathfrak{p}^-}(Hom_{U(\mathfrak{p})}, \delta M_{\pi'}(0))$ is injective in $\mathcal{O}^-_{\pi'}$.

References

[B] B. Broer, *Decomposition varieties in semisimple Lie algebras*, Canad. J. Math. **50** (1998), 929–971.

[Ca] P. Caldero, *On harmonic elements for semisimple Lie algebras*, Adv. Math. **166** (2002), 73–99.

[Co] N. Conze-Berline, *Algèbres d'opérateurs différentiels et quotients des algèbres enveloppantes*, Bull. Soc. Math. France **102** (1974), 379–415.

[D] M. Duflo, *Sur la classification des idéaux primitifs dans l'algèbre enveloppante d'une algèbre de Lie semi-simple*, Ann. Math. **105** (1977), 107–120.

[FJ1] F. Fauquant-Millet, A. Joseph, *Sur les semi-invariants d'une sous-algèbre enveloppante d'une algèbre enveloppante quantifiée*, Transf. Groups **6** (2001), 125–142.

[FJ2] F. Fauquant-Millet, A. Joseph, *Semi-centre de l'algèbre enveloppante d'une sous-algèbre parabolique d'une algèbre de Lie semisimple*, Ann. Sci. École Norm Sup. (4) **38** (2005), 155–191.

[GK] I. M. Gelfand, A. A. Kirillov, *The structure of the Lie field connected with a semisimple decomposable Lie algebra*, Funct. Anal. Appl. **3** (1969), 7–26.

[HJ] I. Heckenberger, A. Joseph, *On the left and right Brylinski–Kostant filtrations,* prerpint, Weizmann Institute, 2005.

[Ja] J.-C. Jantzen, *Moduln mit einem höchsten Gewicht*, Lecture Notes in Math. 750, Springer, Berlin, 1979.

[JLT] A. Joseph, G. Letzter, D. Todoric, *On the Kostant–Parthasarathy–Ranga Rao–Varadarajan determinants*, I–III, J. Algebra **241** (2001), 27–45, 46–66, 67–88.

[JLZ] A. Joseph, G. Letzter, S. Zelikson, *On the Brylinski-Kostant filtration*, JAMS **13** (2000), 945–970.

[Jo1] A. Joseph, *Quantum groups and their primitive ideals*, Springer, Berlin, 1995.

[Jo2] A. Joseph, *On the graded injectivity of the Conze embedding*, J. Algebra **265** (2003), 358–378.

[JT] A. Joseph, D. Todoric, *On the quantum KPRV determinants for semisimple and affine Lie algebras*, Algebras and Rep. Theory **5** (2002), 57–99.

[K] B. Kostant, *Clifford algebra analogue of the Hopf–Koszul–Samelson theorem*, Adv. Math. **125** (1997), 275–350.

Weizmann Institute of Science, Rehovot 76100, Israel

E-mail address: joseph@wisdom.weizmann.ac.il

Amer. Math. Soc. Transl.
(2) Vol. **217**, 2006

Asymptotic Behaviour in the Time Synchronization Model

Vadim Malyshev and Anatoli Manita

ABSTRACT. There are two types $i = 1, 2$ of particles on the line $\mathbf{R}$, with N_i particles of type i. Each particle of type i moves with constant velocity v_i. Moreover, any particle of type $i = 1, 2$ jumps to any particle of type $j = 1, 2$ with rates $N_j^{-1}\alpha_{ij}$. We find phase transitions in the clusterization (synchronization) behaviour of this system of particles on different time scales $t = t(N)$ relative to $N = N_1 + N_2$.

1. The model

It is a pleasure to contribute to this volume devoted to the half-life birthday of Anatoly Vershik. He contributed a lot to probability theory and asymptotic analysis and these are exactly the topics of this paper.

The simplest formulation of the model which we consider here is in terms of the particle system. On the real line there are N_1 particles of type 1 and N_2 particles of type 2, $N = N_1 + N_2$. Each particle of type $i = 1, 2$ performs two independent movements. First of all, it moves with constant speed v_i in the positive direction. We assume further that v_i are constant and different, thus we can assume without loss of generality that $0 \le v_1 < v_2$. The degenerate case $v_1 = v_2$ is different and will be considered separately.

Secondly, at any time interval $[t, t + dt]$ each particle of type i independently of the others with probability $\alpha_{ij}dt$ decides to make a jump to some particle of type j and chooses the coordinate of the j-type particle, where to jump, among the particles of type j, with probability $\frac{1}{N_j}$. Here α_{ij} are given nonnegative parameters for $i, j = 1, 2$. Further on, unless otherwise stated, we assume that $\alpha_{11} = \alpha_{22} = 0, \alpha_{12}, \alpha_{21} > 0$.

After such an instantaneous jump, the particle of type i continues the movement with the same velocity v_i. This defines the continuous time Markov chain $\left\{x_k^{(i)}(t)\right\}$, $i = 1, 2$ and $k = 1, \ldots, N_i$, where $x_k^{(i)}(t)$ is the coordinate of k-th particle of type i at time t. We assume that the initial coordinates $x_k^{(i)}(0)$ of the particles at time

---

2000 *Mathematics Subject Classification.* Primary 60K35; Secondary 60J27, 60F99.
Key words and phrases. Markov process, stochastic particles system, synchronization model.
The work is supported by the Russian Foundation of Basic Research (grants 02-01-00945, 05-01-22001, 06-01-00662).

0 are given. We are interested in the long time evolution of this system on various scales with $N \to \infty$, $t = t(N) \to \infty$.

In different terms, this can be interpreted as the time synchronization problem. In general, the time synchronization problem can be presented as follows. There are N systems (processors, units, persons, etc.). There is an absolute (physical) time t, but each processor j fulfills a homogeneous job in its own proper time $t_j = v_j t, v_j > 0$. Proper time is measured by the amount v_j of the job, accomplished by the processor for the unit of the physical time, if it is disjoint from other processors. However, there is a communication between each pair of processors, which should lead to a drastic change of their proper times. In our case the coordinates $x_k^{(i)}(t)$ can be interpreted as the modified proper times of the particles-processors, the nonmodified proper time being $x_k^{(i)}(0) + v_i t$.

There can be many variants of the exact formulation of such a problem; see [**GMP, MSh1, MiMi, BT, MSi**]. We will call the model considered here the basic model, because there are no restrictions on the jump process. Many other problems include such restrictions; for example, only jumps to the left are allowed. Due to the absence of restrictions, this problem, as we will see below, is a "linear problem" in the sense that after scalings it leads to linear equations. Despite this, it has nontrivial behaviour; one sees a different picture at different time scales.

There are, however, other interesting interpretations of this model, related to psychology, biology and physics. For example, in social psychology, perception of time and life tempo strongly depends on the social contacts and intercourse. We will not give the details here.

2. Main results

We show that the process consists of three consecutive stages: initial desynchronization up to the critical scale, critical slow down of desynchronization and final stabilization.

Final stabilization. The first theorem shows that for N_i fixed and $t \to \infty$ there is a synchronization: all particles asymptotically, as $t \to \infty$, move with the same constant velocity v, that is, like vt. However it does not say how fluctuations depend on N_i.

Put

$$m(t) = \min_{i,k} x_k^{(i)}(t)\,.$$

THEOREM 2.1. *For any fixed N_1, N_2 there exists $v = v(N_1, N_2) > 0$ such that for any $i = 1, 2$ and any $k = 1, ..., N_i$ a.s.*

$$\lim_{t\to\infty} \frac{x_k^{(i)}(t)}{t} = v\,.$$

Moreover, the distribution of the vector $\left\{x_k^{(i)}(t) - m(t),\ i = 1, 2 \text{ and } k = 1, \ldots, N_i\right\}$ *tends to a stationary distribution.*

The velocity v will be written down explicitly in terms of this distribution; it depends of course on α_{ij} and v_i. Note that both the velocity and the distribution do not depend on the initial coordinates.

Initial desynchronization. Now we consider the case when $N \to \infty$ but t is fixed. More exactly, we consider a sequence of pairs (N_1, N_2) such that $N_1, N_2 \to \infty$ so that $\frac{N_i}{N} \to c_i$, where $c_1 + c_2 = 1, c_i > 0$. It is convenient here to consider positive measures or generalized functions

$$m^{(N_i)}(t,x) = \frac{1}{N_i} \sum_k \delta(x - x_k^{(i)}(t)), \quad x \in \mathbf{R}_+,$$

defined by the coordinates of N_i particles of type i at time t. We assume that at time $t = 0$ for any bounded C^1-functions $\phi_i(x)$ on $\mathbf{R}$ the sequence $\langle m_i^{(N_i)}(0,.), \phi_i \rangle$ converges to some number.

THEOREM 2.2. *Then for any t there are weak deterministic limits*

$$\lim_{N\to\infty} \frac{1}{N} m_i^{(N_i)}(t,x) = m_i(t,x)$$

where $m_i(t,x)$ satisfy the equations

$$\frac{\partial m_1}{\partial t} + v_1 \frac{\partial m_1}{\partial x} = \alpha_{12}(m_2 - m_1), \tag{2.1}$$

$$\frac{\partial m_2}{\partial t} + v_2 \frac{\partial m_2}{\partial x} = \alpha_{21}(m_1 - m_2)\,. \tag{2.2}$$

Now we want to study the asymptotic behaviour of $m_i(x,t)$ for $t \to \infty$. Denote

$$a_i(t) = \int x m_i(x,t) dx, \qquad d_i(t) = \int (x - a_i(t))^2 m_i(t,x)\, dx\,.$$

THEOREM 2.3. *There exist constants $v, d > 0$ such that as $t \to \infty$*

$$a_i(t) = vt + a_{i0} + o(1),$$

$$d_i(t) = dt + d_{i0} + o(1)$$

for some constants a_{i0}, d_{i0}. Moreover,

$$\Delta_i(x,t) = \frac{m_i(x,t) - a_i(t)}{\sqrt{d_i(t)}}$$

tends to $\frac{1}{\sqrt{2\pi}} \exp(-\frac{x^2}{2})$ pointwise as $t \to \infty$.

Critical point and uniform estimates. Here we assume that $N_1 = [c_1 N]$, $N_2 = [c_2 N]$ for some $c_i > 0, c_1 + c_2 = 1$. Introduce the empirical means (mass centres) for types 1 and 2

$$\overline{x^{(i)}}(t) = \frac{1}{N_i} \sum_{k=1}^{N_i} x_k^{(i)}(t),$$

the empirical variances

$$S_i^2(t) = \frac{1}{N_i} \sum_{k=1}^{N_i} \left(x_k^{(i)}(t) - \overline{x^{(i)}}(t) \right)^2$$

and their means

$$\boldsymbol{\mu}_i(t) = \mathsf{E}\overline{x^{(i)}}(t), \quad \boldsymbol{l}_{12}(t) = \boldsymbol{\mu}_1(t) - \boldsymbol{\mu}_2(t), \quad R_i(t) = \mathsf{E}S_i^2(t)\,.$$

The following asymptotic results hold for any sequence of pairs (N,t) with $N \to \infty$ and $t = t(N) \to \infty$.

THEOREM 2.4. *We have the following asymptotical results as* $t \to \infty$*:*

$$\boldsymbol{l}_{12}(t) \to \frac{v_1 - v_2}{\alpha_{12} + \alpha_{21}}, \qquad \frac{\boldsymbol{\mu}_i(t)}{t} \to \frac{\alpha_{12}v_2 + \alpha_{21}v_1}{\alpha_{12} + \alpha_{21}}.$$

Assume now that $N_i = c_i N$, where $c_i > 0, c_1 + c_2 = 1$.

THEOREM 2.5. *There are the following three regions of asymptotic behaviour, uniform in* $t(N)$ *for sufficiently large* N*:*

- *if* $\frac{t(N)}{N} \to 0$*, then* $R_i(t(N)) \sim h\varkappa_2 t(N)$*,*
- *if* $t = t(N) = sN$ *for some* $s > 0$*, then* $R_i(t(N)) \sim h\,(1 - e^{-\varkappa_2 s})N$*,*
- *if* $\frac{t(N)}{N} \to \infty$*, then* $R_i(t(N)) \sim hN$*,*

where the constant $\varkappa_2 > 0$ *can be explicitly calculated and*

$$h = \frac{2\alpha_{12}\alpha_{21}\,(v_1 - v_2)^2}{\varkappa_2(\alpha_{12} + \alpha_{21})^3}.$$

The proofs for uniform estimate results will be published elsewhere; see also [**MaMa1, MaMa2**].

3. Limit $t \to \infty$

In this section we will prove Theorem 2.1.

Two particles. It is useful to consider first the case when $N_1 = N_2 = 1$. Thus consider the process $(x^{(1)}(t), x^{(2)}(t))$. We will prove that there exist deterministic limits

$$\lim_{t\to\infty} \frac{x^{(i)}(t)}{t} = v$$

for $i = 1, 2$ and some $v > 0$; moreover the distribution of the random variable $\rho(t) = x^{(2)}(t) - x^{(1)}(t)$ tends to some distribution on $\mathbf{R}_+$.

We can assume that $v_1 = 0, v_2 > 0$. The Markov chain $\rho(t) = x^{(2)}(t) - x^{(1)}(t)$ on $\mathbf{R}_+$ satisfies the Doeblin condition; that is, from any $x \in \mathbf{R}_+$ there is a jump rate to 0, bounded away from zero. Here it equals $\alpha_{12} + \alpha_{21}$. It follows that $\rho(t)$ is ergodic. Then as $t \to \infty$ there exists the limiting (invariant) distribution $F(x)$ for $\rho(t)$. Let

$$t_1 < t_2 < \cdots$$

be the time moments when $x^{(1)}(t) = x^{(2)}(t)$. It is clear that $t_k - t_{k-1}$ are independent random variables, exponentially distributed with parameter $\alpha_{12} + \alpha_{21}$. It follows that $F(x)$ is exponential with the density

$$p(x) = \lambda \exp(-\lambda x), \quad \lambda = \frac{\alpha_{12} + \alpha_{21}}{v_2 - v_1}.$$

Thus, if the limits $\lim_{t\to\infty} \frac{x_i(t)}{t}$ exist, then they are equal. Let us prove that they exist and

$$v = v_1 + \alpha_{12} \int x p(x) dx. \tag{3.1}$$

In fact, the particle 1 moves with constant speed v_1 and performs on the time interval $[0, T]$ independent exponentially distributed jumps in the positive direction. As $T \to \infty$, the number of these jumps asymptotically equals $\alpha_{12}T$, and the mean jump asymptotically is $\int xp(x)dx$.

Similarly one can get

$$v = v_2 - \alpha_{21} \int xp(x)dx \,. \tag{3.2}$$

From this and (3.1) we have

$$v = \frac{\alpha_{21}v_1 + \alpha_{12}v_2}{\alpha_{21} + \alpha_{12}} \,.$$

General case. Let us prove first the second statement of the theorem. We can put $v_1 = 0$ and change the coordinate system putting $m(t) = 0$. Consider a configuration of particles at time t. Denote the particle, which has coordinate $m(t) = 0$ at time t, as particle 0. Let $p(t+2)$ be the probability that at time $t+2$ each particle will be inside the interval $[0, 2v_2]$. This probability can be (very roughly) estimated from below as

$$p(t+2) \geq \min(p_{01}p_2p_1, p_{02}p_3p_4) \,.$$

To prove this, consider first the case when the particle 0 has type 1. Under this condition $p(t+2)$ can be estimated from below as $p_{01}p_2p_1$, where p_{01} is the probability that particle 0 does not do any jumps in the time interval $(t, t+2)$, p_2 is the probability that each particle of type 2 jumps at least once to the particle 0 in the time interval $(t, t+1)$ and does not do any more jumps in the time interval $(t, t+2)$, p_1 is the probability that each particle of type 1 jumps to some particle of type 2 in the time interval $(t+1, t+2)$. Similarly, under the condition that the particle 0 has type 2, $p(t+2)$ can be estimated from below as $p_{02}p_3p_4$, where p_{02} is the probability that the particle 0 does not do any jumps in the time interval $(t, t+2)$, p_3 is the probability that each particle of type 1 jumps at least once to the particle 0 in the time interval $(t, t+1)$ and does not do any more jumps in the time interval $(t, t+2)$, p_4 is the probability that each particle of type 2 jumps to some particle of type 1 in the time interval $(t+1, t+2)$.

This means that the Markov chain $\mathcal{L} = \left\{x_k^{(i)}(t) - m(t), i = 1, 2 \text{ and } k = 1, \dots, N_i\right\}$ satisfies the Doeblin condition. Then it is ergodic and has some stationary distribution. We will now write the formula for v, assuming however that $\alpha_{ii} = 0$. For this we need some marginals of this stationary distribution.

Let $A_i(t)$ be the event that at time t at the point $m(t)$ there is a particle of type i, and let $q_i = \lim_{t\to\infty} P(A_i(t))$ be the stationary (limiting) probability of A_i. Let $p_i(y)$ be the stationary conditional (under the condition A_i) probability density of the distance from m to the nearest particle. In the time interval $[T, T+dt]$ the particle in $m(t)$ moves with the speed v_i and moreover can make one jump. This gives, for example under the condition A_1, constant movement $v_1 dt$ of m and the jump of m to the nearest point with rate $\alpha_{12}dt$. Thus as $T \to \infty$ we have

$$\begin{aligned} \mathsf{E}(m(T+dt) \,|\, m(T)\,) \;-\; m(T) \;\; &= \;\; q_1 \left(v_1 + \alpha_{12} \int yp_1(y)dy\right) dt \\ &+ \;\; q_2 \left(v_2 + \alpha_{21} \int yp_2(y)dy\right) dt \;+\; o(1) \end{aligned}$$

and then

$$v = q_1 \left(v_1 + \alpha_{12} \int yp_1(y)dy\right) + q_2 \left(v_2 + \alpha_{21} \int yp_2(y)dy\right).$$

About Doeblin chains. In the standard theory of Doeblin chains, see [**D**], it is assumed that transition probabilities are absolutely continuous with respect to some positive measure μ on the state space.

If at time 0 all $x_k^{(i)}$ are different, then for any t it is true that all $x_k^{(i)}$ are different a.s. Thus transition probabilities (for example, for the embedded chain at times $0,1,2,\ldots$) are absolutely continuous with respect to Lebesgue measure on $(\mathbf{R}_+^{N_1-1}\times\mathbf{R}_+^{N_2})\cup(\mathbf{R}_+^{N_1}\times\mathbf{R}_+^{N_2-1})$. If at time 0 some coordinates coincide, then a.s. in finite time τ they become all different.

4. Limit $N\to\infty$

It is very intuitive to introduce the following continuous model. Let $m_i(0,x)$, $x\in\mathbf{R}$, $i=1,2$, be positive smooth functions, $M_i=\int m_i(0,x)dx=1$. We call them continuous mass distributions of type i at time $t=0$. The dynamics of the masses is deterministic — during time dt from each element dm_1 of the mass the part $\alpha_{12}\,dt\,dm_1$ goes out and distributes correspondingly to the mass $m_2(x)$, namely it becomes the mass distribution with density $m_2(x)\,\alpha_{12}\,dt\,dm_1$, and vice-versa, interchanging 1 and 2. Moreover each mass element moves with velocities v_1 and v_2, respectively. From this we easily get linear equations (2.1)–(2.2) for mass distribution $m_i(t,x)$ at time t with the initial conditions

$$m_i(0,x)=f_i(x).$$

Now we will prove the convergence of the N particle model to the continuous model.

4.1. Convergence: the martingale problem. Here we prove Theorem 2.2. We consider the continuous time Markov process

$$\xi_{N_1,N_2}(t)=\left(x_1^{(1)}(t),\ldots,x_{N_1}^{(1)}(t);x_1^{(2)}(t),\ldots,x_{N_2}^{(2)}(t)\right) \tag{4.1}$$

with the state space $\mathbf{R}^{N_1+N_2}$. Its generator

$$\begin{aligned}(L_{N_1,N_2}f)\left(x^{(1)};x^{(2)}\right) &= \left[v_1\sum_{i=1}^{N_1}\frac{\partial}{\partial x_i^{(1)}}+v_2\sum_{j=1}^{N_2}\frac{\partial}{\partial x_j^{(2)}}\right]f\left(x^{(1)};x^{(2)}\right)\\ &\quad+\sum_{i=1}^{N_1}\sum_{j=1}^{N_2}\frac{\alpha_{12}}{N_2}\left[f\left(\left(x^{(1)};x^{(2)}\right)_{i\to j}\right)-f\left(x^{(1)};x^{(2)}\right)\right]\\ &\quad+\sum_{j=1}^{N_2}\sum_{i=1}^{N_1}\frac{\alpha_{21}}{N_1}\left[f\left(\left(x^{(1)};x^{(2)}\right)_{i\leftarrow j}\right)-f\left(x^{(1)};x^{(2)}\right)\right],\end{aligned}$$

where the notation

$$\begin{aligned}\left(x^{(1)};x^{(2)}\right) &= \left(x_1^{(1)},\ldots,x_{N_1}^{(1)};x_1^{(2)},\ldots,x_{N_2}^{(2)}\right),\\ \left(x^{(1)};x^{(2)}\right)_{i\to j} &= \left(x_1^{(1)},\ldots,x_{i-1}^{(1)},x_j^{(2)},x_{i+1}^{(1)},\ldots,x_{N_1}^{(1)};x_1^{(2)},\ldots,x_{N_2}^{(2)}\right),\\ \left(x^{(1)};x^{(2)}\right)_{i\leftarrow j} &= \left(x_1^{(1)},\ldots,x_{N_1}^{(1)};x_1^{(2)},\ldots,x_{j-1}^{(2)},x_i^{(1)},x_{j+1}^{(2)},\ldots,x_{N_2}^{(2)}\right)\end{aligned}$$

used is defined on bounded C^1-functions.

We will consider the limiting behaviour of this process when $t = const$ and $N_1, N_2 \to \infty$. It is not convenient to deal with the sequence $\xi_{N_1,N_2}(t)$ of processes because the dimension of the state space changes with N_1, N_2.

Denote

$$M_{N_1,N_2}(t) = \left(\frac{1}{N_1} \sum_{i=1}^{N_1} \delta(\cdot - x_i^{(1)}(t)), \frac{1}{N_2} \sum_{j=1}^{N_2} \delta(\cdot - x_j^{(2)}(t)) \right),$$

where $\delta(x)$, $x \in \mathbf{R}$, is the δ-function. One can see that the generalized functions

$$\frac{1}{N_1} \sum_{i=1}^{N_1} \delta(\cdot - x_i^{(1)}(t)), \qquad \frac{1}{N_2} \sum_{j=1}^{N_2} \delta(\cdot - x_j^{(2)}(t))$$

represent empirical "densities" or masses of (type 1 and 2, respectively) particles at time t. Thus, if $\phi(x) = (\phi_1(x), \phi_2(x))$, where $\phi_i \in S(\mathbf{R})$, then for fixed particle positions $x_1^{(1)}(t), \ldots, x_{N_1}^{(1)}(t)$ and $x_1^{(2)}(t), \ldots, x_{N_2}^{(2)}(t)$ the vector function $M_{N_1,N_2}(t)$ is a linear functional on the vector test functions ϕ; that is,

$$\langle M_{N_1,N_2}(t), \phi \rangle = \frac{1}{N_1} \sum_{i=1}^{N_1} \phi_1(x_i^{(1)}(t)) + \frac{1}{N_2} \sum_{j=1}^{N_2} \phi_2(x_j^{(2)}(t)).$$

Fix some $T > 0$. Then $(M_{N_1,N_2}(t),\, 0 \le t \le T)$ can be considered as a Markov process taking its values in the space of tempered distributions $S'(\mathbf{R}) \times S'(\mathbf{R})$. In the sequel we consider $S'(\mathbf{R}) \times S'(\mathbf{R})$ as a topological space equipped with the strong topology (see subsection 5.2). Without loss of generality one can assume that the trajectories of the process $M_{N_1,N_2}(t)$ are right continuous functions with left limits. So it is natural to consider the Skorohod space $\Pi^T = D([0,T], S'(\mathbf{R}) \times S'(\mathbf{R}))$ of functions on $[0,T]$ with values in $S'(\mathbf{R}) \times S'(\mathbf{R})$ as a coordinate space of the process $M_{N_1,N_2}(t)$. Subsection 5.2 explains how to introduce topology on this space. Let $\mathcal{B}(\Pi^T)$ be the corresponding Borel σ-algebra. Denote by $P^T_{N_1,N_2}$ the probability measure on $\big(\Pi^T, \mathcal{B}(\Pi^T)\big)$, induced by the process $(M_{N_1,N_2}(t),\, 0 \le t \le T)$.

Our assumption for the theorem is that for any test function $\phi(x)$ the sequence $\langle M_{N_1,N_2}(0), \phi \rangle$ weakly converges as $N_1, N_2 \to \infty$.

We want to prove that as $N_1, N_2 \to \infty$ the sequence of probability distributions $P^T_{N_1,N_2}$ has a weak limit, and this limit is a one-point measure, that is the only trajectory $(m_1(t), m_2(t))$, $0 \le t \le T$, which is the classical solution of the system (2.1)–(2.2). We split the proof of this result into the next two propositions.

PROPOSITION 4.1. *The family of probability distributions* $\{P^T_{N_1,N_2}\}_{N_1,N_2}$ *on* $\big(\Pi^T, \mathcal{B}(\Pi^T)\big)$ *is tight.*

PROPOSITION 4.2. *Limit points of the family of distributions* $P^T_{N_1,N_2}$ *are concentrated on the weak solutions of the system* (2.1)–(2.2).

4.1.1. *Tightness.* Before proving Proposition 4.1, we start with some preliminary lemmas. We want to prove that the family of distributions $P^T_{N_1,N_2}$ of the random process $(M_{N_1,N_2}(t),\, 0 \le t \le T)$, with values in the space of generalized functions, is tight. By Theorem 4.1 of [**M**] (see also subsection 5.2), it is sufficient to prove that for any test function $\psi = (\psi_1(x), \psi_2(x))$ the family of random processes $(\langle M_{N_1,N_2}(t), \psi \rangle,\, 0 \le t \le T)$, with values in $\mathbf{R}^1$, is tight. This will be done in Proposition 4.5 below.

Fix some test function $\psi = (\psi_1(x), \psi_2(x))$ and consider the random process

$$\begin{aligned} F_{\psi,N_1,N_2}\left(x^{(1)}(t); x^{(2)}(t)\right) &= \langle M_{N_1,N_2}(t), \psi\rangle \\ &= \frac{1}{N_1}\sum_{i=1}^{N_1} \psi_1(x_i^{(1)}(t)) + \frac{1}{N_2}\sum_{j=1}^{N_2} \psi_2(x_j^{(2)}(t)) . \end{aligned}$$

This is a function of the Markov process $\xi_{N_1,N_2}(t)$; thus (see [**KL**, Lemma 5.1, p. 330], for example) the following two processes are martingales:

$$\begin{aligned} W_{\psi,N_1,N_2}(t) &= F_{\psi,N_1,N_2}\left(x^{(1)}(t); x^{(2)}(t)\right) - F_{\psi,N_1,N_2}\left(x^{(1)}(0); x^{(2)}(0)\right) \\ &\quad - \int_0^t L_{N_1,N_2} F_{\psi,N_1,N_2}\left(x^{(1)}(s); x^{(2)}(s)\right) ds , \qquad (4.2) \\ V_{\psi,N_1,N_2}(t) &= \left(W_{\psi,N_1,N_2}(t)\right)^2 - \int_0^t L_{N_1,N_2} F^2_{\psi,N_1,N_2}\left(x^{(1)}(s); x^{(2)}(s)\right) ds \\ &\quad + 2\int_0^t F_{\psi,N_1,N_2}\left(x^{(1)}(s); x^{(2)}(s)\right) L_{N_1,N_2} F_{\psi,N_1,N_2}\left(x^{(1)}(s); x^{(2)}(s)\right) ds. \end{aligned}$$

For shortness we will write $F(x^{(1)}; x^{(2)})$ instead of $F_{\psi,N_1,N_2}(x^{(1)}; x^{(2)})$.

LEMMA 4.3. *The following estimates hold:*

i) $\left|L_{N_1,N_2} F\left(x^{(1)}; x^{(2)}\right)\right| \le C_1(\psi, v_1, v_2, \alpha_{12}, \alpha_{21})$ *uniformly in* N_1, N_2 *and* $\left(x^{(1)}; x^{(2)}\right)$;

ii) *uniformly in* $x^{(1)}$, $x^{(2)}$

$$\begin{aligned} \left|L_{N_1,N_2} F^2(x^{(1)}; x^{(2)}) - F(x^{(1)}; x^{(2)}) L_{N_1,N_2} F(x^{(1)}; x^{(2)})\right| \\ \le \frac{C_{12}(\alpha_{12}, \psi_1)}{N_1} + \frac{C_{21}(\alpha_{21}, \psi_2)}{N_2}. \end{aligned} \qquad (4.3)$$

PROOF. Note that

$$\begin{aligned} F\left(\left(x^{(1)}; x^{(2)}\right)_{i\to j}\right) - F\left(x^{(1)}; x^{(2)}\right) &= \frac{1}{N_1}\left(\psi_1\left(x_j^{(2)}\right) - \psi_1\left(x_i^{(1)}\right)\right), \\ F\left(\left(x^{(1)}; x^{(2)}\right)_{i\leftarrow j}\right) - F\left(x^{(1)}; x^{(2)}\right) &= \frac{1}{N_2}\left(\psi_2\left(x_i^{(1)}\right) - \psi_2\left(x_j^{(2)}\right)\right). \end{aligned}$$

Thus

$$\begin{aligned} L_{N_1,N_2} F\left(x^{(1)}; x^{(2)}\right) &= \frac{v_1}{N_1}\sum_{i=1}^{N_1} \psi_1'\left(x_i^{(1)}\right) + \frac{v_2}{N_2}\sum_{j=1}^{N_2} \psi_2'\left(x_j^{(2)}\right) \\ &\quad + \sum_{i=1}^{N_1}\sum_{j=1}^{N_2} \frac{\alpha_{12}}{N_2}\cdot\frac{1}{N_1}\left(\psi_1\left(x_j^{(2)}\right) - \psi_1\left(x_i^{(1)}\right)\right) \\ &\quad + \sum_{j=1}^{N_2}\sum_{i=1}^{N_1} \frac{\alpha_{21}}{N_1}\cdot\frac{1}{N_2}\left(\psi_2\left(x_i^{(1)}\right) - \psi_2\left(x_j^{(2)}\right)\right). \qquad (4.4) \end{aligned}$$

Then

$$\begin{aligned} \left|L_{N_1,N_2} F\left(x^{(1)}; x^{(2)}\right)\right| &\le |v_1|\,\|\psi_1'\|_C + |v_2|\,\|\psi_2'\|_C \\ &\quad + 2\alpha_{12}\,\|\psi_1\|_C + 2\alpha_{21}\,\|\psi_2\|_C , \end{aligned}$$

and the assertion **i)** of the lemma is proved. To prove assertion **ii)**, it is convenient to represent $L_{N_1,N_2} = L^0_{N_1,N_2} + L^1_{N_1,N_2}$ as the sum of the "differential" $L^0_{N_1,N_2}$ and "jump" $L^1_{N_1,N_2}$ parts.

It is easy to see that

$$L^0_{N_1,N_2} F^2(x^{(1)};x^{(2)}) - 2F(x^{(1)};x^{(2)}) L^0_{N_1,N_2} F(x^{(1)};x^{(2)}) = 0.$$

Let us prove that uniformly in $x^{(1)}$, $x^{(2)}$

$$\left| L^1_{N_1,N_2} F^2(x^{(1)};x^{(2)}) - F(x^{(1)};x^{(2)}) L^1_{N_1,N_2} F(x^{(1)};x^{(2)}) \right| \\ \leq \frac{4\alpha_{12} \left\|\psi_1\right\|_C^2}{N_1} + \frac{4\alpha_{21} \left\|\psi_2\right\|_C^2}{N_2}. \tag{4.5}$$

In fact,

$$\begin{aligned} F^2 &\left(\left(x^{(1)};x^{(2)} \right)_{i\to j} \right) - F^2\left(x^{(1)};x^{(2)}\right) \\ &= \left(F\left(\left(x^{(1)};x^{(2)} \right)_{i\to j} \right) - F\left(x^{(1)};x^{(2)}\right) \right) \\ &\quad \times \left(2F\left(x^{(1)};x^{(2)}\right) + \frac{1}{N_1} \left(\psi_1\left(x_j^{(2)}\right) - \psi_1\left(x_i^{(1)}\right) \right) \right) \\ &= 2F\left(x^{(1)};x^{(2)}\right) \left[F\left(\left(x^{(1)};x^{(2)} \right)_{i\to j} \right) - F\left(x^{(1)};x^{(2)}\right) \right] \\ &\quad + \left[\frac{1}{N_1} \left(\psi_1\left(x_j^{(2)}\right) - \psi_1\left(x_i^{(1)}\right) \right) \right]^2 \end{aligned}$$

and similarly for expressions with $\left(x^{(1)};x^{(2)}\right)_{i\leftarrow j}$. Thus,

$$\begin{aligned} L^1_{N_1,N_2} &F^2(x^{(1)};x^{(2)}) \\ &= 2F\left(x^{(1)};x^{(2)}\right) \sum_{i=1}^{N_1} \sum_{j=1}^{N_2} \frac{\alpha_{12}}{N_2} \cdot \left[F\left(\left(x^{(1)};x^{(2)} \right)_{i\to j} \right) - F\left(x^{(1)};x^{(2)}\right) \right] \\ &\quad + \sum_{i=1}^{N_1} \sum_{j=1}^{N_2} \frac{\alpha_{12}}{N_2} \cdot \left[\frac{1}{N_1} \left(\psi_1\left(x_j^{(2)}\right) - \psi_1\left(x_i^{(1)}\right) \right) \right]^2 \\ &\quad + 2F\left(x^{(1)};x^{(2)}\right) \sum_{j=1}^{N_2} \sum_{i=1}^{N_1} \frac{\alpha_{21}}{N_1} \cdot \left[F\left(\left(x^{(1)};x^{(2)} \right)_{i\leftarrow j} \right) - F\left(x^{(1)};x^{(2)}\right) \right] \\ &\quad + \sum_{j=1}^{N_2} \sum_{i=1}^{N_1} \frac{\alpha_{21}}{N_1} \cdot \left[\frac{1}{N_2} \left(\psi_2\left(x_i^{(1)}\right) - \psi_2\left(x_j^{(2)}\right) \right) \right]^2 \\ &= 2F L^1_{N_1,N_2} F + \frac{\alpha_{12}}{N_1} \sum_{i=1}^{N_1} \sum_{j=1}^{N_2} \frac{1}{N_2 N_1} \left(\psi_1\left(x_j^{(2)}\right) - \psi_1\left(x_i^{(1)}\right) \right)^2 \\ &\quad + \frac{\alpha_{21}}{N_2} \sum_{j=1}^{N_2} \sum_{i=1}^{N_1} \frac{1}{N_1 N_2} \left(\psi_2\left(x_i^{(1)}\right) - \psi_2\left(x_j^{(2)}\right) \right)^2, \end{aligned}$$

and the estimate (4.5) follows from this. The lemma is proved. □

COROLLARY 4.4.

$$\sup_{t\le T} \mathsf{E}\left(W_{\psi,N_1,N_2}(t)\right)^2 \to 0, \qquad N_1, N_2 \to \infty\,.$$

PROOF. As V_{ψ,N_1,N_2} is a martingale with mean zero, it is sufficient to prove that the expectation of

$$\int_0^t \Big[L_{N_1,N_2}F^2\left(x^{(1)}(s); x^{(2)}(s)\right) \\ -2F\left(x^{(1)}(s); x^{(2)}(s)\right) L_{N_1,N_2}F\left(x^{(1)}(s); x^{(2)}(s)\right)\Big]\, ds$$

tends to zero. This follows from estimate (4.3) of the lemma. □

PROPOSITION 4.5. *The sequence of distributions of the following real-valued random processes*

$$F_{\psi,N_1,N_2}\left(x^{(1)}(t); x^{(2)}(t)\right), \; t \in [0,T],$$

is tight.

PROOF OF PROPOSITION 4.5. Recall that the following representation holds:

$$F_{\psi,N_1,N_2}\left(x^{(1)}(t); x^{(2)}(t)\right) = F_{\psi,N_1,N_2}\left(x^{(1)}(0); x^{(2)}(0)\right) + W_{\psi,N_1,N_2}(t) \\ + \int_0^t L_{N_1,N_2}F_{\psi,N_1,N_2}\left(x^{(1)}(s); x^{(2)}(s)\right) ds\,.$$

Note that our initial assumption is that the sequence $F_{\psi,N_1,N_2}\left(x^{(1)}(0); x^{(2)}(0)\right)$ weakly converges as $N_1, N_2 \to \infty$.

We shall now prove that the sequence

$$\left\{\eta^{N_1,N_2}(t) = \int_0^t L_{N_1,N_2}F\left(x^{(1)}(s); x^{(2)}(s)\right) ds,\; t \in [0,T]\right\}_{N_1,N_2}$$

is tight. We use subsection 5.1 of the Appendix. By assertion **i)** of the lemma

$$\left|\int_0^t L_{N_1,N_2}F\left(x^{(1)}(s); x^{(2)}(s)\right) ds\right| \le C_1(\psi, v_1, v_2, \alpha_{12}, \alpha_{21}) \cdot T\,;$$

thus, condition 1) of Theorem 5.1 of the Appendix holds. Condition 2) also holds, as one can prove that

$$w'(\eta^{N_1,N_2}, \gamma) \le 2\gamma \cdot C_1(\psi, v_1, v_2, \alpha_{12}, \alpha_{21})\,.$$

We shall prove that the sequence $\{W_{\psi,N_1,N_2}(t),\, t \in [0,T]\}_{N_1,N_2}$ is tight. Using Kolmogorov's inequality for submartingales with right continuous trajectories (see [**D**]), we have the following estimate, uniform in N_1, N_2:

$$P\left(\sup_{t\le T} |W_{\psi,N_1,N_2}(t)| > C\right) \le \frac{\sup_{t\le T} \mathsf{E}\left(W_{\psi,N_1,N_2}(t)\right)^2}{C^2}\,.$$

Then from Corollary 4.4 condition 1) of the Appendix holds. Thus,

$$\begin{aligned} & P\left(|W_{\psi,N_1,N_2}(\tau+\theta) - W_{\psi,N_1,N_2}(\tau)| > \varepsilon\right) \\ & \qquad \le \frac{\mathsf{E}\left(W_{\psi,N_1,N_2}(\tau+\theta) - W_{\psi,N_1,N_2}(\tau)\right)^2}{\varepsilon^2} \\ & \qquad = \frac{\mathsf{E}\int_\tau^{\tau+\theta} V_{\psi,N_1,N_2}(s)\,ds}{\varepsilon^2} \\ & \qquad \le \frac{\theta\cdot\left(C_{12}(\alpha_{12},\psi_1)/N_1 + C_{21}(\alpha_{21},\psi_2)/N_2\right)}{\varepsilon^2}\,. \end{aligned}$$

Using this estimate, one can check the sufficient condition of Aldous. Then Proposition 4.5 is proved. □

This also concludes the proof of Proposition 4.1.

4.1.2. *Weak solutions.*

DEFINITION 4.6. We say that the pair of functions $M(t) = (m_1(t,x), m_2(t,x))$ is a weak solution of the system (2.1)–(2.2), if for any pair $\phi_1(x), \phi_2(x) \in S(\mathbf{R})$ the identities

$$\begin{aligned} \langle M(t),\phi\rangle &= \langle M(0),\phi\rangle \\ &\quad + \int_0^t \langle M(s), (v_1\phi_1' - \alpha_{12}\phi_1 + \alpha_{21}\phi_2, v_2\phi_2' - \alpha_{12}\phi_1 + \alpha_{21}\phi_2)\rangle\,ds \end{aligned}$$

hold, where $\phi(x) = (\phi_1(x),\phi_2(x))$, and the action of $G(x) = (g_1(x), g_2(x))$ on the test function $\phi(x)$ can be written as

$$\langle G,\phi\rangle = \int g_1(x)\phi_1(x)\,dx + \int g_2(x)\phi_2(x)\,dx\,.$$

Note that from the representation (4.2) and the identity (4.4) it follows that

$$\begin{aligned} &\langle M_{N_1,N_2}(t),\phi\rangle = W_{\phi,N_1,N_2}(t) + \langle M_{N_1,N_2}(0),\phi\rangle \\ &\quad + \int_0^t \langle M_{N_1,N_2}(s), (v_1\phi_1' - \alpha_{12}\phi_1 + \alpha_{21}\phi_2, v_2\phi_2' - \alpha_{12}\phi_1 + \alpha_{21}\phi_2)\rangle\,ds. \end{aligned}$$

Let $h = h(t) \in \Pi^T = D([0,T], S'(\mathbf{R}) \times S'(\mathbf{R}))$. For fixed ϕ define the functional

$$\begin{aligned} J_{\phi,T}(h) = \sup_{t\le T}\Big| &\langle h(t),\phi\rangle - \langle h(0),\phi\rangle \\ & - \int_0^t \langle h(s), (v_1\phi_1' - \alpha_{12}\phi_1 + \alpha_{21}\phi_2, v_2\phi_2' - \alpha_{12}\phi_1 + \alpha_{21}\phi_2)\rangle\,ds\Big|. \end{aligned}$$

In particular,

$$\sup_{t\le T}|W_{\phi,N_1,N_2}(t)| = J_{\phi,T}(M_{N_1,N_2}).$$

The rest of the proof is standard (see [**KL**]) and consists of three steps.

Step 1. From the definition of the topology on Π^T it follows that $J_{\phi,T}(\cdot) : \Pi^T \to \mathbf{R}_+$ is a continuous functional.

Step 2. Note that

$$\forall\varepsilon > 0 \quad P\{J_{\phi,T}(M_{N_1,N_2}) > \varepsilon\} \equiv P^T_{N_1,N_2}\{h : J_{\phi,T}(h) > \varepsilon\} \to 0 \qquad (N_1, N_2 \to \infty)$$

by Kolmogorov's inequality and Corollary 4.4.

Step 3. As $J_{\phi,T}(\cdot)$ is continuous, the set $\{h:\ J_{\phi,T}(h)>0\}$ is open in Π^T. It follows now that for any limit point P^T_∞ of the family $\left\{P^T_{N_1,N_2}\right\}_{N_1,N_2}$ we have

$$P^T_\infty\left\{h:\ J_{\phi,T}(h)>\varepsilon\right\}\le\limsup_{N_1,N_2}P^T_{N_1,N_2}\left\{h:\ J_{\phi,T}(h)>\varepsilon\right\}.$$

That is, for any $\varepsilon>0$ we have $P^T_\infty\left\{h:\ J_{\phi,T}(h)>\varepsilon\right\}=0$. In other words, all limit points P^T_∞ of the family $\left\{P^T_{N_1,N_2}\right\}_{N_1,N_2}$ have support on the set

$$\left\{h:\ J_{\phi,T}(h)=0\right\},$$

which consists of weak solutions of (2.1)–(2.2).

This completes the proof of Proposition 4.2.

The problem of uniqueness of the weak solution of (2.1)–(2.2) is quite simple because the system (2.1)–(2.2) is *linear*. In subsection 4.2 we shall see that this system of first-order differential equations has a unique classical solution which can be obtained in an explicit way.

4.2. Time asymptotics for the continuous model. We prove here Theorem 2.3.

Define the mean (mass centrum) $a_i(t)=\displaystyle\int x m_i(t,x)\,dx$ and the variance (momentum of inertia) $d_i(t)=\displaystyle\int (x-a_i(t))^2 m_i(t,x)\,dx$.

From (2.1)–(2.2) we get the following equations for the means:

$$\begin{aligned}\dot{a_1}&=v_1+\alpha_{12}\left(a_2-a_1\right),\\ \dot{a_2}&=v_2+\alpha_{21}\left(a_1-a_2\right).\end{aligned}$$

It follows that the equation for $a_2(t)-a_1(t)$ is closed and has the following solution:

$$a_2(t)-a_1(t)=\frac{v_2-v_1}{\alpha_{12}+\alpha_{21}}\left(1-e^{-(\alpha_{12}+\alpha_{21})t}\right)+\left(a_2(0)-a_1(0)\right)e^{-(\alpha_{12}+\alpha_{21})t}.$$

Thus

$$a_2(t)-a_1(t)\to\frac{v_2-v_1}{\alpha_{12}+\alpha_{21}}\qquad(t\to+\infty),$$

and similarly

$$\frac{d}{dt}a_i(t)\to\frac{\alpha_{21}v_1+\alpha_{12}v_2}{\alpha_{12}+\alpha_{21}}\qquad(t\to+\infty).$$

The equations for variances are

$$\begin{aligned}\dot{d_1}&=\alpha_{12}\left(d_2-d_1\right)+\alpha_{12}\left(a_2(t)-a_1(t)\right)^2,\\ \dot{d_2}&=\alpha_{21}\left(d_1-d_2\right)+\alpha_{21}\left(a_1(t)-a_2(t)\right)^2.\end{aligned}$$

Or, equivalently,

$$\begin{aligned}\frac{d}{dt}\left(\alpha_{21}d_1+\alpha_{12}d_2\right)&=2\alpha_{12}\alpha_{21}\left(a_2(t)-a_1(t)\right)^2,\\ \frac{d}{dt}\left(d_2-d_1\right)&=-\left(\alpha_{12}+\alpha_{21}\right)\left(d_2-d_1\right)+\left(\alpha_{21}-\alpha_{12}\right)\left(a_2(t)-a_1(t)\right)^2.\end{aligned}$$

From this we get

$$d_2(t)-d_1(t)\to const=\left(\frac{v_2-v_1}{\alpha_{12}+\alpha_{21}}\right)^2\cdot\frac{\alpha_{21}-\alpha_{12}}{\alpha_{12}+\alpha_{21}}$$

and

$$\frac{d}{dt}(\alpha_{21}d_1 + \alpha_{12}d_2) \to 2\alpha_{12}\alpha_{21}\left(\frac{v_2 - v_1}{\alpha_{12} + \alpha_{21}}\right)^2 .$$

Thus the growth of variances is asymptotically linear. Moreover, both are asymptotically equal.

Now we come to the solution of the equations. Define the Fourier transforms

$$m_i(x,t) = \int \exp(ixp) g_i(p,t) dp .$$

We get

$$\frac{\partial g_1}{\partial t} + v_1 ipg_1 = \alpha_{12}(g_2 - g_1) ,$$
$$\frac{\partial g_2}{\partial t} + v_2 ipg_2 = \alpha_{21}(g_1 - g_2)$$

with the initial conditions $m_i(0,x) = m_i(x)$, $i = 1,2$. We write this system in the vector form

$$\frac{dg}{dt} = Ag ,$$

where

$$A = \begin{pmatrix} -iv_1p - \alpha_{12} & \alpha_{12} \\ \alpha_{21} & -iv_2p - \alpha_{21} \end{pmatrix} .$$

For the eigenvalues we have

$$\lambda_\pm = -\frac{a}{2} \pm \sqrt{\frac{a^2}{4} - b} ,$$

where

$$a = i(v_1 + v_2)p + \alpha_{12} + \alpha_{21}, \qquad b = -v_1v_2p^2 + ip(v_1\alpha_{21} + v_2\alpha_{12}) .$$

One can write the solution as

$$g = C_+\phi_+ \exp(t\lambda_+) + C_-\phi_- \exp(t\lambda_-) ,$$

where $\phi_\pm$ are eigenfunctions. Note that for small p there are two roots. One has $\operatorname{Re}\lambda_- < 0$, thus a strongly decreasing term. The other is

$$\lambda_+ = c_1p + c_2p^2 + O(p^3), \qquad c_2 \neq 0, \tag{4.6}$$

for small p.

Let ξ_t be a random variable with density $m(x,t)$, and let $g(k)$ be its characteristic function. We are interested in $\frac{1}{\sqrt{t}}(\xi_t - a), a = \mathsf{E}\xi_t$, its characteristic function being

$$\exp\left(-ia\frac{k}{\sqrt{t}}\right) g\left(\frac{k}{\sqrt{t}}\right) .$$

Using (4.6), we get the result.

REMARK 4.7. One can see that there is no solution of the type

$$m_i(t,x) = f_i(x - vt)$$

as then f_i would be exponents.

REMARK 4.8. For the singular initial conditions, that is, when $x_k^{(i)}(0) = 0$ for $k = 1, \dots, N_i$ and $i = 1,2$, one can get the same asymptotic results.

5. Appendix

5.1. Probability measures on the Skorohod space: tightness. Let us consider a sequence $\{(\xi_t^n, t \in [0,T])\}_{n\in\mathbf{N}}$ of real random processes whose trajectories are right-continuous and admit left-hand limits for every $0 < t \le T$. We will consider ξ^n as random elements with values in the Skorohod space $D_T(\mathbf{R}) := D\left([0,T], \mathbf{R}^1\right)$ with the standard topology. Denote byP_T^n the distribution of ξ^n, defined on the measurable space $(D_T(\mathbf{R}), \mathcal{B}\left(D_T(\mathbf{R})\right))$. The following result can be found in [**B**].

THEOREM 5.1. *The sequence of probability measures* $\{P_T^n\}_{n\in\mathbf{N}}$ *is tight iff the following two conditions hold:*

1) for any $\varepsilon > 0$ *there is* $C(\varepsilon) > 0$ *such that*

$$\sup_n P_T^n \left(\sup_{0\le t\le T} |\xi_t^n| > C(\varepsilon) \right) \le \varepsilon\,;$$

2) for any $\varepsilon > 0$

$$\lim_{\gamma\to 0} \limsup_n P_T^n \left(\xi_\cdot : w'(\xi;\gamma) > \varepsilon\right) = 0\,,$$

where for any function $f : [0,T] \to \mathbf{R}$ *and any* $\gamma > 0$ *we define*

$$w'(f;\gamma) = \inf_{\{t_i\}_{i=1}^r} \max_{i<r} \sup_{t_i\le s<t<t_{i+1}} |f(t) - f(s)|\,;$$

moreover the inf *is taken over all partitions of the interval* $[0,T]$ *such that*

$$0 = t_0 < t_1 < \cdots < t_r = T, \qquad t_i - t_{i-1} > \gamma, \quad i = 1,\ldots,r.$$

The following theorem is known as the sufficient condition of Aldous [**KL**].

THEOREM 5.2. *Condition 2) of the previous theorem follows from the condition*

$$\forall\varepsilon > 0 \qquad \lim_{\gamma\to 0} \limsup_n \sup_{\tau\in\mathcal{R}_T,\,\theta\le\gamma} P_T^n\left(|\xi_{\tau+\theta} - \xi_\tau| > \varepsilon\right) = 0\,,$$

where $\mathcal{R}_T$ *is the set of Markov moments (stopping times) not exceeding* T *.*

5.2. Strong topology on the Skorohod space. Mitoma theorem. Recall that the Schwartz space $S(\mathbf{R})$ is a Frechet space (complete locally convex space, the topology of which is generated by a countable family of seminorms, which implies metrizability; see [**RS**]). In the dual space $S'(\mathbf{R})$ of tempered distributions there are at least two ways to define topology (both not metrizable):

1) *weak topology* on $S'(\mathbf{R})$, where all functionals

$$\langle\,\cdot\,,\phi\rangle\,, \quad \phi \in S(\mathbf{R}),$$

are continuous;

2) *strong topology* on $S'(\mathbf{R})$, which is generated by the set of seminorms

$$\left\{\rho_A(M) = \sup_{\phi\in A} |\langle M,\phi\rangle| \;:\; A \subset S(\mathbf{R}) \text{ is bounded}\right\}.$$

We shall consider $S'(\mathbf{R})$ as equipped with the strong topology. Details can be found in [**RS**].

The problem of introducing the Skorohod topology on the space $D_T(S') := D([0,T], S'(\mathbf{R}))$ was studied in [**M**] and [**J**]. The topology on this space is defined as follows. Let $\{\rho_A\}$ be a family of seminorms, which generates the strong topology in $S'(\mathbf{R})$. For each seminorm ρ_A define a pseudometric

$$d_A(y,z) = \inf_{\lambda\in\Lambda}\left\{\sup_t \left|y_t - z_{\lambda(t)}\right| + \sup_{t\neq s}\left|\log\frac{\lambda(t)-\lambda(s)}{t-s}\right|\right\}, \quad y,z\in D_T(S'),$$

where the inf is taken over the set $\Lambda = \{\lambda = \lambda(t),\, t\in[0,T]\}$ of all strictly increasing maps of the interval $[0,T]$ into itself. Equipped with the topology of the projective limit for the family $\{d_A\}$, the set $D_T(S')$ becomes a completely regular topological space.

Let $\mathcal{B}(D_T(S'))$ be the corresponding Borel σ-algebra. Let $\{P_n\}$ be a sequence of probability measures on $(D_T(S'),\mathcal{B}(D_T(S')))$. For each $\phi\in S(\mathbf{R})$ consider a map $\mathcal{I}_\phi : y\in D_T(S') \to y_{\cdot}(\phi)\in D_T(\mathbf{R})$. The following result belongs to I. Mitoma [**M**].

THEOREM 5.3. *Suppose that for any $\phi\in S(\mathbf{R})$ the sequence $\left\{P_n\mathcal{I}_\phi^{-1}\right\}$ is tight in $D_T(\mathbf{R})$. Then the sequence $\{P_n\}$ itself is tight in $D_T(S')$.*

References

[B] P. Billingsley, *Convergence of probability measures,* John Wiley & Sons, 1968.

[BT] D. P. Bertsekas, J. N. Tsitsiklis, *Parallel and Distributed Computation: Numerical Methods,* Athena Scientific, Belmont, MA, 1997.

[D] J. L. Doob, *Stochastic Processes,* John Wiley & Sons, New York.

[GMP] A. Greenberg, V. Malyshev, S. Popov, *Stochastic Model of Massively Parallel Simulation,* Markov Proc. Rel. Fields **1** (1995), 473–490.

[J] A. Jakubowski, *On the Skorohod topology,* Ann. Inst. H. Poincaré: Probabilités et Statistiques **22** (1986), 263–285.

[K] T. Kato, *Perturbation Theory for Linear Operators,* Springer, 1976.

[KL] C. Kipnis, C. Landim, *Scaling Limits of Interacting Particle Systems,* Springer, 1998.

[M] I. Mitoma, *Tightness of probabilities on $C([0,1];S')$ and $D([0,1];S')$,* Ann. Probab. **11** (1983), 989–999.

[MaMa1] V. Malyshev, A. Manita, *Time synchronization model,* Preprint INRIA no. RR-5204, 2004.

[MaMa2] V. A. Malyshev, A. D. Manita, *Phase transitions in the time synchronization model,* Probability Theory and Applications **50** (2005), 150–158.

[MiMi] D. Mitra, I. Mitrani, *Analysis and optimum performance of two message-passing parallel processors synchronized by rollback,* Performance Evaluation **7** (1987), 111–124.

[MSh1] A. Manita, V. Shcherbakov, *Asymptotic analysis of a particle system with mean-field interaction,* Markov Processes Relat. Fields **11** (2005), 489–518.

[MSh2] A. Manita, V. Shcherbakov, *Stochastic particle system with non-local mean-field interaction,* Proceedings of Int. Conf. "Kolmogorov and Contemporary Mathematics", Moscow, June 16–21, 2003, pp. 549–550.

[MSi] A. Manita, F. Simonot, *Clustering in Stochastic Asynchronous Algorithms for Distributed Simulations,* Lecture Notes in Computer Science **3777** (2005), 26–37.

[RS] M. Reed, B. Simon, *Methods of Modern Mathematical Physics. Vol. 1: Functional Analysis,* Academic Press, 1972.

INRIA – ROCQUENCOURT, B.P. 105, 78153 LE CHESNAY CEDEX, FRANCE
E-mail address: vadim.malyshev@inria.fr

FACULTY OF MATHEMATICS AND MECHANICS, MOSCOW STATE UNIVERSITY, 119992, MOSCOW, RUSSIA
E-mail address: manita@mech.math.msu.su

Amer. Math. Soc. Transl.
(2) Vol. **217**, 2006

Stable Densities and Operators of Fractional Differentiation

Yuri A. Neretin

ABSTRACT. Let $D(s)$ be a fractional derivation of order s. For real $\alpha \neq 0$, we construct an integral operator $A(\alpha)$ in an appropriate functional space such that $A(\alpha)D(s)A(\alpha)^{-1} = D(\alpha s)$ for all s. The kernel of the operator $A(\alpha)$ is expressed in terms of a function similar to the stable densities.

Introduction

0.1. Definition of functions $\mathbb{L}_{\alpha,\beta}$. This paper contains several simple observations concerning the special function

$$\mathbb{L}_{\alpha,\beta}(z) = \sum_{n=0}^{\infty} \frac{(-1)^n \Gamma(\alpha n + \beta)}{n!} z^n , \tag{0.1}$$

where $0 < \alpha < 1$, $\operatorname{Re}\beta > 0$, and $z \in \mathbb{C}$. We can also represent this function in the form

$$\mathbb{L}_{\alpha,\beta}(z) = \int_0^{\infty} x^{\beta-1} \exp(-zt^{\alpha} - t)\, dt , \tag{0.2}$$

$$\mathbb{L}_{\alpha,\beta}(z) = \frac{1}{2\pi i} \int_{-i\infty}^{+i\infty} \Gamma(\beta - \alpha s)\, \Gamma(s) z^{-s}\, ds . \tag{0.3}$$

The integrals (0.2), (0.3) also make sense for $\alpha > 1$. The definition of functions $\mathbb{L}_{\alpha,\beta}$ is discussed in detail below in Section 1.

The function $\mathbb{L}_{\alpha,\beta}$ is one of the simplest examples of the so-called *H-functions* (or *Fox functions*); see [**M**]. In a strange way, the function $\mathbb{L}_{\alpha,\beta}$ has no official name. Obviously, for rational $\alpha = p/q$ the function $\mathbb{L}_{\alpha,\beta}$ can be expressed in the terms of higher hypergeometric functions. But for $q > 4$ such expressions do not seem to be very useful.

2000 *Mathematics Subject Classification.* Primary 33C60; Secondary 26A33, 44A10, 60E07.
Key words and phrases. Fox function, H-function, Mellin-Barnes integral, Wright function, stable distribution, fractional derivative, Laplace transform, operation calculus, Hardy space.

0.2. Results of the paper. Integral operators with functions $\mathbb{L}_{\alpha,\beta}$ in kernels.

A. We consider the space $\mathcal{K}$ of functions holomorphic in the half-plane $\operatorname{Re} z > 0$, smooth up to the line $\operatorname{Re} z = 0$, and satisfying the following condition:

- for each $k > 0$ and $N > 0$ there exists M such that

$$|f^{(k)}(z)| \leqslant M(1+|z|)^{-N} . \tag{0.4}$$

We define the operators of *fractional differentiation* D_h in the space $\mathcal{K}$ by

$$D_h f(z) = \frac{\Gamma(h+1)}{2\pi} \int_{-\infty}^{+\infty} \frac{f(it)\,dt}{(-it+z)^{h+1}} .$$

For a positive integer n, we have $D_n = (-1)^n d^n/dz^n$; the operator D_{-n} is the indefinite integration iterated n times. Also $D_{h+r} = D_h D_r$. See Section 2 below for details.

Next, for $\alpha > 0$ we define the kernel

$$K_\alpha(u,v) = \int_0^\infty \exp(-ux^\alpha - vx)\,dx = v^{-1}\mathbb{L}_\alpha(u/v^\alpha),$$

and the operator in the space $\mathcal{K}$ given by

$$A_\alpha f(v) = \frac{1}{2\pi}\int_{-\infty}^{+\infty} K_\alpha(-it, v) f(it)\,dt . \tag{0.5}$$

The operators A_α form a one-parameter group (see subsection 3.5),

$$A_\alpha A_\beta = A_{\alpha\beta}. \tag{0.6}$$

We also show that they satisfy the property

$$A_\alpha D_h A_\alpha^{-1} = D_{\alpha h} . \tag{0.7}$$

REMARK. We emphasize the following particular cases of (0.7):

$$A_{m/n}\frac{d^n}{dz^n}A_{m/n}^{-1} = (-1)^{m-n}\frac{d^m}{dz^m},$$

$$A_k \frac{d}{dz} A_k^{-1} = (-1)^{k-1}\frac{d^k}{dz^k}$$

for integers m, n, k.

REMARK. Also

$$A_\alpha z A_\alpha^{-1} f(z) = \frac{1}{\alpha} D_{1-\alpha}(zf(z)) . \tag{0.8}$$

REMARK. It seems that (0.7), (0.8) and formula (0.12) below give a possibility to strange transformations of partial differential equations and their solutions.

Furthermore, the operators of dilatation

$$R_a g(z) = a^{-1} g(z/a), \qquad a > 0 ,$$

satisfy

$$A_\alpha^{-1} R_a A_\alpha = R_{a^\alpha} . \tag{0.9}$$

The generator of the one-parameter group R_a is $(z\,d/dz + 1)$. Hence (0.9) can be written in the form

$$A_\alpha^{-1}\Big(z\frac{d}{dz}+1\Big)A_\alpha = \alpha\Big(z\frac{d}{dz}+1\Big) .$$

B. In Section 3 we consider the group G of integral operators in $\mathcal{K}$ generated by the operators A_α, the fractional derivations D_h, and the dilatations R_a. We observe that G is a 6-dimensional solvable Lie group with 2-dimensional center and kernels of all elements of this group admit simple expressions in the terms of the functions $\mathbb{L}_{\alpha,\beta}$ (Theorems 3.1, 3.2).

C. In Section 4 we consider the usual Riemann–Liouville fractional integrations J_h in the space of functions on the half-line $x \geqslant 0$; see (4.1). For $0 < \alpha < 1$, we consider the Zolotarev operators [**Z1, Z2**] defined by the formula

$$B_\alpha f(x) = \frac{1}{\pi x}\int_0^\infty \operatorname{Im}\Big\{\mathbb{L}_{\alpha,1}(x^{-\alpha} y e^{i\pi\alpha})\Big\} f(y)\,dy\,. \tag{0.10}$$

We have

$$B_\alpha B_\beta = B_{\alpha\beta}\,, \tag{0.11}$$

$$B_\alpha J_h = J_{h\alpha} B_\alpha \tag{0.12}$$

(but we cannot represent the identity (0.12) in the form (0.7), since the operators B_α are not invertible).

These operators can be included into a 7-dimensional *semigroup* of integral operators on the half-line; this semigroup has a 2-dimensional center (Theorem 4.2).

0.3. Some references on functions $\mathbb{L}_{\alpha,\beta}$.

1) Barnes in 1906 [**Ba**] evaluated asymptotics of several H-functions and, in particular, for $\mathbb{L}_{\alpha,\beta}$. But it seems that he had no reasons to investigate $\mathbb{L}_{\alpha,\beta}$ in detail; in the following years this function (as far as I know) had not attracted specialists in special functions.

2) The functions

$$\mathbb{W}_{\alpha,\beta}(z) = \sum_{n=0}^\infty \frac{z^n}{n!\,\Gamma(\alpha n+\beta)}, \qquad \alpha > 0, \tag{0.13}$$

('*Wright functions*', '*Bessel–Maitland functions*')[1] were discussed more; see [**Wr, A**] and references in "Higher transcendental functions" [**EMOT**, Section "Mittag-Leffler function"]. The functions $\mathbb{L}_{\alpha,\beta}(z)$, $\mathbb{W}_{\alpha,\beta}(z)$ quite often appear in formulas in similar cases (for instance, see subsection 1.5). Another 'relative' of the function $\mathbb{L}_{\alpha,\beta}$ is the Mittag-Leffler function $\sum z^n/\Gamma(\alpha n+1)$ that appears in the literature quite often.

3) The functions $\mathbb{L}_{\alpha,\beta}$ appear (see Feller [**F1**]) if we solve the Cauchy problem for the partial pseudo-differential equation

$$\left[\frac{d}{dt} - \frac{d^\alpha}{dx^\alpha}\right] f(x,t) = 0, \qquad f(x,0) = \psi(x)\,,$$

where d^α/dx^α is some fractional derivative. Sometimes it is possible to write

$$f(x,t) = \int K(t,x,y) f(y)\,dy\,,$$

[1] Apparently, Wright and Maitland are the same person (Edward Maitland Wright), which also coincides with the author of the well-known book "An introduction to the theory of numbers" by Hardy and Wright.

where the kernel K can be expressed in the terms of the function $\mathbb{L}_{\alpha,\beta}$. The work of Feller generated a wide literature on diffusions generated by pseudo-differential operators.

4) Now, we recall the most important situation, where the functions $\mathbb{L}_{\alpha,\beta}$ arise in a natural way.

Consider a sequence ξ_j of independent random variables and its partial sums $S_n = \xi_1 + \dots + \xi_n$. Consider the distribution μ_n of S_n. Let us center and normalize μ_n in some way, $\widetilde{\mu}_n(t) := \mu_n(a_n t + b_n)$, where $a_n > 0$, $b_n \in \mathbb{R}$ are some constants. Which distributions can appear as limits of sequences $\widetilde{\mu}_n$? In the most common cases, we obtain a normal (Gauss) distribution. Nevertheless, there are other possible limits ([**L3**], see also [**F2**]); they are called *stable distributions*. Densities of these distributions admit a simple expression (0.15) in terms of the functions $\mathbb{L}_{\alpha,1}$.

A logical possibility of non-normal distributions in limit theorems of this kind was observed by Cauchy in 1853; see [**C**]. He claimed that the distribution whose densities are given by

$$\varphi_\alpha(x) = \int_0^\infty \exp(-t^\alpha)\cos(tx)\,dt \tag{0.14}$$

can appear in limit theorems for sums of independent random variables. Firstly, it was necessary to verify positivity of the functions φ_α. They are indeed positive for $0 < \alpha \leqslant 2$, but Cauchy could not prove this except for several simple cases ($\alpha = 1, 1/2, 2$). In 1922 P. Levy[2] attracted attention to the problem [**L1**], and in 1923 Polya [**P**] proved positivity of (0.14) for $0 < \alpha < 1$.

After the appearance of the Kolmogorov–Levy–Hinchin integral representation for infinitely divisible laws, a complete description of stable distributions became a solvable problem, and the final result is presented in the books of P. Levy [**L2**], 1937, and A. Hinchin (another spelling is 'Khintchine') [**H**], 1938. The stable densities can be represented in the form

$$p(x;\alpha,\gamma) = \frac{1}{\pi x}\operatorname{Im}\mathbb{L}_{\alpha,1}\left(x^{-\alpha}e^{i(\gamma-\alpha)\pi/2}\right), \tag{0.15}$$

where $0 < \alpha < 2$, $\gamma \in \mathbb{R}$, and $|\gamma| < \min(\alpha, 2-\alpha)$ (in this formula, we omit the exceptional and simple case $\alpha = 1$).

It was clear that the integrals of the form (0.2), (0.14) have no expression in terms of classical special functions, but they were important for probabilists and attracted their interest; see [**F1, F2, Z1**]. The basic text on this subject is Zolotarev's book [**Z2**] published in 1986; also see the bibliography in this book.

Levy also introduced stable stochastic processes (see [**L3**]). Non-explicitness of stable densities makes stable processes difficult for investigations; nevertheless some collection of explicit formulae is known (see Dynkin [**D**], Neretin [**N1**], Pitman–Yor [**PY**]).

In this paper, expression (0.15) appears in formulas (0.10), (1.10). Also formulas (0.11), (0.6) are variants of the "multiplication theorem for the stable laws" [**Z2**, Theorem 3.3.1]; there are many other places where we touch formulas from Zolotarev's book [**Z2**], and I do not try to mention all similarities in formulas.

---

[2]Some references between 1853 and 1922 can be found in [**Wi**]. Also, there was a work of Holtsmark (1919) on the distribution of the gravitation force in the universe; see its exposition in [**F2, Z2**].

5) The functions $\mathbb{L}_{\alpha,\beta}$ arise in a relatively natural way in the theory of the Laplace transform (the 'operation calculus'); see below. The tables of McLachlan, Humbert, Poli [**MHP**], 1950, contain 18 partial cases of the integral transformations defined below; also the transformations (0.10) are contained in Zolotarev [**Z1**], and a similar construction with Wright functions is a subject of Agarwal [**A**].

6) It is known (see [**MR**]) that pseudo-differential equations with constant coefficients of the type

$$\Big(\sum_{k=0}^{n} a_k D_{k\alpha}\Big) f = 0$$

admit explicit analysis. Apparently this phenomena is related to the identities (0.7), (0.12).

0.4. Structure of the paper. In Section 1, we discuss various definitions of the functions $\mathbb{L}_{\alpha,\beta}$, theirs integral representations, and also some integrals containing products of two functions $\mathbb{L}_{\alpha,\beta}$.

In Section 2, we discuss the space $\mathcal{K}$ of holomorphic functions defined above; also we introduce the standard scale H_μ of Hilbert spaces of holomorphic functions in the half-plane. The latter spaces are well known in the representation theory of $SL_2(\mathbb{R})$.

For our purposes, the space $\mathcal{K}$ and the Hardy space $\mathcal{H}^2$ are almost sufficient.

In Section 3, we introduce a simple construction in the spirit of the Vilenkin–Klimyk book [**VK**]. We consider the 6-dimensional solvable Lie group of operators

$$f(x) \mapsto \lambda x^h f(ax^\alpha), \qquad \lambda \in \mathbb{C}^*,\ h \in \mathbb{C}, a > 0,\ \alpha \in \mathbb{R} \setminus 0,$$

acting in the space of functions on the half-line and consider the image of this group under the Laplace transform. As a result, we obtain a group of continuous operators, whose kernels are expressed in terms of $\mathbb{L}_{\alpha,\beta}$. The most interesting property of these operators is the identity (0.7) given above.

In Section 4, we consider a similar construction. We start from a 7-dimensional semigroup (4.2) of operators acting in the space of holomorphic functions on the half-plane and consider its image under the inverse Laplace transform. As a result, we obtain a semigroup of integral operators acting in an appropriate space of functions on the half-line.

1. Some properties of the functions $\mathbb{L}_{\alpha,\beta}$.

1.1. Definition. We define the function $\mathbb{L}_{\alpha,\beta}$ as the Barnes integral

$$\mathbb{L}_{\alpha,\beta}(z) = \frac{1}{2\pi i}\int_{-i\infty}^{+i\infty} \Gamma(s)\Gamma(\beta - \alpha s)\, z^{-s} ds\,. \tag{1.1}$$

We must explain the meaning of elements of this formula.

1) Our indices are in the domain $\alpha \in \mathbb{R}$, $\beta \in \mathbb{C}$. Assume also

$$\beta + \alpha m + n \neq 0 \qquad \text{for all } n, m = 0, 1, 2, \ldots. \tag{1.2}$$

2) Our integral is convergent if $|\arg z| < (1 + |\alpha|)\pi/2$.

3) Our integrand has poles at the points

$$s = 0, -1, -2, \ldots \qquad \text{and} \qquad s = \beta/\alpha, (\beta+1)/\alpha, (\beta+2)/\alpha, \ldots.$$

We describe the integration contour separately in two cases $\alpha > 0$ and $\alpha < 0$.

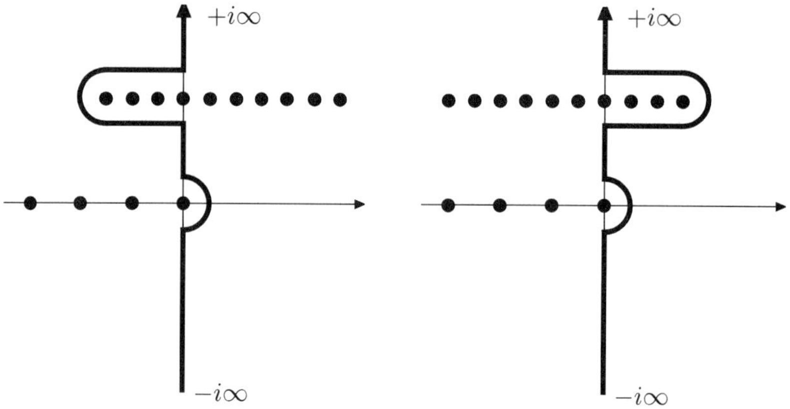

FIGURE 1. The integration contour for (1.1). We show its positions for $\alpha > 0$ and $\alpha < 0$, respectively. For $0 < \alpha < 1$ the Barnes method gives a sum of the residues at the points $s = -n$. For $\alpha > 1$ the Barnes method gives a sum of the residues at the points $s = (\beta+n)/\alpha$. For $\alpha < 0$ we obtain a sum of the residues at all the poles. If $\alpha > 0$ and (1.2) is not satisfied, then we cannot draw a contour separating the left and the right series of poles. For $\alpha < 0$ condition (1.2) is not important for defining $\mathbb{L}$, but expansion (1.5) is valid under condition (1.2) only.

First, let $\alpha > 0$. For $\beta > 0$, we can assume that the contour of the integration is the imaginary axis $i\mathbb{R}$ and we leave the pole $s = 0$ on the left-hand side from the contour (denote this contour by $+0 + i\mathbb{R}$). Otherwise, we consider the contour L coinciding with $i\mathbb{R}$ near $\pm i\infty$ and separating the left series of the poles ($s = -n$) and the right series $s = (\beta + n)/\alpha$ of the poles. Such a contour exists due to condition (1.2).

We can also transform this contour integral to the form

$$\int_L = \int_{+0+i\mathbb{R}} - \sum_{n:\,(\beta+n)/\alpha<0} \operatorname{res}_{s=(\beta+n)} \,.$$

Second, let $\alpha < 0$. Then we consider an arbitrary contour L coinciding with the imaginary axis near $\pm i\infty$ and leaving all the poles of the integrand on the left-hand side. If $\beta > 0$, then we can choose L to be $+0 + i\mathbb{R}$.

REMARK. For fixed β, $z > 0$, the function $\mathbb{L}_{\alpha,\beta}(z)$ as a function of the parameter α is C^∞-smooth at $\alpha = 0$ [3] but it is not real analytic in α at this point (compare (1.4) and (1.5)). Thus it is not quite clear, whether it would be natural to consider $\mathbb{L}_{\alpha,\beta}$ as one function or as two functions defined for $\alpha > 0$ and $\alpha < 0$. For local purposes of this paper, the first variant is more convenient.

[3] See the integral representation (1.9); we differentiate it in α and apply the Lebesgue dominant convergence theorem.

1.2. Expansion of $\mathbb{L}_{\alpha,\beta}$ into power series. We write expansions of $\mathbb{L}_{\alpha,\beta}$ in series applying the standard Barnes method; see [**S, M**].

a) Let $0 < \alpha < 1$. Then the integral (1.1) is the sum of residues at the points $s = -n$, i.e.,

$$\mathbb{L}_{\alpha,\beta}(z) = \sum_{n=0}^{\infty} \frac{(-1)^n \Gamma(\alpha n + \beta)}{n!} z^n . \tag{1.3}$$

This function is well defined on the whole complex plane $z \in \mathbb{C}$. Due to (1.2), the Γ-functions in the numerators have no poles.

b) Let $\alpha > 1$. Then (1.1) is the sum of residues at the points $s = (\beta + n)/\alpha$, i.e.,

$$\mathbb{L}_{\alpha,\beta}(z) = -\frac{1}{\alpha} \sum_{n=0}^{\infty} \frac{(-1)^n \Gamma\big((n+\beta)/\alpha\big)}{n!} z^{(n+\beta)/\alpha} . \tag{1.4}$$

Here we assume $z^\nu = \exp(\nu \ln z)$ and $\ln z \in \mathbb{R}$ for $z > 0$. The series is convergent in the domain $|\arg z| < \infty$ (i.e., our function is defined on the universal covering surface $\widetilde{\mathbb{C}}^*$ of the punctured complex plane $\mathbb{C}^* = \mathbb{C} \setminus 0$).

c) For $\alpha < 0$, the integral (1.1) is the sum of the residues at all the poles, i.e.,

$$\mathbb{L}_{\alpha,\beta}(z) = \sum_{n=0}^{\infty} \frac{(-1)^n \Gamma(\alpha n + \beta)}{n!} z^n - \frac{1}{\alpha} \sum_{n=0}^{\infty} \frac{(-1)^n \Gamma\big((n+\beta)/\alpha\big)}{n!} z^{(n+\beta)/\alpha} . \tag{1.5}$$

This expression is valid if the poles are simple, i.e., $\beta + \alpha n + m \neq 0$ for $n, m = 0, 1, 2, 3, \ldots$. But the points $\beta = -\alpha n - m$ are not really singular, for in these cases some of the poles of the integrand have order 2, and we must apply the formula for a residue in a non-simple pole (or remove singularities in (1.5)).

The domain of convergence of (1.5) is $|\arg z| < \infty$.

1.3. A symmetry.

LEMMA 1.1.

(a) *For* $\alpha > 0$, $\beta > 0$,

$$\mathbb{L}_{\alpha,\beta}(z) = \alpha^{-1} z^{-\beta/\alpha} \mathbb{L}_{1/\alpha,\beta/\alpha}\left(z^{-1/\alpha}\right) . \tag{1.6}$$

(b) *For* $0 < \alpha < 1$,

$$\mathbb{L}_{\alpha,\alpha}(z) = -\frac{1}{\alpha z} \left[\mathbb{L}_{\alpha,1}(z) - 1\right] .$$

(c) *For* $\alpha > 0$,

$$\mathbb{L}_{\alpha,1}(z) = 1 - \mathbb{L}_{1/\alpha,1}(z^{-1/\alpha}) . \tag{1.7}$$

PROOF. (a) Substituting $t = \beta - \alpha s$ into (1.1), we obtain

$$\mathbb{L}_{\alpha,\beta}(z) = \frac{1}{2\pi i \alpha} \int_{-i\infty}^{+i\infty} \Gamma\big((\beta - t)/\alpha\big)\Gamma(t)\, z^{-(\beta-t)/\alpha} dt \tag{1.8}$$

as was required.

Statement (b) follows from

$$\mathbb{L}_{\alpha,\alpha}(z) = \sum_{n=0}^{\infty} \frac{(-1)^n \Gamma(\alpha n + \alpha)}{n!} z^n = \frac{1}{\alpha} \sum_{n=0}^{\infty} \frac{(-1)^n \Gamma(\alpha n + \alpha)(\alpha n + \alpha)}{(n+1)!} z^n .$$

(c) Substitute $\beta = 1$ into (1.6) and assume $\alpha > 1$. Applying (b), we obtain the required statement for $\alpha > 1$. But the identity (1.6) is symmetric with respect to the transformation $\alpha \mapsto 1/\alpha$, $z \mapsto z^{-1/\alpha}$. □

REMARK. The statement (c) is a well-known symmetry in the theory of stable distributions; see [**F2**, (17.6.10)] and also [**Z2**, Section 2.3].

1.4. Some integral representations of $\mathbb{L}_{\alpha,\beta}$.

LEMMA 1.2. *Let* $\alpha \in \mathbb{R}$, $\operatorname{Re} u > 0$, $\operatorname{Re} v > 0$, $\operatorname{Re} h > 0$. *Then*

$$\int_0^\infty x^{h-1} \exp(-ux^\alpha - vx)\, dx = v^{-h} \mathbb{L}_{\alpha,h}(u/v^\alpha) . \tag{1.9}$$

PROOF. It is very easy to verify this for $\alpha > 0$.

1) For $0 < \alpha < 1$: we expand the factor $\exp(-ux^\alpha)$ in (1.9) in Taylor series and integrate termwise

$$\sum_{n=0}^\infty \frac{(-u)^n}{n!} \int_0^\infty x^{\alpha n + h - 1} e^{-vx} dx = \sum_{n=0}^\infty \frac{(-u)^n}{n!} \cdot \frac{\Gamma(\alpha n + h)}{v^{\alpha n + h}} .$$

2) Similarly, for $\alpha > 1$, we expand the factor $\exp(-vx)$ into a Taylor series

$$\sum_{n=0}^\infty \frac{(-v)^n}{n!} \int_0^\infty x^{h+n-1} \exp(-ux^\alpha)\, dx$$

$$= \frac{1}{\alpha} \sum_{n=0}^\infty \frac{(-u)^n}{n!} \cdot \frac{\Gamma((h+n)/\alpha)}{u^{(h+n)/\alpha}} = \frac{1}{\alpha} u^{-h/\alpha} \mathbb{L}_{1/\alpha, h/\alpha}(vu^{-1/\alpha})$$

and apply the symmetry (1.6).

3) The case $\alpha < 0$ is not obvious, and we give a calculation that is valid for all $\alpha \in \mathbb{R}$. Consider the space L^2 on $\mathbb{R}_+$ with respect to the measure dx/x. The left-hand side of (1.9) is the L^2-inner product of the functions Φ_1, Φ_2 given by

$$\Phi_1(x) = \exp(-ux^\alpha), \qquad \Phi_2(x) = x^{\overline{h}} \exp(-\overline{v}x) .$$

The Mellin transform of Φ_1 is

$$\widetilde{\Phi}_1(\lambda) = \int_0^\infty x^{\lambda-1} \exp(-ux^\alpha)\, dx = \frac{\operatorname{sgn}(\alpha)}{\alpha} \int_0^\infty \exp(-uy) y^{\lambda/\alpha - 1}\, dy$$
$$= \frac{\operatorname{sgn}(\alpha)\Gamma(\lambda/\alpha)}{\alpha u^{\lambda/\alpha}} .$$

The Mellin transform of Φ_2 is

$$\widetilde{\Phi}_2(\lambda) = \int_0^\infty x^{\lambda + \overline{h} - 1} \exp(-\overline{v}x)\, dx = v^{-(\overline{h}+\lambda)}\Gamma(\overline{h}) .$$

By the Plancherel formula for the Mellin transform, we have

$$\int_0^\infty \Phi_1(x)\overline{\Phi_2(x)}\, \frac{dx}{x} = \frac{1}{2\pi} \int_{-\infty}^\infty \widetilde{\Phi}_1(is)\overline{\widetilde{\Phi}_2(is)}\, ds ,$$

i.e.,

$$\int_0^\infty x^{h-1} \exp(-ux^\alpha - vx)\, dx = \frac{\operatorname{sgn}(\alpha)}{2\pi\alpha v^h} \int_{-\infty}^{+\infty} \Gamma(is/\alpha)\Gamma(h+is)\, u^{-is/\alpha} v^{is}\, ds .$$

Then we introduce the new variable $t = s/\alpha$. □

1.5. Integral representations. Variants. Now let x, $y > 0$. Let $\alpha < 1$, $\theta > 0$. Then

$$(1.10)\quad \frac{1}{2i}\int_{-i\infty}^{+i\infty} p^{\theta-1}\exp(-p^\alpha x + py)\,dp = -\operatorname{Im}\bigl[y^{-\theta}e^{\pi i\theta}\mathbb{L}_{\alpha,\theta}(xy^{-\alpha}e^{\pi i\alpha})\bigr]$$
$$= -\sum_{n=0}^{\infty}\frac{(-1)^n\Gamma(n\alpha+\theta)}{n!}x^n y^{-n\alpha-\theta}\sin(n\alpha+\theta)\,,$$

where the integral is taken over the imaginary axis and p^α is a positive real for $p > 0$.

Indeed, we represent the integral in the form

$$\frac{1}{2i}e^{\pi i\theta/2}\int_0^\infty \exp(-t^\alpha x e^{\pi i\alpha/2} + ity)\,t^{\theta-1}dt$$
$$-\frac{1}{2i}e^{-\pi i\theta/2}\int_0^\infty \exp(-t^\alpha x e^{-\pi i\alpha/2} - ity)\,t^{\theta-1}dt$$

and apply (1.9) with $u = x\exp(\pm\pi i\alpha/2)$, $v = y\exp(\mp\pi i/2)$.

REMARK. For $0 < \alpha < 1, \theta = 1$ expression (1.10) is the density of a stable subordinator.

REMARK. The calculation given above survives in the case $\alpha < 0$. The factor $\exp(-t^\alpha x e^{\pi i\alpha/2})$ is flat at $t = 0$ and hence we can consider $\theta < 0$. We transform

$$\Gamma(\alpha n+\theta)\sin[(\alpha n+\theta)\pi] = \pi/\Gamma(1-\theta-\alpha n)\,,$$

and we reduce (1.10) to the form $y^{-\theta}\mathbb{W}_{-\alpha,1-\theta}(x/y^\alpha)$, where $\mathbb{W}$ is the Wright function.

1.6. Remark. An $L^2(\mathbb{R})$-inner product. Let $\alpha > 0$, $\beta > 0$. Let x, $y > 0$. Consider the function

$$(1.11)\qquad \Psi_{\alpha,\beta,y}(x) := x^{\beta-1}\exp\left(-yx^\alpha\right)\,.$$

By (1.9), its Laplace transform is

$$(1.12)\qquad \widehat{\Psi}_{\alpha,\beta,y}(\xi) = \int_0^\infty x^{\beta-1}\exp\{-yx^\alpha - \xi x\}\,dx = \xi^{-\beta}\mathbb{L}_{\alpha,\beta}(y/\xi^\alpha)\,.$$

Evaluating the $L^2(\mathbb{R})$-inner product of $\Psi_{\alpha_1,\beta_1,y_1}$ and $\Psi_{\alpha_2,\beta_2,y_2}$, we obtain

$$\int_0^\infty x^{\beta_1-1}\exp(-y_1x^{\alpha_1})x^{\beta_2-1}\exp(-y_2x^{\alpha_2})\,dx$$
$$(1.13)\qquad = \alpha_2^{-1}y_2^{-(\beta_1+\beta_2-1)/\alpha_2}\mathbb{L}_{\alpha_1/\alpha_2,(\beta_1+\beta_2-1)/\alpha_2}(y_1y_2^{-\alpha_1/\alpha_2})$$

(we substitute $t = x^{\alpha_2}$ and apply Lemma 1.2). The expression in the right-hand side is symmetric with respect to $(\alpha_1,\beta_1)\leftrightarrow(\alpha_2,\beta_2)$ by Lemma 1.1.

By the Plancherel formula for the Fourier transform, the same expression can be written in the form

$$(1.14)\qquad \frac{1}{2\pi}\int_{-\infty}^{+\infty}(it)^{-\beta_1}\mathbb{L}_{\alpha_1,\beta_1}(y_1/(it)^{\alpha_1})(-it)^{-\beta_2}\mathbb{L}_{\alpha_2,\beta_2}(y_2/(-it)^{\alpha_2})\,dt\,.$$

Thus, the expression (1.14) equals (1.13).

REMARK. The integral (1.14) looks like a kernel of a product of two integral operators; moreover (1.13) shows that this product has the same form, i.e., we obtain a family of integral operators closed with respect to multiplication. Below we propose two ways to give a precise sense for this observation; apparently, there are other possibilities.

1.7. Remark. Convolutions. We keep the notation (1.11), (1.12). We have

$$\Psi_{\alpha,\beta,y}(x)\Psi_{\alpha,\beta',y'}(x) = \Psi_{\alpha,\beta+\beta',y+y'}(x) .$$

Hence

$$\frac{1}{2\pi i}\int_{-i\infty}^{i\infty} \widehat{\Psi}_{\alpha,\beta,y}(u)\widehat{\Psi}_{\alpha',\beta',y'}(z-u)du = \widehat{\Psi}_{\alpha,\beta+\beta',y+y'}(z) .$$

2. Spaces of holomorphic functions. Preliminaries

2.1. Spaces $L^2_\mu(\mathbb{R}_+)$. Fix $\mu > 0$. Denote by $L^2_\mu(\mathbb{R}_+)$ the space L^2 on the half-line $\mathbb{R}_+$, $x > 0$, with respect to the weight $\Gamma(\mu)^{-1}x^{\mu-1}dx$, i.e., the Hilbert space with the inner product

$$\langle f, g\rangle_{[\mu]} := \frac{1}{\Gamma(\mu)}\int_0^\infty f(x)\overline{g(x)}x^{\mu-1}dx .$$

For instance,

$$\langle \exp(-zx), \exp(-ux)\rangle_{[\mu]} = (z+\overline{u})^{-\mu} \tag{2.1}$$

for arbitrary complex u, z satisfying $\operatorname{Re} z > 0$, $\operatorname{Re} u > 0$.

2.2. Hilbert spaces of holomorphic functions on a half-plane. Let Π be the right half-plane $\operatorname{Re} z > 0$ on the complex plane. We consider the Hardy space $\mathcal{H}^2$ on Π. Recall that this space consists of functions holomorphic in the half-plane, whose boundary values on the imaginary axis $\operatorname{Re} z = 0$ exist and are contained in $L^2(\mathbb{R})$.

The inner product in $\mathcal{H}^2$ is given by

$$\langle f, g\rangle = \frac{1}{2\pi}\int_{-\infty}^{+\infty} f(it)\overline{g(it)}\,dt := \lim_{\varepsilon\to+0}\frac{1}{2\pi}\int_{-\infty}^{+\infty} f(\varepsilon+it)\overline{g(\varepsilon+it)}\,dt .$$

The Hardy space belongs to the following one-parametric scale H_μ, $\mu > 0$, of spaces of holomorphic functions.

Fix $\mu > 1$. Consider the space $H_\mu = H_\mu(\Pi)$ consisting of functions $f(z)$ holomorphic in Π and satisfying the condition

$$\int_\Pi |f(z)|^2(\operatorname{Re} z)^{\mu-2}dz\,d\overline{z} < \infty ,$$

where $dz\,d\overline{z}$ denotes the Lebesgue measure on Π.

We define an inner product in H_μ by

$$\langle f, g\rangle_\mu = \frac{\mu-1}{\pi}\int_\Pi f(z)\overline{g(z)}(\operatorname{Re} z)^{\mu-2}dz\,d\overline{z} . \tag{2.2}$$

The space H_μ is a Hilbert space with respect to this inner product.

The reproducing kernel[4] of this space is

$$K_\mu(z,u) = (z+\overline{u})^{-\mu} .$$

[4] For machinery of reproducing kernels, see, for instance, [**Be**], [**N2**].

This means that the function $\Xi_u(z)$ given by

$$\Xi_u(z) = (z + \overline{u})^{-\mu}$$

satisfies the reproducing property

$$\langle f, \Xi_u \rangle_\mu = f(u) \qquad \text{for all } f \in H_\mu. \tag{2.3}$$

In particular,

$$\langle \Xi_u, \Xi_w \rangle_\mu = (w + \overline{u})^{-\mu} . \tag{2.4}$$

The space H_μ can be defined by (2.3), (2.4) without reference to the explicit formula (2.2) for the inner product. Indeed, consider an abstract Hilbert space H with a system of vectors Ξ_u, where $u \in \Pi$, and assume that their inner products are given by (2.4). Such a Hilbert space exists; see formula (2.1). Assume also that linear combinations of Ξ_u are dense in H. Then for each $h \in H$ we consider the holomorphic function on Π given by

$$f_h(u) := \langle h, \Xi_u \rangle_H , \tag{2.5}$$

and thus we have identified our space H with some space of holomorphic functions on Π.

But the last construction survives for arbitrary $\mu > 0$ (since the existence of H_μ is provided by formula (2.1) and this formula is valid for $\mu > 0$).

REMARK. For $\mu = 1$ we obtain the Hardy space $\mathcal{H}^2$.

REMARK. For $0 < \mu < 1$, it is possible to write an integral formula for the inner product in H_μ involving derivatives. But it is more convenient to use the definition (2.3)–(2.4) or to consider the analytic continuation of the integral (2.2) with respect to μ.

We define the *weighted Laplace transform* $\mathcal{L}_\mu$ by

$$\mathcal{L}_\mu f(z) = \frac{1}{\Gamma(\mu)} \int_0^\infty f(x) \exp(-ux)\, x^{\mu-1} dx .$$

For $\mu = 1$ we obtain the usual Laplace transform $\mathcal{L} = \mathcal{L}_1$. The following statement is well known[5].

LEMMA 2.1. *The weighted Laplace transform is a unitary operator*

$$\mathcal{L}_\mu : \; L^2_\mu(\mathbb{R}_+) \to H_\mu(\Pi) .$$

PROOF. Consider the function

$$\xi_u(x) = \exp(-ux)$$

in L^2_μ. Its image under $\mathcal{L}_\mu$ is Ξ_u. It remains to compare (2.1) and (2.4). □

REMARK. The transform $\mathcal{L}_\mu$ is precisely the operator defined by formula (2.5). Indeed, we can assume $H = L_\mu$. Then the corresponding space of functions f_h is H_μ.

[5]The case $\mu = 1$ is the Paley–Wiener theorem.

2.3. Operators in the spaces H_μ. Recall a standard general trick, apparently discovered by Berezin [**Be**] (formulas (2.6), (2.9) given below are valid in arbitrary Hilbert space defined by a reproducing kernel).

Let A be a *bounded* operator $H_\mu \to H_\mu$. Define the function

$$M(z,u) = A\Xi_u(z) \,. \tag{2.6}$$

Then $M(z,u)$ is the *kernel* of the operator A. For $\mu > 1$ let us consider the formal expression

$$Af(z) = \frac{\mu - 1}{\pi} \int_\Pi M(z,u) f(u) (\operatorname{Re} u)^{\mu-2}\, du\, d\overline{u}\,. \tag{2.7}$$

Respectively, for $\mu = 1$,

$$Af(z) = \frac{1}{2\pi} \int_{-\infty}^{\infty} M(z,-it) f(it)\, dt\,. \tag{2.8}$$

For general $\mu > 0$, we can write as

$$Af(z) = \langle f, \overline{M(z,u)} \rangle_\mu \tag{2.9}$$

the inner product in the space H_μ of functions depending on the variable u. Formulae (2.7)–(2.8) are partial cases of this formula. In particular, the integrals (2.7)–(2.8) are convergent for each $z \in \Pi$ and $f \in H_\mu$.

2.4. Space of rapidly decreasing functions. We also consider the space $\mathcal{K} = \mathcal{H}^2 \cap \mathcal{S}(i\mathbb{R})$, where $\mathcal{S}(i\mathbb{R})$ is the Schwartz space (consisting of functions on the imaginary axis rapidly decreasing with all derivatives).

We can say that $\mathcal{K}$ is the space of functions holomorphic in $\operatorname{Re} z > 0$ and continuous in $\operatorname{Re} z \geqslant 0$ such that

a) $f(it)$ is C^∞-smooth;

b) for each $k > 0$ and $N > 0$ there exists M such that

$$|f^{(k)}(z)| \leqslant M(1 + |z|)^{-N}\,. \tag{2.10}$$

Consider the space $\mathcal{S}(\mathbb{R}_+)$ consisting of smooth functions f on $[0,\infty)$ such that

a) $f^{(k)}(0) = 0$ for all $k \geqslant 0$;

b) $\lim_{x\to+\infty} f^{(k)}(x) x^N = 0$ for all $N > 0$, $k \geqslant 0$.

In other words, $\mathcal{S}(\mathbb{R}_+)$ is the intersection of the Schwartz space $\mathcal{S}(\mathbb{R})$ on $\mathbb{R}$ and the space $L^2(\mathbb{R}_+)$.

LEMMA 2.2. *The space $\mathcal{K}$ is the image of $\mathcal{S}(\mathbb{R}_+)$ under the Laplace transform.*

PROOF. Let $f \in \mathcal{S}(\mathbb{R}_+)$. Integrating by parts, we obtain

$$\int_0^\infty f^{(k)}(x) e^{-px} dx = p^k \int_0^\infty f(x) e^{-px} dx\,.$$

The left-hand side is a bounded function in p; looking at the right-hand side, we observe that $(\mathcal{L}f)(p)$ is rapidly decreasing for $\operatorname{Re} p \geqslant 0$.

Conversely, a function F satisfying (2.10) is an element of $\mathcal{H}^2$. Hence $f = \mathcal{L}^{-1} F$ is supported by $\mathbb{R}_+$. Since (2.10) is valid for $z \in i\mathbb{R}$, we have $f \in \mathcal{S}(\mathbb{R})$. ☐

2.5. Fractional derivations. We define the operators of *fractional differentiation* D_h in $\mathcal{K}$ by

$$D_h f(z) = \frac{\Gamma(h+1)}{2\pi} \int_{-\infty}^{+\infty} \frac{f(it)\,dt}{(-it+z)^{h+1}} . \tag{2.11}$$

A branch of $\theta(z,it) = (-it+z)^{h+1}$ is determined from the condition $\theta(x,0) > 0$ for $x > 0$.

LEMMA 2.3.
(a) *D_h is an operator $\mathcal{K} \to \mathcal{K}$ for each $h \in \mathbb{C}$.*
(b) *For integer $n > 0$,*

$$D_n f(z) = (-1)^n \frac{d^n}{dz^n} f(z) .$$

(c) *For positive integer m,*

$$\begin{aligned} D_{-m} f(z) &:= \lim_{s \to m} \frac{\Gamma(-s+1)}{2\pi i} \int_{-\infty}^{+\infty} (-it+z)^{s-1} f(t)\,du \\ &= (-1)^m \int_{-i\infty}^{z} dz_1 \int_{-i\infty}^{z_1} dz_2 \ldots \int_{-i\infty}^{z_{n-1}} f(z_n)\,dz_n . \end{aligned} \tag{2.12}$$

(d) $D_{h_1} D_{h_2} = D_{h_1+h_2}$.

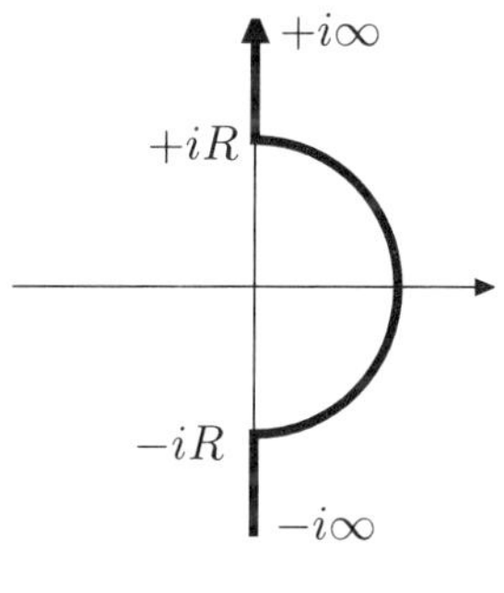

FIGURE 2. The contour for the proof of Lemma 2.3.

PROOF. (a) Convergence of the integral for $\operatorname{Re} u > 0$ is obvious. Let us show rapid decreasing of $g := D_h f$ at $z \to \infty$.

Let $\operatorname{Re} h < 0$. We represent (2.11) as a contour integral

$$\frac{\Gamma(h+1)}{2\pi i} \int_{-i\infty}^{+i\infty} \frac{f(u)\,du}{(-u+z)^{h+1}} . \tag{2.13}$$

Then we replace a part $(-iR, +iR)$ of the contour of the integration $i\mathbb{R}$ by the semi-circle $R\exp(i\varphi)$, where $\varphi \in (-\pi, \pi)$. By (2.10), all three summands of the integral rapidly tend to 0 as R tend to ∞.

If $\operatorname{Re} h \geqslant 0$, we integrate expression (2.13) by parts and obtain

$$\frac{(-1)^k \Gamma(h-k+1)}{2\pi i} \int_{-i\infty}^{+i\infty} (z-u)^{k-h-1} f^{(k)}(u)\,du .$$

We choose $k > \operatorname{Re}(h+1)$ and repeat the same consideration.

Also,

$$\frac{d}{dz} D_h f = -D_{h+1} f \ ,$$

and this implies rapid decreasing (2.10) of derivatives.

(b) This is the Cauchy integral representation for derivatives.

(c) First, we give a remark that formally is not necessary. The factor $\Gamma(1+h)$ has a pole at $h = -m$. Let us show that the integral vanishes at this point. Indeed, we have the expression

$$\int_{-i\infty}^{i\infty} (z-u)^{m-1} f(u)\, du \ .$$

We replace the part $(-iR, +iR)$ of the contour of the integration $i\mathbb{R}$ by the semicircle $R\exp(i\varphi)$ as above and let R go to ∞.

Now we give a proof of (c). Consider the operator of indefinite integration

$$If(z) = \int_{-i\infty}^{z} f(u)\, du \ . \tag{2.14}$$

Changing the contour as above, we obtain $If \in \mathcal{K}$.

For $f \in \mathcal{K}$ we have

$$\Gamma(1-s) \int_{-i\infty}^{+i\infty} (z-u)^{s-1} f(u)\, du$$
$$= (-1)^m \Gamma(1-s+m) \int_{-i\infty}^{+i\infty} (z-u)^{s-m-1} (I^m f)(u)\, du \ .$$

Now we can substitute $s = m$.

(d) This is valid for $\operatorname{Re} h_1 = \operatorname{Re} h_2 = 0$ by the statement (b) of the following Lemma 2.4. Then we consider the analytic continuation in h. □

LEMMA 2.4.

(a) *For $s \in \mathbb{R}$, the operator D_{is} is a unitary operator in each H_μ. Its kernel (in the sense of H_μ) is*

$$\frac{\Gamma(\mu+is)}{\Gamma(is)} (z+\overline{u})^{-\mu-is} \ .$$

(b) $D_{is_1} D_{is_2} = D_{is_1+is_2}$.

PROOF. (a) Consider the operator U_{is} in $L^2_\mu(\mathbb{R}_+)$ given by

$$U_{is} f(x) = f(x)\, x^{is} \ .$$

Let us evaluate the kernel of the operator

$$\mathcal{L}_\mu U_{is} \mathcal{L}_\mu^{-1} : H_\mu \to H_\mu \ .$$

By (2.6), we must evaluate the function $\mathcal{L}_\mu U_{is} \mathcal{L}_\mu^{-1} \Xi_u$. We have

$$\mathcal{L}_\mu^{-1} \Xi_u(x) = \exp(-\overline{u}x),$$
$$U_{is} \mathcal{L}_\mu^{-1} \Xi_u(x) = \exp(-\overline{u}x) x^{is},$$
$$\mathcal{L}_\mu U_{is} \mathcal{L}_\mu^{-1} \Xi_u(z) = (z+\overline{u})^{-\mu-is} \Gamma(\mu+is)/\Gamma(is) \ .$$

For $\mu = 1$ our operator coincides with the operator D_{is} defined above. In fact, all the operators $\mathcal{L}_\mu U_{is} \mathcal{L}_\mu^{-1} : H_\mu \to H_\mu$ induce the same operator D_{is} in $\mathcal{K}$. Indeed, the operator $L_\mu^{-1}L_\nu$ is the operator of multiplication by $x^{\mu-\nu}$, and this operator commutes with U_{is}.

Part (b) corresponds to the identity $x^{i(s_1+s_2)} = x^{is_1}x^{is_2}$ after the Laplace transform. □

3. Operators in spaces of holomorphic functions

3.1. Some operators acting in $\mathcal{S}(\mathbb{R}_+)$. We consider the following one-parameter groups of operators in the space $\mathcal{S}(\mathbb{R}_+)$ (see Section 2.4):

$$U_\alpha f(x) = f(x^\alpha), \qquad \alpha > 0; \tag{3.1}$$

$$V_a(x)f(x) = f(ax), \qquad a > 0; \tag{3.2}$$

$$W_h f(x) = x^h f(x), \qquad h \in \mathbb{C}\,. \tag{3.3}$$

The last group is a complex one-parameter group, i.e., a real two-parameter group. The infinitesimal generators of these groups are, respectively,

$$E_1 f(x) = x \ln x \frac{d}{dx} f(x); \qquad E_2 f(x) = x \frac{d}{dx} f(x); \qquad E_3 f(x) = (\ln x) f(x)\,.$$

They satisfy the commutation relations

$$[E_1, E_2] = -E_2; \qquad [E_1, E_3] = E_3; \qquad [E_2, E_3] = 1\,. \tag{3.4}$$

Thus we obtain a real 6-dimensional Lie algebra $\mathfrak{g}$ spanned by the operators

$$E_1, \quad E_2, \quad E_3, \quad iE_3, \quad 1, \quad i\,.$$

The algebra $\mathfrak{g}$ is solvable and it contains a 2-dimensional center $\mathbb{R}\cdot 1 + \mathbb{R}\cdot i$.

Also $\mathfrak{g}$ is a real subalgebra (but not a real form) in the 4-dimensional complex Lie algebra

$$\mathbb{C}\cdot E_1 + \mathbb{C}\cdot E_2 + \mathbb{C}\cdot E_3 + \mathbb{C}\cdot 1\,.$$

Obviously,

$$U_\alpha W_h U_\alpha^{-1} = W_{\alpha h}; \tag{3.5}$$

$$U_\alpha V_a U_\alpha^{-1} = V_{a^{1/\alpha}}\,. \tag{3.6}$$

Consider the group G generated by the one-parameter groups (3.1)–(3.3). A general element of this group is an operator of the form

$$\lambda \cdot R(h, \alpha, a)\,,$$

where $\lambda \in \mathbb{C}^*$ and

$$R(h,\alpha,a)f(x) = x^h f(ax^\alpha) = W_h U_\alpha V_a f(x)\,. \tag{3.7}$$

The product is given by

$$R(g,\beta,b)R(h,\alpha,a) = b^h R(g + h\beta, \alpha\beta, ab^\alpha)\,. \tag{3.8}$$

We also can add the operator

$$U_{-1} f(x) = f(1/x)\,,$$

or equivalently we can allow $\alpha \in \mathbb{R}\setminus 0$ in (3.1), (3.7), (3.8). Then we obtain a Lie group G° of operators containing two connected components; the group G defined above is the connected component containing 1.

3.2. Operators in $L^2_\mu(\mathbb{R}_+)$. Now, let us fix $\mu > 0$. Then the operators

$$|\alpha|^{1/2} a^{1/2} R\left((\alpha-\mu)/2 + is, \alpha, a\right)$$

are unitary in $L^2_\mu(\mathbb{R}_+)$. Such operators form a 4-dimensional solvable Lie group with a 1-dimensional center; denote this group by G°_μ.

REMARK. All other operators $R(h,\alpha,a)$ are unbounded in $L^2_\mu(\mathbb{R})$.

3.3. Operators in H_μ. Let us evaluate the kernel in H_μ of the operator $\mathcal{L}_\mu R(h,\alpha,a)\mathcal{L}_\mu^{-1}$ using 2.3.

We have

$$\mathcal{L}_\mu^{-1}\Xi_u(x) = \exp(-\overline{u}x);$$

$$R(h,\alpha,a)L_\mu^{-1}\Xi_u(x) = x^h\exp(-a\overline{u}x^\alpha);$$

$$\mathcal{L}_\mu R(h,\alpha,a)\mathcal{L}_\mu^{-1}\Xi_u(z) = \frac{1}{\Gamma(\mu)}\int_0^\infty x^{h+\mu-1}\exp(-a\overline{u}x^\alpha - zx)\,dx \tag{3.9}$$

$$= z^{-\mu-h}\mathbb{L}_{\alpha,h+\mu}(\overline{u}a/z^\alpha)\,, \tag{3.10}$$

and we obtain (for $\mu > 1$) the integral operator

$$\widetilde{R}(h,\alpha,a)F(z) = \frac{1}{\pi\Gamma(\mu-1)} z^{-\mu-h}\int_\Pi \mathbb{L}_{\alpha,h+\mu}(\overline{u}a/z^\alpha)F(u)\,(\operatorname{Re}u)^{\mu-2}du\,d\overline{u}\,. \tag{3.11}$$

For $\mu = 1$ we understand (3.11) as (2.8), and for $\mu < 1$, as (2.9).

Formally, our algorithm of evaluating the kernel is valid only for bounded operators. Thus we have proved the following theorem:

THEOREM 3.1. *Let*

$$\operatorname{Re}h = (\alpha-\mu)/2\,.$$

Then the operator $\widetilde{R}(h,\alpha,a)$ defined by (3.11) *is unitary in H_μ up to a scalar factor. The product of two operators $\widetilde{R}$ is given by the formula*

$$\widetilde{R}(g,\beta,b)\widetilde{R}(h,\alpha,a) = b^h\widetilde{R}(g+h\beta,\alpha\beta,ab^\alpha)\,. \tag{3.12}$$

These operators generate a 4*-dimensional solvable Lie group isomorphic to the group G°_μ described in* 3.2.

3.4. Operators in the space $\mathcal{K}$. The group G° defined in subsection 3.1 acts in the space $\mathcal{S}(\mathbb{R}_+)$. Since the weighted Laplace transform $\mathcal{L}_\mu$ identifies $\mathcal{S}(\mathbb{R}_+)$ and $\mathcal{K}$, this group also acts in $\mathcal{K}$. For the subgroup $G^\circ_\mu \subset G^\circ$, the action was constructed in the previous subsection, but formula (3.11) for the kernel almost survives for a general element of G°. We consider separately the cases $\alpha > 0$ and $\alpha < 0$.

a) Let $\alpha > 0$. We substitute $y = x^\alpha$ into the expression (3.9) and obtain

$$\frac{1}{|\alpha|}\int_0^\infty y^{(h+\mu)/\alpha-1}\exp(-a\overline{u}y - zy^{1/\alpha})\,dy\,. \tag{3.13}$$

If h satisfies the condition

$$(\operatorname{Re}h+\mu)/\alpha > 0\,, \tag{3.14}$$

then our integral is convergent (otherwise, we would have a non-integrable singularity at 0). The expression (3.13) is the Laplace transform of the function $y^{(h+\mu)/\alpha-1}\exp(-zy^{1/\alpha})$. Since this function is an element of L^1, its Laplace transform is a bounded function in Π. Hence, for $F \in \mathcal{K}$, the integral (3.11) is convergent and is holomorphic in h. Thus the formula (3.11) defines the $\mathcal{L}_\mu$-image of $R(h,\alpha,a)$ for all triples h, α, a satisfying (3.14).

Next, we write

$$R(h,\alpha,a) = W_{-n} \circ R(h+n,\alpha,a)\,. \tag{3.15}$$

For sufficiently large n, the $\mathcal{L}_\mu$-image of $R(h+n,\alpha,a)$ is defined by formula (3.11); the $\mathcal{L}_\mu$-image of W_{-n} is the iterated indefinite integration (2.12).

b) $\alpha < 0$. We again transform the integral to the form (3.13). Now the integrand is smooth at 0; for convergence at infinity we require the condition $(\operatorname{Re} h+\mu)/\alpha < 0$. If it is not satisfied, then we write[6]

$$R(h,\alpha,a) = W_n \circ R(h-n,\alpha,a)\,.$$

Thus we obtain the following theorem.

THEOREM 3.2. *Fix $\mu > 0$. Let $h \in \mathbb{C}$, $\alpha \in \mathbb{R}\setminus 0$, and $a > 0$. If $\operatorname{sgn}(\alpha)\cdot(\operatorname{Re} h + \mu) > 0$, then we define the integral operator $\widetilde{R}(h,\alpha,a)$ in $\mathcal{K}$ by* (3.11).

— *For $\alpha > 0$, we consider n such that $\operatorname{Re} h + n > 0$ and define*

$$\widetilde{R}(h,\alpha,a) = (-1)^n I^n \circ \widetilde{R}(h+n,\alpha,a)\,,$$

where I is the operator of indefinite integration (2.14).

— *For $\alpha < 0$, we write*

$$\widetilde{R}(h,\alpha,a) = \left(-\frac{d}{dz}\right)^n \circ \widetilde{R}(h-n,\alpha,a)$$

for sufficiently large n.

Then all these operators are bounded in $\mathcal{K}$ and their product is given by (3.12). *The group generated by $\widetilde{R}(h,\alpha,a)$ is isomorphic to the group G° defined in* 3.1.

REMARK. For different μ we obtain the same group of operators in $\mathcal{K}$; but the identification of this group with G° depends on μ.

3.5. Statements formulated in the Introduction. Let $\mu = 1$. The operator A_α given by (0.5) is the $\mathcal{L}_1$-image of U_α, the fractional derivation D_h is the $\mathcal{L}_1$-image of W_h, and the operator R_a is the $\mathcal{L}_1$-image of V_a. Now (0.7), (0.9) follow from (3.5), (3.6).

The operator $F \mapsto zF$ is the $\mathcal{L}$-image of d/dx, and this implies (0.8).

3.6. Hankel type transforms. The kernel of the operator $\widetilde{R}(h,-1,1)$ is

$$K(z,u) = \text{const}\cdot \int_0^\infty x^{h+\mu-1}\exp(-\overline{u}/x - zx)\,dx\,.$$

This expression is a modified Bessel function of Macdonald; see [**EMOT**]. The corresponding integral transform is similar to the Hankel transform.

[6]This is not really necessary, since for $\operatorname{Re} u > 0$ integral (3.13) is convergent. But for the case $\mu = 1$ it is nice to have an expression for $\operatorname{Re} u = 0$.

3.7. Another group of symmetries. Now we consider the group of unitary operators in $L^2(\mathbb{R}_+)$ generated by

$$U'_\alpha f(x) = |\alpha|^{1/2} f(x^\alpha) x^{(\alpha-1)/2};$$
$$V'_a f(x) = a^{1/2} f(ax);$$
$$T_\beta(is) f(x) = \exp(isx^\beta) f(x), \qquad s \in \mathbb{R}, \beta \in \mathbb{R}\,.$$

This group is infinite dimensional since it contains all the operators having the form

$$f(x) \mapsto f(x) \exp\left(i \sum_{j=1}^{N} s_j x^{\beta_j} \right).$$

Obviously, we have

$$U_\alpha T_\beta(is) U_\alpha^{-1} f(x) = T_{\alpha\beta}(is); \qquad V_a T_\beta(is) V_a^{-1} = T_\beta(isa^\beta)\,. \tag{3.16}$$

Consider the image of our group of operators under the standard Laplace transform $\mathcal{L}$. The operators U'_α, V'_a are contained in the group G°, and their images $\widetilde{U}'_\alpha$, $\widetilde{V}'_a$ are described above. The image of $T_\beta(is)$ is the convolution operator

$$\widetilde{T}_\beta(is) F(z) = \frac{1}{2\pi i} \int_{-i\infty}^{i\infty} M(z-u) F(u)\, du\,,$$

where

$$M(z) = \int_0^\infty \exp(isx^\beta - zx)\, dx = \mathbb{L}_{\beta,1}(is/z^\beta)\,.$$

In particular, we obtain the identity

$$\widetilde{U}_\alpha \widetilde{T}_\beta(is) \widetilde{U}_\alpha^{-1} f(x) = \widetilde{T}_{\alpha\beta}(is)$$

for our operators with $\mathbb{L}$-kernels.

4. Operators in the space of functions on the half-line

4.1. Spaces of functions. Consider the space $\mathcal{P}$ consisting of C^∞-functions on the half-line $x \geqslant 0$ satisfying

a) $f^{(k)}(0) = 0$ for all $k \geqslant 0$;

b) $\lim_{x\to+\infty} f^{(m)}(x) \exp(-\varepsilon x) = 0$ for all $\varepsilon > 0$, $m \geqslant 0$.

Consider also the space $\mathcal{F}$, whose elements are the functions holomorphic in the half-plane $\operatorname{Re} z > 0$ and satisfying the condition

- for each $\varepsilon > 0$, $N > 0$ there exists C such that

$$|f(z)| < C/|z|^{-N} \qquad \text{for } \operatorname{Re} z > \varepsilon\,.$$

LEMMA 4.1. *The image of the space $\mathcal{P}$ under the Laplace transform is $\mathcal{F}$.*

PROOF. a) For $f \in \mathcal{P}$ denote $F = \mathcal{L}f$. Fix small $\delta > 0$.

$$F(p) = \int_0^\infty f(x) e^{-px} dx = \int_0^\infty \left[f(x) e^{-\delta x} \right] e^{-(p-\delta)x} dx\,.$$

Since $f(x)e^{-\delta x}$ is an element of the Schwartz space, for each N we have an estimate

$$|F(p)| \leqslant C|p-\delta|^{-N}$$

for $\operatorname{Re} p \geqslant \delta$. For $\operatorname{Re} p > 2\delta$, we can write

$$|F(p)| \leqslant 2^N C|p|^{-N} .$$

Thus, $F \in \mathcal{F}$.

b) Let $F \in \mathcal{F}$. The inversion formula for $\mathcal{L}$ gives

$$f(x) = \frac{1}{2\pi i} \int_{a-i\infty}^{a+i\infty} e^{px} F(p)\, dp = \frac{1}{2\pi} e^{xa} \int_{-i\infty}^{i\infty} e^{itx} F(a+it)\, dt .$$

Since $F(a+it)$ is an element of Schwartz space $\mathcal{S}(\mathbb{R})$, the function

$$e^{-ax} f(x) = \frac{1}{2\pi} \int_{-i\infty}^{i\infty} e^{itx} F(a+it)\, dt$$

is an element of $\mathcal{S}(\mathbb{R})$. By the Paley–Wiener theorem this function is supported by $\mathbb{R}^+$. Since $a > 0$ is arbitrary, we obtain the required statement. □

4.2. Fractional derivations. Consider the Riemann–Liouville operators of fractional integration (see [**SKM**])

$$J_r f(x) = \frac{1}{\Gamma(r)} \int_0^x f(y)(x-y)^{r-1}\, dy \tag{4.1}$$

in the space $\mathcal{P}$. The integral is convergent for $r > 0$. For a positive integer $r = n$ we have

$$J_r f(x) = \int_0^x dx_1 \int_0^{x_1} dx_2 \ldots \int_0^{x_{n-1}} f(x_n)\, dx_n .$$

For fixed f and x, the function $J_r f(x)$ admits a holomorphic continuation to the whole plane $r \in \mathbb{C}$, and for a negative integer $r = -n$ we have

$$J_{-n} f(x) = \frac{d^n}{dx^n} f(x) .$$

We also have the identity

$$J_r J_p = J_{r+p}$$

for all $r,\, p \in \mathbb{C}$.

The Laplace transform $\mathcal{L}$ identifies the Riemann–Liouville fractional integrations J_r with the operators in $\mathcal{F}$ given by

$$F(z) \mapsto z^{-r} F(z) .$$

4.3. Operators. Now we consider the semigroup Γ of operators in $\mathcal{F}$ consisting of transformations

$$\lambda Q(\theta, \alpha, a) ,$$

where $\lambda \in \mathbb{C}^*$,

$$Q(\theta, \alpha, a) F(z) = z^\theta F(a z^\alpha) , \tag{4.2}$$

and the parameters satisfy the conditions

$$0 < \alpha < 1, \qquad \arg a + \alpha\pi/2 < \pi/2, \qquad \arg a - \alpha\pi/2 > -\pi/2, \qquad \theta \in \mathbb{C} . \tag{4.3}$$

REMARK. The restrictions (4.3) mean that $z \in \Pi$ implies $a z^\alpha \in \Pi$.

Obviously, we have

$$Q(\theta', \alpha', a') Q(\theta, \alpha, a) = (a')^\theta Q(\theta' + \theta\alpha', \alpha\alpha', a(a')^\alpha) . \tag{4.4}$$

Thus Γ is a 7-dimensional semigroup with 2-dimensional center.

REMARK. The semigroup Γ can be embedded into a 7-dimensional Lie group. The parameters of this group are $\lambda \in \mathbb{C}^*$, $a \in \mathbb{C}^*$, $\theta \in \mathbb{C}$, $\alpha > 0$. The multiplication in this group is determined by formula (4.4), where $\alpha > 0$, and θ, $a \in \mathbb{C}$. But the corresponding operators (4.2) are not well defined in the space $\mathcal{F}$.

The $\mathcal{L}^{-1}$-image of the operators $Q(\theta, \alpha, a)$ is given by the formula

$$\widetilde{Q}(\theta, \alpha, a) f(x) = \int_0^\infty N(x, y) f(y)\, dy\,, \tag{4.5}$$

where the kernels $N(x, y)$ are given by

$$N(x, y) = \frac{1}{2\pi} \int_{-i\infty}^{i\infty} z^\theta \exp(-az^\alpha y + zx)\, dz\,.$$

For $\theta > -1$, we transform this integral in the same way as in 1.5 and obtain

$$N(x, y) = \frac{1}{2\pi i y^{\theta+1}} \Big\{ e^{i(\theta+1)\pi/2} \mathbb{L}_{\alpha,\theta+1}(yax^{-\alpha} e^{i\pi\alpha/2}) \\ - e^{-i(\theta+1)\pi/2} \mathbb{L}_{\alpha,\theta+1}(yax^{-\alpha} e^{-i\pi\alpha/2}) \Big\}\,.$$

For real $a > 0$, $\theta > -1$ we have an expression of the form (1.10).

If $\theta < -1$, we write $z^\theta F(az^\alpha) = z^{-n} z^{\theta+n} F(az^\alpha)$, and

$$\widetilde{Q}(\theta, \alpha, a) := J_n \circ \widetilde{Q}(\theta + n, \alpha, a) \tag{4.6}$$

for sufficiently large n.

THEOREM 4.2. *For α, a satisfying* (4.3), *the operators* $\widetilde{Q}(\theta, \alpha, a)$ *defined by* (4.5), (4.6) *are bounded in* $\mathcal{P}$ *and their product is given by* (4.4).

Formulas (0.11)–(0.12) given in the Introduction are particular cases of this statement.

References

[A] R. P. Agarwal, *Sur une generalisation de la transformation de Hankel,* Ann. Soc. Sci. Bruxelles. Sér. I. **64** (1950), 164–168.

[Ba] E. W. Barnes, *The asymptotic expansions of integral functions defined by Taylor series,* Phil. Trans. Roy. Soc. London **206A** (1906), 249-297.

[Be] F. A. Berezin, *Wick and anti-Wick symbols of operators,* Mat. Sb. (N.S.) **86(128)** (1971), 578–610; English translation in Sbornik Math. (1971).

[C] A. Cauchy, *Sur les résultats moyens d'observations de meme nature, et sur les résultates les plus probables,* Compt. Rend. Acad. Sci. Paris **37** (1853), 198–206; reprinted in Oeuvres completes, Ser 1., vol. 12, Paris, 1900, pp. 94–104.

[D] E. B. Dynkin, *Some limit theorems for sums of independent random variables with infinite mathematical expectations,* Izv. Akad. Nauk SSSR. Ser. Mat. **19** (1955), 247–266; English transl.: Select. Transl. Math. Statist. and Probability, vol. 1, Inst. Math. Statist. and Amer. Math. Soc., Providence, R.I., 1961, pp. 171–189.

[EMOT] A. Erdelyi, W. Magnus, F. Oberhettinger, F. G. Tricomi, *Higher transcendental functions,* vols. II, III, McGraw-Hill, 1955.

[F1] W. Feller, *On a generalization of Marcel Riesz' potentials and the semi-groups generated by them,* Comm. Sem. Math. Univ. Lund [Medd. Lunds Univ. Mat. Sem.] Tome Supplementaire (1952), 72–81.

[F2] W. Feller, *An introduction to probability theory and its applications,* vol. II, J. Wiley, 1966.

[H] A. Ya. Hinchin, *Limit laws for sums of independent random variables,* ONTI, Moscow, 1938 (Russian).

[L1] P. Levy, *Sur le rôle de la loi de Gauss dans la théorie des erreurs,* Compt. Rend. **174** (1922), 855–857.

[L2] P. Levy, *Theorie de l'addition des variables aleatoires,* 2nd ed., Gauthier-Villars, Paris, 1954.

[L3] P. Levy, *Processus stochastiques et mouvement brownien,* 2nd ed., Gauthier-Villars, Paris, 1965.

[M] O. I. Marichev, *A method of calculating integrals from special functions (theory and tables of formulas),* Nauka i Tekhnika, Minsk, 1978; English transl.: O. I. Marichev, *Handbook of integral transforms of higher transcendental functions. Theory and algorithmic tables,* J. Wiley and Sons, 1983.

[MHP] N. W. McLachlan, P. Humbert, L. Poli, *Supplement au formulaire pour le calcul symbolique,* Memor. Sci. Math., no. 113, Gauthier-Villars, Paris, 1950.

[MR] K. S. Miller, B. Ross, *An introduction to the fractional calculus and fractional differential equations,* J. Wiley and Sons, Inc., New York, 1993.

[N1] Yu. A. Neretin, *The group of diffeomorphisms of a ray, and random Cantor sets,* Mat. Sb. **187**:6 (1996), 73–84; English translation Sb. Math. **187** (1996), 857–868.

[N2] Yu. A. Neretin, *Plancherel formula for Berezin deformation of L^2 on Riemannian symmetric space,* J. Funct. Anal. **189** (2002), 336–408.

[P] G. Polya, *Herleitung des Gaussschen Fehlergesetzes aus ainer Funktionalgleichung,* Math. Z. **18** (1923), 96–108.

[PY] J. Pitman, M. Yor, *The two-parameter Poisson-Dirichlet distribution derived from a stable subordinator,* Ann. Probab. **25** (1997), 855–900.

[S] L. J. Slater, *Generalized hypergeometric functions,* Cambridge University Press, Cambridge, 1966.

[SKM] S. G. Samko, A. A. Kilbas, O. I. Marichev, *Integrals and derivatives of fractional order and some of their applications,* Nauka i Tekhnika, Minsk, 1987; English translation: Gordon and Breach, 1993.

[VK] N. Ja. Vilenkin, A. U. Klimyk, *Representation of Lie groups and special functions. Vol. 1. Simple Lie groups, special functions and integral transforms,* Kluwer, Dordrecht, 1991.

[Wi] A. Wintner, *Stratifications of Cauchy's "stable" transcendents and of Mittag-Leffler's entire functions,* Amer. J. Math. **80** (1958), 111–124.

[Wr] E. M. Wright, *The asymptotic expansion of the generalized Bessel function,* Proc. London Math. Soc. (2) **38** (1934), 257–270.

[Z1] V. M. Zolotarev, *On analytic properties of stable distribution laws,* Vestnik Leningrad. Univ. **11**:1 (1956), 49–52 (in Russian).

[Z2] V. M. Zolotarev, *One-dimensional stable distributions,* Nauka, Moscow, 1983; English translation: American Mathematical Society, Providence, RI, 1986.

Mathematical Physics Group, Institute of Theoretical and Experimental Physics, B. Cheremushkinskaya 25, Moscow 117259, Russia

Current address: University of Vienna, Mathematics Department, Nordbergstrasse 15, Vienna 1090, Austria

E-mail address: `neretin@mccme.ru`

Amer. Math. Soc. Transl.
(2) Vol. **217**, 2006

Skew Products and Livsic Theory

William Parry and Mark Pollicott

ABSTRACT. In this article we consider skew products of compact groups over hyperbolic maps. In particular, we describe approaches to studying their ergodic decomposition and stability of ergodicity based on applications of the Livsic Theorem.

1. Introduction

This is a short exposition of various aspects of skew products from the point of view of Livsic theory — an essay, as it were, focussed on compact group extensions only and hence partial in at least two respects. We have taken the liberty of expanding recent results (our real topic) so that placed in context, they might interest a wider readership. Experts, of course, will pass over much of this material.

Skew products date back (at least) to von Neumann who proposed their study in connection with the classification theory of ergodic measure-preserving transformations of a probability space. Von Neumann provided examples of such transformations which were non-isomorphic, but which shared the same spectral characteristics. This represented a first step away from the well-understood theory of transformations with discrete spectrum (due to his work done with Halmos). Subsequently Anzai, Abramov and Furstenberg laid the foundations for later research in this area. At that time the emphasis was on skew products over translations of compact abelian groups and so involved algebraic structure at two levels — the base and fibre — giving rise to noticeable rigidity of behaviour.

We shall be interested in skew products where the base is *chaotic*, or more precisely hyperbolic, and where motion in the fibres is determined by compact group translations. The systems therefore incorporate random and deterministic features.

The purpose of this paper is to give an exposition of three aspects of the theory:

(1) ergodic decompositions of skew products (heavily indebted to [**Br**]);
(2) cohomology theory (Livsic theory proper [**Li**]);
(3) ergodic stability problems (motivated initially in [**HPS**] and recently revived in [**AKS, PP1, FP, FMT, D**]).

2000 *Mathematics Subject Classification.* Primary 37D20; Secondary 37A25.
Key words and phrases. Hyperbolic maps, skew products, Livsic theorem.

The amount of detail provided varies with each section in accordance with our estimate of what is and what is not 'well-known'.

Let (X, d) be a compact metric space. For $0 < \alpha < 1$, we say that a continuous function $f : X \longrightarrow \mathbb{C}^d$ is Hölder continuous (with exponent α) if

$$|f(x) - f(y)| \leq Cd(x,y)^\alpha$$

for some constant C and for all $x, y \in X$. Here, as elsewhere, $|\cdot|$ denotes the Euclidean norm. Likewise $f : X \to M(d)$, ($d \times d$ complex matrices) is Hölder if $|f(x) - f(y)| \leq Cd(x,y)^\alpha$ for some constant C and for all $x, y \in X$. (Here $|\cdot|$ represents the Euclidean operator norm.) The least such C above is denoted $|f|_\alpha$ and we define $C^\alpha(X, \mathbb{C}^d), C^\alpha(X, M(d))$ to be the spaces of Hölder continuous functions, which with respect to the norm $\|f\| = |f|_\infty + |f|_\alpha$ are Banach spaces.

Let $\phi : X \longrightarrow X$ be a continuous surjective map. Then two functions $f, g \in C^\alpha(X, \mathbb{C}^d)$ are said to be *cohomologous* (with respect to ϕ) if there exists $h \in C^\alpha(X, \mathbb{C}^d)$ such that

$$f - g = h\phi - h\,.$$

If f is cohomologous to 0, then f is called a *coboundary*. Similarly $f, g \in C^\alpha(X, G)$, where $G \subset U(d) \subset M(d)$, are said to be cohomologous if there exists $h \in C^\alpha(X, G)$ such that $f = h^{-1}gh\phi$, and when f is cohomologous to 1 (the constant function assuming the identity value 1 in G), we say f is a *coboundary*.

These notions (which can be extended to other categories of functions) arise as follows:

For $f \in C^\alpha(X, G)$ we define the *skew product* $\phi_f : X \times G \longrightarrow X \times G$ by

$$\phi_f(x, y) = (\phi x, yf(x))\,.$$

For $g_0 \in G$, the map $g_0(x, y) = (x, g_0 y)$ defines an action of G on $X \times G$ which commutes with ϕ_f (ϕ_f is equivariant). Cohomological considerations arise when we look for a G-equivariant isomorphism ψ between two skew products ϕ_f, ϕ_g with the restriction that ψ factors to the identity on X, i.e., ψ is a skew product of the form $\psi(x, y) = (x, yh(x))$. The existence of such an isomorphism is precisely the requirement that f, g are cohomologous.

2. Measure-preserving transformations

Let ϕ be a measure-preserving transformation of a probability space $(X, \mathcal{B}, m)$. ϕ is said to be *ergodic* if $\phi^{-1}B = B$ ($B \in \mathcal{B}$) implies $m(B) = 0$ or 1. A stronger condition is *(weak-) mixing* which can be defined in several ways. In particular ϕ is weak-mixing if the equation

$$F \circ \phi = \alpha F \quad (\alpha \in \mathbb{C},\ F \in L^2(X))$$

implies that F is constant almost everywhere (i.e., apart from the one-dimensional space of constants, the *induced* isometry of $L^2(X)$ has no discrete spectrum).

Keynes and Newton [**KN**], generalising earlier work of Anzai and Furstenberg, established the following criteria for the ergodicity or weak-mixing of skew products in the measure-category.

THEOREM 2.1. *Let ϕ be an ergodic measure-preserving transformation of the probability space $(X, \mathcal{B}, m)$ and let $\phi_f : X \times G \longrightarrow X \times G$ be a skew product*

$$\phi_f(x, y) = (\phi x, yf(x))\,,$$

where $f : X \longrightarrow G$ *is measurable. Then* ϕ_f *is ergodic with respect to the measure* $m \times m_G$ *(*m_G *is the Haar measure) if and only if the equation*

$$F(\phi x) = R_\chi(f(x))F(x)$$

a.e. (where $F \in L^2(X, \mathbb{C}^d)$ *and* R_χ *is an irreducible unitary representation of* $G \subset U(d)$*) has only the trivial solutions:* $d = 1, R_\chi$ *is trivial and* F *is constant a.e.*

THEOREM 2.2. *Let* ϕ *be a weak-mixing measure-preserving transformation of the probability space* $(X, \mathcal{B}, m)$ *and let* $\phi_f : X \times G \longrightarrow X \times G$ *be a skew product. Then* ϕ_f *is weak-mixing if in addition to being ergodic (see above theorem), the equation*

$$F(\phi x) = \alpha\gamma(f(x))F(x)$$

(where $F \in L^2(X, \mathbb{C}), \alpha \in \mathbb{C}, \gamma$ *is a one-dimensional unitary representation) has only the trivial solution:* F *is constant.*

The equations in Theorems 2.1 and 2.2 are suggestive of the cohomology conditions of the previous section. This will be made explicit later when we consider measure-spaces which are also metric spaces. More precisely we shall be looking exclusively at so-called hyperbolic sets of Axiom A diffeomorphisms [**Sm**]. The interplay of measure and topology will be a recurrent theme in our exposition.

3. Hyperbolicity and shifts of finite type

Let $\phi : M \longrightarrow M$ be a C^∞ diffeomorphism of a compact Riemannian manifold whose metric we denote by d.

A closed ϕ-invariant set X is said to have *hyperbolic structure* if the tangent bundle $T_X M$ over X can be written as a direct sum $T_X M = E^u \oplus E^s$ (continuously split) such that

(1) $D\phi(E^u) = E^u$ and $D\phi(E^s) = E^s$;
(2) there exist constants $c > 0, 0 < \lambda < 1$ such that

$$\begin{aligned} \|D\phi^n v\| &\leq c\lambda^n \|v\| \quad \text{for } v \in E^s , \\ \|D\phi^{-n} v\| &\leq c\lambda^n \|v\| \quad \text{for } v \in E^u . \end{aligned}$$

The restriction $\phi = \phi\,|_X$ is called a *hyperbolic map* if

(1) X has hyperbolic structure;
(2) ϕ is *transitive*, i.e., there exists $x \in X$ such that $\{\phi^n x : n \in \mathbb{Z}\}$ is dense in X;
(3) the periodic points of ϕ are dense in X;
(4) there exists an open set $U \supset X$ such that

$$\bigcap_{n=-\infty}^{\infty} \phi^n U = X .$$

Very often we shall take ϕ to be *topologically mixing*, which means that condition (2) is strengthened to

(2′) for all open U, V there exists N such that

$$\phi^n U \cap V \neq \emptyset \quad \text{for all } n > N .$$

If we simply assume that M itself has hyperbolic structure (without the other assumptions), then $\phi : M \longrightarrow M$ is called an Anosov diffeomorphism.

The prototype of a diffeomorphism which is Anosov is an automorphism of a torus whose defining matrix does not have eigenvalues of modulus 1.

Examples of hyperbolic systems (which are not Anosov) are provided by *shifts of finite type*, since Williams [**W**] has shown that every such shift can be embedded as a hyperbolic system in S^3, and Bowen [**Bo3**] proved that every zero-dimensional hyperbolic system is (homeomorphic to) a shift of finite type.

Shifts of finite type (topological Markov shifts) are defined as follows: Let A be a $k \times k$ $0-1$ irreducible matrix and let

$$\Sigma = \Sigma_A = \{x \in (1, \ldots, k)^{\mathbb{Z}} : A(x_n, x_{n+1}) = 1 \text{ for all } n \in \mathbb{Z}\}.$$

(We shall always consider *aperiodic* shifts, i.e., assume that there exists N such that $A^N(i,j) > 0$ for all i, j.) The shift transformation σ (which is defined on doubly infinite sequences $x = \{x_n\}$ by $(\sigma x)_n = x_{n+1}$) leaves Σ invariant. Its restriction $\sigma = \sigma \mid_\Sigma$ is a shift of finite type.

The underlying space Σ is compact (with respect to the Tychonov topology), zero-dimensional, and metrizable. A convenient metric is given by $d(x, y) = 1/2^N$, where $x_n = y_y$for $|n| \leq N$ and either $x_N \neq y_N$ or $x_{-N} \neq y_{-N}$.

Shifts of finite type are connected with hyperbolic systems in another, perhaps more fundamental, way. They provide symbolic models for all hyperbolic maps in the following sense [**Bo1**]:

THEOREM 3.1 (Bowen). *Let $\phi : X \longrightarrow X$ be a hyperbolic map. Then there exists a shift of finite type, $\sigma : \Sigma \to \Sigma$, and a Hölder continuous surjective map $\pi : \Sigma \longrightarrow X$ such that $\pi\sigma = \phi\pi$. Moreover, except for points in a set of first category, each $x \in X$ has a unique inverse image, and in any case there exists a bound b on the number of inverse images for all $x \in X$.*

Briefly, the theorem comes from the hyperbolic structure of X as follows: for each $x \in X$ define the stable set through x as $W^s(x) = \{y \in X : d(\phi^n x, \phi^n y) \longrightarrow 0\}$ and the unstable set as $W^u(x) = \{y \in X : d(\phi^{-n} x, \phi^{-n} y) \longrightarrow 0\}$. The stable sets foliate (partition) X as do the unstable sets. Rectangles (with precise properties) can be defined by these sets and X can be covered by a finite union of rectangles, $R_1, \ldots, R_k$, in such a way that the interiors of these sets are pairwise disjoint, each R_i is the closure of its interior, and for each $i = 1, \ldots, k$

$$\begin{aligned} \phi(W^s(x) \cap R_i) &\subset W^s(\phi x) \cap R_j \text{ if } x \in \operatorname{int}(R_i \cap \phi^{-1} R_j)\,, \\ \phi^{-1}(W^u(x) \cap R_i) &\subset W^u(\phi^{-1} x) \cap R_j \text{ if } x \in \operatorname{int}(R_i \cap \phi R_j)\,. \end{aligned}$$

Such a partition (strictly speaking, covering) into rectangles is called a Markov partition.

The existence of arbitrary small Markov partitions was proved in increasing order of generality by Adler and Weiss [**AW**](for ergodic automorphisms of a 2-torus), Sinai [**Si1**] (for Anosov diffeomorphisms) and Bowen [**Bo1**](for hyperbolic maps). From this data one can now produce a shift of finite type (Σ, σ) as follows: Define $A(i,j) = 1$ if $R_i \cap \phi^{-1} R_j$ has non-empty interior and $A(i,j) = 0$ otherwise. When the Markov partition is small enough, there is at most one point x for which

$\phi^n x \in \text{int} R_{x_n}$ for all $n \in \mathbb{Z}$ which implies that the map

$$\pi(x) = \bigcap_{n=-\infty}^{\infty} \phi^{-n} R_{x_n} \quad (x = \{x_n\})$$

is well defined and maps $\Sigma_A = \Sigma$ onto X with the exceptional points y with more than one inverse image being those for which there exists $n \in \mathbb{N}$ such that $\phi^n y$ belongs to the boundary of some rectangle. These exceptional points form a set of first category.

4. Equilibrium states and one-sided shifts

Let $\phi : X \longrightarrow X$ be a hyperbolic map and let $f \in C^\alpha(X, \mathbb{R})$. The *pressure* of f is defined by

$$P(f) = \sup \left\{ h(\mu) + \int f \, d\mu \right\},$$

where the supremum is taken over all ϕ invariant probabilities μ and $h(\mu)$ denotes the μ measure-theoretic entropy of ϕ. If $P(f) = h(m) + \int f \, dm$, we say that m is an *equilibrium state* associated with f. A key fact for hyperbolic systems is that for each Hölder continuous function there is one and only one equilibrium state [**Bo2**]. In much of what is to follow there will be a (Hölder) equilibrium state in the background, which frequently need not be specified. If m is the equilibrium state of f and $\pi : \Sigma \longrightarrow X$ is the projection defined in Bowen's theorem, then $m = \mu\pi^{-1}$, where μ is the equilibrium state of $f \circ \pi$. Moreover π is a measure-theoretic isomorphism between (Σ, σ) and (X, ϕ) with respect to these measures, a fact which frequently allows problems for hyperbolic systems to be transferred to problems for shifts.

In addition to the *two-sided* shifts of finite type, defined above, there are *one-sided* versions. Again we begin with the matrix A and define

$$\Sigma^+ = \Sigma_A^+ = \{x \in (1, \ldots, k)^{\mathbb{Z}^+} : A(x_n, x_{n+1}) = 1 \text{ for all } n \in \mathbb{Z}^+\}$$

and $\sigma(x)_n = x_{n+1}$. (We retain the same notation for the one-sided shift.) The shifts $\sigma : \Sigma \longrightarrow \Sigma$ and $\sigma : \Sigma^+ \longrightarrow \Sigma^+$ are related by the obvious projection which deletes negatively indexed coordinates. Using this projection, it is not difficult to see that any probability measure which is invariant for the one-sided shift gives rise to a corresponding invariant probability for the two-sided shift — and vice versa. Moreover the corresponding systems enjoy the same ergodic or mixing properties.

Let $\sigma : \Sigma \longrightarrow \Sigma$ be a shift of finite type and let $f \in C^\alpha(\Sigma, \mathbb{C}^d)$. The following theorem enables one to transfer problems to a one-sided setting.

THEOREM 4.1 ([**Si2**]). *If $f \in C^\alpha(X, \mathbb{C}^d)$, then there exist functions $f^+, h \in C^{\alpha/2}(X, \mathbb{C}^d)$ such that $f^+ = f + h\sigma - h$, where f^+ is independent of the past, i.e., $f^+(x) = f'(\pi x)$ where $f' \in C^{\alpha/2}(X^+, \mathbb{C}^d)$.*

To establish properties of f, it is frequently useful to prove them for f' in the context of one-sided shifts and then to invoke the theorem.

The proof of the following theorem mimics the proof of Sinai. (In fact the crucial property used is the bi-invariance of the group metric.)

THEOREM 4.2. *If $f \in C^\alpha(X, G)$, $(G \subset U(d))$, then there exist functions $f^+, h \in C^{\alpha/2}(X, G)$ such that $f^+ = h^{-1} f h\sigma$, where f^+ is independent of the past.*

5. The Ruelle transfer operator

In this section we concentrate on a one-sided shift of finite type $\sigma : \Sigma^+ \longrightarrow \Sigma^+$. For $u \in C^\alpha(\Sigma^+, \mathbb{R})$, we define the transfer operator,

$$L_u : C^\alpha(\Sigma^+, \mathbb{R}) \longrightarrow C^\alpha(\Sigma^+, \mathbb{R})$$

by

$$(L_u w)(x) = \sum_{\sigma y = x} e^{u(y)} w(y)\,,$$

which is a continuous linear operator.

THEOREM 5.1 (Ruelle). *If $P(u)$ is the pressure function of Section 4, then $e^{P(u)}$ is the maximum eigenvalue of L_u; it is simple and its eigenfunctions $h \in C^\alpha(\Sigma^+, \mathbb{R})$ may be taken to be strictly positive. The rest of the spectrum of L_u is contained in a disc of radius strictly less than $e^{P(u)}$.*

As in [**R**], for $L_u h = e^{P(u)} h$ we define $u' = u + \log h - \log h\sigma - P(u)$, so that $u' - u$ is cohomologous to a constant. Then this eigenfunction equation transforms to $L_{u'} 1 = 1$, and we say that u' is *normalised* (the normalisation of u).

The equilibrium state m defined by u (or any other function which is cohomologous to it plus a constant) can be defined using the normalised operator $L_{u'}$. It is the unique measure satisfying

$$\int (L_{u'} w)\, dm = \int w\, dm$$

for all continuous w (i.e., $L_{u'}^* m = m$). In view of Ruelle's theorem, $L_{u'}$ has 1 as a simple eigenvalue corresponding to the constant function 1, and the rest of the spectrum is contained in a disk of radius less than 1. This latter fact plays a crucial role in establishing mixing properties of both the one-sided and two-sided shifts. In particular it is used to show that σ is *exponentially mixing*, meaning that there exists $c > 0, 0 < \lambda < 1$ such that for all $v, w \in C^\alpha(\Sigma^+, \mathbb{R})$,

$$\left| \int v \circ \sigma^n w\, dm - \int v\, dm \int w\, dm \right| \le c.\lambda^n \|v\| \|w\|\,.$$

We can define a transfer operator for complex-valued functions $u + iv \in C^\alpha(\Sigma^+, \mathbb{C})$ in the same way as above,

$$\begin{aligned} (L_{u+iv} w)(x) &= \sum_{\sigma y = x} e^{u(y)} e^{iv(y)} w(y) \\ &= L_u(e^{iv} w). \end{aligned}$$

Or, more generally, if $f \in C^\alpha(\Sigma^+, G)$, where $G \subset U(d)$, we can define

$$L_f : C^\alpha(\Sigma^+, \mathbb{C}^d) \longrightarrow C^\alpha(\Sigma^+, \mathbb{C}^d)$$

by $L_f w = L_u(fw)$. (Here u is taken to be a fixed element of $C^\alpha(\Sigma^+, \mathbb{R})$.)

Proposition 5.4 to follow was first proved by Pollicott (for the case $d = 1$, which amounts to the complex operator case). We shall present a generalised version. Crucial to the proof is the inequality given by

LEMMA 5.2. *If L_u is normalised and $f \in C^\alpha(\Sigma^+, U(d))$, then there exists a constant C such that for all $w \in C^\alpha(\Sigma^+, \mathbb{C}^d)$,*

$$\left| L_f^n w \right|_\alpha \le C |w|_\infty + \alpha^n |w|_\alpha\,.$$

(See [**PP2**] for a proof when $d = 1$. The general case is similar.)
We shall also need

LEMMA 5.3. *The function* $w \in C^\alpha(\Sigma^+, \mathbb{C}^d)$ *satisfies* $L_f w = \alpha w$ *(with* $|\alpha| = 1$*) if and only if* $fw = \alpha w \circ \sigma$.

PROOF. The second equation implies the first equation by a simple application of the operator L_u. Assuming $L_f w = \alpha w$, we see that $L_u|w| \geq |w|$, and integration then shows that we must have $L_u|w| = |w|$. This means that $|w|$ takes a constant value, which we may take to be 1. For each $x, \alpha w(x)$ is a point of the unit sphere in $\mathbb{C}^d$, and

$$\sum_{\sigma y = x} e^{u(y)} f(y) w(y)$$

is a convex combination of points $f(y)w(y)$ on the unit sphere, because

$$\sum_{\sigma y = x} e^{u(y)} = 1 \, .$$

The equation therefore implies that for each $y \in \sigma^{-1}x$, $f(y)w(y) = \alpha w(x)$, i.e., $f(y)w(y) = \alpha w(\sigma y)$ for all $y \in \Sigma^+$. □

PROPOSITION 5.4. *Let* $V \subset C^\alpha(X^+, \mathbb{C}^d)$ *be a closed subspace such that* $L_f V \subset V$ *and* $L_f = L_f \mid_V$ *has no eigenvalues of modulus* 1. *Then* $\rho(L_f \mid_V) < 1$, *where* ρ *is the spectral radius.*

PROOF. By Lemma 5.2, we see that for $M = C+1$, $L_f^n B_1 \subset B_M$ for all $n \in \mathbb{N}$, where B_1, B_M are the closed balls of radius 1 and M, respectively. In particular, this means that

$$A_N(w) = \frac{1}{N} \sum_{n=0}^{N-1} L_f^n w \in B_M \quad \text{for all } w \in B_1 \, ,$$

and since B_M is uniformly compact, $A_N(w)$ converges uniformly through some subsequence to, say, $w^* \in V$. Moreover, $L_f w^* = w^*$ so that $w^* = 0$ (for otherwise 1 would be an eigenvalue). The sequence $|A_{2^N}(w)|_\infty$ is decreasing since

$$\left| \frac{1}{2^{N+1}} \sum_{n=0}^{2^{N+1}-1} L_f^n w \right|_\infty \leq \frac{\left| \sum_{n=0}^{2^N-1} L_f^n w \right|_\infty}{2^{N+1}} + \frac{\left| L_f^{2^N} \left(\sum_{n=0}^{2^N-1} L_f^n w \right) \right|_\infty}{2^{N+1}} \, ,$$

i.e., $|A_{2^{N+1}}(w)|_\infty \leq |A_{2^N}(w)|_\infty$. Since B_1 is uniformly compact, we therefore conclude that for every $\varepsilon > 0$ there exists N such that $|A_N(w)|_\infty < \varepsilon$ for *all* $w \in B_1$. The basic inequality gives

$$\left| L_f^k A_N(w) \right|_\alpha \leq |A_N(w)|_\infty . C + \alpha^k \, |A_N(w)|_\alpha$$

and

$$\begin{aligned} |A_N(w)|_\alpha = \left| \frac{1}{N} \sum_{n=0}^{N-1} L_f^n w \right|_\alpha &\leq \frac{1}{N} \sum_{n=0}^{N-1} \left| L_f^n w \right|_\alpha \\ &\leq \frac{1}{N} \sum_{n=0}^{N-1} (|w|_\infty C + \alpha^n |w|_\alpha) \\ &\leq \|w\|(C+1) \, . \end{aligned}$$

Thus,

$$\left|L_f^k A_N(w)\right|_\alpha \le C.\left|A_N(w)\right|_\infty + \alpha^k(C+1)\|w\|\ ,$$

and

$$\left|L_f^k A_N(w)\right|_\infty \le \left|A_N(w)\right|_\infty\ ,$$

which added gives

$$\left\|L_f^k A_N(w)\right\| \le (C+1)\left(\left|A_N(w)\right|_\infty + \alpha^k\|w\|\right)\ .$$

If we choose $\varepsilon = 1/(2C+2)$ and N such that $|A_N(w)|_\infty < \varepsilon$ for all $w \in B_1$, and k such that $\alpha^k < 1/(2C+2)$, we will have $\|L_f^k A_N(w)\| < 1$ for all $w \in B_1$. Thus 1 is not an eigenvalue of L_f, and hence $\rho(L_f^k \circ A_N) = \sup\{|p(\lambda)| : \lambda \in \mathrm{sp}L_f\} < 1$, where

$$p(\lambda) = \frac{1}{N}\sum_{n=k}^{k+N-1} \lambda^n\ .$$

Therefore $1 \notin \mathrm{sp}L_f$.

Repeating this argument with f replaced by αf with $|\alpha| = 1$, we conclude that α is not an eigenvalue of L_f. Hence $\mathrm{sp}L_f$ is contained in the open unit disc. □

THEOREM 5.5. *The space $C^\alpha(X^+, \mathbb{C}^d) = C^\alpha$ decomposes as $C^\alpha = V_0 + V_1$, where $L_f : V_1 \longrightarrow V_1$, $L_f : V_0 \longrightarrow V_0$, and*

(1) $\rho(L_f\,|_{V_0}) < 1$,
(2) $\dim(V_1) \le d$,
(3) *V_1 is spanned by (generalized) eigenvectors corresponding to eigenvalues of modulus* 1.

PROOF. Let V_1 be the linear space of finite linear combinations of eigenvectors whose eigenvalues have modulus 1. We equip this space with the inner product $\langle\langle v, w\rangle\rangle = \int \langle v(x), w(x)\rangle\, dm$, where $\langle v(x), w(x)\rangle$ is the Euclidean inner product on $\mathbb{C}^d$. If $L_f v = \alpha v$ and $L_f w = \beta w$, where $|\alpha| = |\beta| = 1$, then $fv = \alpha v \circ \sigma$ and $fw = \beta w \circ \sigma$, so that (using the fact that f is $U(d)$-valued), $\langle v(x), w(x)\rangle = \alpha\overline{\beta}\langle v(\sigma x), w(\sigma x)\rangle$. Since σ is mixing, we conclude that $\langle v(x), w(x)\rangle$ is constant, and $\alpha = \beta$ if this constant is non-zero. Choosing a fixed point $x_0 = \sigma x_0$, we therefore have $\langle v(x), w(x)\rangle = \langle v(x_0), w(x_0)\rangle$ for all $x \in X^+$. Since V_1 is spanned by L_f eigenfunctions, we see that the map $V_1 \to \mathbb{C}^d$ given by $v \to v(x_0)$ is an isometry. Hence $\dim(V_1) \le d$.

Now let $V_0 = \{v \in C^\alpha : v \perp V_1\}$. It is clear that V_0 is a closed subspace of C^α. If $v \in V_0$ and $L_f v_0 = \alpha v_0 \in V_1$, where $|\alpha| = 1$ (so that $fv_0 = \alpha v_0 \sigma$), then $\langle L_f v, v_0\rangle = \langle L(fv), v_0\rangle = \langle fv, v_0\sigma\rangle = \langle v, \overline{f} v_0 \sigma\rangle = \langle v, \overline{\alpha} v_0\rangle = 0$, i.e., $L_f V_1 \subset V_1$. Furthermore, $C^\alpha = V_0 + V_1$ since V_1 is a closed subspace of $L^2(m)$. By Proposition 5.4, $\rho(L_f|V_0) < 1$. □

6. Regularisation of functions

In this section we shall show how the spectral properties of transfer operators discussed in the last section can be used to prove Livsic type theorems. Typically, one wants to show that measurable solutions to certain functional equations are necessarily (Hölder) continuous.

THEOREM 6.1. *Let $L_f w = \alpha w$ a.e., where $|\alpha| = 1, w \in L^2(\Sigma^+, \mathbb{C}^d)$, $f \in C^\alpha(\Sigma^+, U(d))$. Then there exists $w_1 \in C^\alpha(\Sigma^+, U(d))$ such that $w = w_1$ a.e.*

PROOF. Replacing f by $\overline{\alpha}f$, if necessary, we can assume $L_f w = w$. Let $L^2 = V_1^\perp \oplus V_1$. Then $V_1^\perp \supset V_0$ and $C^\alpha = V_0 \oplus V_1$. Choose $\varepsilon > 0$ and $v \in C^\alpha$ such that $\int |v-w|^2\, dm < \varepsilon^2$. Then $v = v_0 + v_1$, $w = w_0 + w_1$, where $v_0 \in V_0$, $v_1 \in V_1$, $w_0 \in V_1^\perp$, $w_1 \in V_1$. Thus, $\varepsilon^2 \geq \int |v-w|^2\, dm = \int |v_0 - w_0|^2\, dm + \int |v_1 - w_1|^2\, dm$, and hence $\int |v_0 - w_0|^2\, dm \leq (\int |v_0 - w_0|^2\, dm)^{1/2} \leq \varepsilon$. Furthermore, $L_f V_1^\perp \subset V_1^\perp$. To see this, suppose $\langle u, u_1 \rangle = 0$ whenever $L_f u_1 = \alpha u_1$ ($|\alpha| = 1$), or, equivalently, $f u_1 = \alpha u_1 \circ \sigma$. Then $\langle u, \alpha f^{-1} u_1 \circ \sigma \rangle = 0$, which implies that $\langle L_f u, \alpha u_1 \rangle = 0$. This shows that $L_f u \perp V_1$ when $u \perp V_1$.

Hence $\int |L_f^n v_0 - L_f^n w_0|\, dm \leq \int |v_0 - w_0|\, dm \leq \varepsilon$, and since $L_f^n v_0 \longrightarrow 0$, we have $\lim_{n\to\infty} \int |L_f^n w_0|\, dm \leq \varepsilon$, i.e., $w_0 = 0$ a.e., and $w = w_1$ a.e. □

For a proof of the following cf. [**PP2**].

LEMMA 6.2. *If $f \in C^\alpha(\Sigma, U(d))$ depends only on future coordinates and $w \in L^2(\Sigma)$ satisfies*

$$fw = \alpha w \circ \sigma \quad \text{a.e. } (|\alpha| = 1)\,,$$

then w depends (essentially) only on future coordinates.

This can now be used in conjunction with Theorem 6.1 and Theorem 4.1 to prove

THEOREM 6.3. *Let $\sigma : \Sigma \longrightarrow \Sigma$ be a (two-sided) shift of finite type and let $f \in C^\alpha(\Sigma, U(d))$. If $w \in L^2(\Sigma, \mathbb{C}^d)$ satisfies*

$$fw = \alpha w \circ \sigma \quad \text{a.e. } (|\alpha| = 1),$$

then there exists $w_1 \in C^{\alpha/2}(\Sigma, \mathbb{C}^d)$ such that $w = w_1$ a.e.

PROOF. By Theorem 4.1 there exists f^+, $h \in C^{\alpha/2}(\Sigma, \mathbb{C}^d)$ such that $f = h(\sigma)^{-1} f^+ h$, where f^+ depends only on future coordinates. Hence

$$f^+(hw) = \alpha(hw) \circ \sigma\,.$$

By Lemma 6.2 it follows that hw depends only on future coordinates, and applying Theorem 6.1, we see that $hw = w_1$ a.e. for some $w_1 \in C^{\alpha/2}(\Sigma, \mathbb{C}^d)$, i.e., $w = h^{-1} w_1$ a.e. □

We are now in a position to prove the following generalisation of a theorem due to Livsic (who considered the case where one of the functions f, g is trivial).

THEOREM 6.4. *Let $f, g \in C^\alpha(\Sigma, G)$ ($G \subset U(d)$) and let $h : \Sigma \longrightarrow G$ be measurable, where*

$$f = h^{-1} g h \sigma \quad \text{a.e.}$$

Then there exists $h' \in C^{\alpha/2}(\Sigma, G)$ such that $h = h'$ a.e. (and $f = h^{-1} g h \sigma$ everywhere).

PROOF. We consider the Euclidean space $M(d)$ endowed with the inner product $\langle A, B \rangle = \operatorname{Trace} AB^*$. The linear map $A \mapsto UAV$ (where $U, V \in U(d)$) is a unitary operator on this d^2-dimensional space since

$$\begin{aligned} \langle UAV, UBV \rangle &= \operatorname{Tr}(UAVV^*B^*U^*) \\ &= \operatorname{Tr}(UAB^*U^{-1}) \\ &= \operatorname{Tr} AB^* \\ &= \langle A, B \rangle\,. \end{aligned}$$

We can therefore apply Theorem 6.1 to the unitary-valued function

$$x \mapsto \phi(x)^{-1} A f(x) \; .$$

In other words, the equation

$$g(x)^{-1} h(x) f(x) = h(\sigma x) \quad \text{a.e.}$$

can be represented in the form required by Theorem 6.1. We conclude that the $M(d)$-valued (in fact G-valued) function h equals a.e. a function $h' \in C^{\alpha/2}(\Sigma, G)$. □

Using Bowen's theorem (Theorem 3.1), the above result can be used to prove the following.

THEOREM 6.5. *If $\phi : X \longrightarrow X$ is hyperbolic and $f, g \in C^\alpha(X, G)$, where $G \subset U(d)$, and if*

$$f = h^{-1} g h \phi \quad \text{a.e.},$$

then there exists $h' \in C^{\alpha/2}(X, G)$ such that $h = h'$ a.e.

The basic idea is to lift the given equation to the setting of a shift of finite type using the projection π in Bowen's theorem. The functions $f \circ \pi, g \circ \pi$ are Hölder continuous on the shift space and $h \circ \pi$ is measurable. Thus $h \circ \pi$ may be replaced by a Hölder continuous function h'' on the shift space. The next step is to show that $h'' = h' \circ \pi$ is the lift of a continuous function on X (cf. [**PP1**]). Hence $f = h'^{-1} g h' \circ \phi$, where h' is continuous. One can now appeal to the use of Markov partitions in Theorem 3.1 to show that $h' \in C^{\alpha/2}(X, G)$.

Theorem 6.5 may be thought of as a 'rigidity' theorem. It has the consequence that if ϕ_f, ϕ_g are equivariant, e.g. isomorphic (in the measure-theoretic sense) via an extension of the identity, then this isomorphism is (essentially) Hölder continuous.

7. Ergodic components

Let $\phi : X \longrightarrow X$ be hyperbolic, and let $f \in C^\alpha(X, G)$, $G \subset U(d)$. In Theorem 2.1 we gave criteria for ϕ_f to be ergodic with respect to $m \times m_G$, where m is an equilibrium state. The situation there was purely measure-theoretic. In view of Theorem 6.5, we can get a clearer picture regarding ergodic components.

THEOREM 7.1. *The skew product is ergodic if and only if the equation*

$$R(f) w = w \phi \quad \text{(everywhere)} \; ,$$

where R is an irreducible d-dimensional representation and $w \in C^\alpha(X, \mathbb{C}^d)$, has only the trivial solution given by the trivial $d = 1$ dimensional representation and constant w.

For the case of a shift of finite type, $\sigma : \Sigma \longrightarrow \Sigma$, we can give a description of the ergodic components of ϕ_f (where ϕ_f is non-ergodic). For, if $R(f) w = w \circ \sigma$ is a non-trivial equation, we see that the function $R(y^{-1}) w(x)$ is a non-constant σ_f invariant function.

Choose $x_0 \in \Sigma$ and define $E = \{(x, y) : R(y^{-1}) w(x) = w(x_0)\}$. If $(x, y) \in E$ and $(x, y g_0) \in E$, then $R(g_0) w(x_0) = w(x_0)$ and vice versa. Let $H = \{g_0 : R(y_0) w(x_0) = w(x_0)\}$. Then H is a closed proper subgroup of G for otherwise R would be the trivial representation. It is clear that $E \cap E_{g_0} = \emptyset$ if $g_0 \notin H$ and $E \equiv E g_0$ if $g_0 \in H$. Furthermore, E is ϕ_f invariant since $w(\sigma x) = R(f(x)) w(x)$.

Choosing $x_0 \in \Sigma$ to be a point with a dense orbit, one sees that the closed set E projects onto Σ.

LEMMA 7.2. *If ϕ_f is not ergodic, then there exists a closed proper subgroup H and a $C^\alpha(\Sigma, H)$ function f' such that f, f' are cohomologous.*

PROOF. Define H as above. Since Σ is zero-dimensional, one can choose a Hölder cross-section in E, i.e., $h : \Sigma \longrightarrow G$ with $(x, h(x)) \in E$ for all $x \in \Sigma$. Hence $(\sigma x, f(x)h(x)) \in E$. This means that there exists $f'(x) \in H$ such that $f(x)h(x) = h(\sigma x)f'(x)$, which was to be proved. □

THEOREM 7.3. *If ϕ_f is not ergodic, then there exists a* minimal *closed subgroup $H \subset G$ for which f is cohomologous to some $f' \in C^\alpha(\Sigma, H)$. Moreover, $\sigma_f : \Sigma \times H \longrightarrow \Sigma \times H$ is ergodic with respect to $m \times m_H$.*

This follows from the lemma on noting that if $\sigma_{f'}$ is not ergodic, then we conclude that there is a closed proper $\sigma_{f'}$ invariant $E' \subset E$, which is H'-invariant, where $H' \subset H$ (proper) and E' projects to Σ. This would contradict the minimality of H.

We have therefore arrived at a complete ergodic decomposition of $\Sigma \times G$. If Γ consists of choices from each H coset above, then the σ_f invariant sets Eg, $g \in \Gamma$, are disjoint and σ_f invariant, and $\Sigma \times G = \bigcup_{g \in \Gamma} Eg$. Furthermore, Eg is $g^{-1}Hg$ invariant and $\phi_f \mid_{Eg}$ is ergodic. These observations in conjunction with Bowen's theorem (Theorem 3.1) lead to

THEOREM 7.4. *Let $\phi_f : X \times G \longrightarrow X \times G$, where $\phi : X \longrightarrow X$ is hyperbolic and $f \in C^\alpha(X, G)$. Then there exists a closed ϕ_f invariant set E' which projects to X and a closed subgroup $H \subset G$ such that $E'g_0 \cap E' = \emptyset$ if $g_0 \notin H$ and $E'g_0 = E'$ if $g_0 \in H$. Furthermore, $\phi_f \mid_{E'}$ is ergodic with respect to the measure $m \times m_H$.*

PROOF. Let $\pi : \Sigma \longrightarrow X$ be the projection of Bowen's theorem and let $\pi(x, g) = (\pi x, g)$. Then it is easy to see that $\pi E = E'$, and H satisfies our requirements, where E, H are defined in the theorem. □

REMARK 7.5. Although this complete description of the ergodic decomposition exists for the general case of a skew product ϕ when ϕ is hyperbolic, we cannot assert (in general) the existence of a function f' corresponding to the one defined in Lemma 7.2 since its construction depended on the existence of a cross-section h to the set E, and this is not always true.

For an orbit segment $\ell = (y, \phi y, \ldots, \phi^n y)$ we define $f(\ell) = f(y) \cdots f(\phi^n y)$, and we define $f_{-n}(x) = f(\phi^{-n}x) \cdots f(\phi^{-1}x)$, $f_n(x) = f(\phi x) \cdots f(\phi^n x)$. Let z be a fixed point, and let $W^u(z) = \{x : \phi^{-n}x \longrightarrow z\}, W^s(z) = \{x : \phi^n x \longrightarrow z\}$ and $W(z) = W^u(z) \cap W^s(z)$. Points in $W(z)$ are called *homoclinic* (with respect to z). Since $f(z)$ is not necessarily 1, we choose and fix a sequence $\{N_n\}$ such that $f(z)^{N_n} \longrightarrow 1$ and define $f_-(x) = \lim_{n\to\infty} f_{-N_n}(x)$, when $x \in W^u(z)$ and $f_+(x) = \lim_{n\to\infty} f_{N_n}(x)$ when $x \in W^s(x)$. These limits exist since both $f(z)^{-n}f_{-n}(x)$ and $f_n(x)f(z)^{-n}$ converge due to the fact that the convergences $\phi^{-n}x \longrightarrow z$ and $\phi^n x \longrightarrow z$ are exponentially fast and f is Hölder continuous. If $x \in W(z)$, then we refer to the homoclinic orbit $\ell(x) = \{\phi^n x\}$ as a loop based at z (generated by x), and we define $f\ell(x) = f_-(x)f(x)f_+(x)$. Note that we must distinguish between the loop generated by x and the loop generated by ϕx since $f\ell(\phi x) = f(z)^{-1}f\ell(x)f(z)$. We also note that $z \in W(z)$ and $\ell(z) = \{z\}$, so that $f\ell(z) = f(z)$.

Let Γ denote the free group generated by loops $\ell(x), x \in W(z)$, a typical element being a formal product $\ell_1^{\pm1}\ell_2^{\pm1}\cdots\ell_k^{\pm1}$ of loops, some of which may be empty, i.e., coincide with the identity element of Γ. With these definitions it is clear that $f:\Gamma \longrightarrow G$ (defined by $f(\ell_1^{\pm1}\ell_2^{\pm1}\cdots\ell_k^{\pm1}) = f(\ell_1)^{\pm1}f(\ell_2)^{\pm1}\cdots f(\ell_k)^{\pm1})$ is a homomorphism.

THEOREM 7.6. *For $f \in C^\alpha(X,G)$ the skew product ϕ_f is ergodic if and only if $f(\Gamma)$ is dense in G.*

PROOF. Suppose ϕ_f is ergodic and let $K = \{f(z)^n\}$. Define $E = \{(x, kf_-(x)) : k \in K \text{ and } x \in W(z)\}$. Then $\overline{E}$ is ϕ_f- invariant and projects to X. Observe that points homoclinic to z are dense. Hence $\overline{E}$ contains a transitive point so that $\overline{E} = X\times G$. Let $g \in G$ be arbitrary, and approximate (z,g) by $(x, kf_-(x)) \in E$ with x so close to z that $f_n(x)f(z)^{-n}$ is close to 1 for all $n > N$, and $f(x)$ is close to $f(z)$. Evidently, $kf_-(x)f(x)f_nf(z)^{-(n+1)}$ is close to g as is $kf_-(x)f(x)f_+(x)f(z)^{-1} = kf\ell(x)f(z)^{-1}$. Thus g is arbitrarily close to $f(\Gamma)$, i.e., $f(\Gamma)$ is dense in G.

Conversely, if Γ is dense and $(z,1) \in E$, an ergodic component, then the orbit of $(x, f_-(x))$ belongs to E since

$$\phi_f^{-N_n}(x, f_-(x)) = \left(\phi^{-N_n}x, f_-(x)\left(f_{-N_n}(x)\right)^{-1}\right) \longrightarrow (z,1)\,.$$

Furthermore,

$$\phi^{N_n}(x, f_-(x)) = (\phi^{N_n}x, f_-(x)f(x)f_{N_n-1}(x)) \longrightarrow (z, f(\ell(x))f(z)^{-1}) \in E.$$

We conclude that $f\ell(x)f(z)^{-1}$ belongs to the closed subgroup H of Theorem 7.3. Hence $f\ell(x) \in H$ for all $x \in W(z)$, i.e., $H = G$ and ϕ_f is ergodic. □

8. Weighted periodic points

Up to now the main theme has been regularisation theorems. In other words, given an equivariant *measure-theoretic* isomorphism between two skew products (with appropriate conditions), we showed that the isomorphism is (essentially) Hölder continuous.

In this section we shall not be provided with an initial isomorphism. Instead, our conditions are given independently for each skew product and are expressed in terms of periodic points.

As before we are interested in the cocycle equation. So let $\phi : X \longrightarrow X$ be hyperbolic and let $f, g \in C^\alpha(X,G)$, $G \subset U(d)$. The question we shall address is when are f, g cohomologous? If there exists $h \in C^\alpha(X,G)$ such that

$$f = h^{-1}gh\circ\phi\,,$$

then for every periodic point x ($\phi^n x = x$) we have

$$f(x)\cdots f(\phi^{n-1}x) = h(x)^{-1}g(x)\cdots g(\phi^{n-1}x)h(x)\,,$$

i.e., the *weights* $f(x)\cdots f(\phi^{n-1}x)$ and $g(x)\cdots g(\phi^{n-1}x)$ of the closed orbit $x, \phi x, \ldots, \phi^{n-1}x$ are conjugate in G.

We shall see in the next section that, conversely, if a function $h \in C^\alpha(X,G)$ given in advance provides such a conjugacy of weights for all periodic orbits, then f and g are cohomologous. By modifying one of these functions appropriately, one may as well take $h \equiv 1$ and require that the weights be equal.

The identity of weights (or equivalently the provision of the above function h) is a strong condition. We shall concentrate on the more general problem: How are the functions f and g related if f-weights and g-weights are *conjugate* in G, i.e., when no function h as above is given?

When f (or g) is trivial (consistently equal to 1, the identity of G), then the above condition is simply that all weights are equal to 1, and in this case Livsic's celebrated theorem says that the non-trivial function is cohomologous to 1, i.e., it is a coboundary. We are interested in the general situation which does not reduce to Livsic's case when G is non-abelian.

Our answer to the above question will be sketched, with less detail than in earlier sections.

Let $\phi z = z$ be fixed throughout. (If no fixed points exist, our argument may be modified using a periodic point.) In what follows, f and g will be $C^\alpha(X, G)$ functions ($G \subset U(d)$), and $f(x) \cdots f(\phi^{n-1}x), g(x) \cdots g(\phi^{n-1}x)$ will be conjugate whenever $\phi^n x = x$, and in particular $f(z), g(z)$ will be conjugate. In fact, by conjugating one of f, g by an element of G, there is no harm in assuming $f(z) = g(z)$. We need the following closing lemmas (cf. [**Bo1**]).

For two orbit segments $\ell = (x, \phi x, \ldots, \phi^n x)$ and $\ell' = (x', \phi x', \ldots, \phi^n x')$ of the same length we define

$$d_\alpha(\ell, \ell') = \sum_{i=0}^{n} d(\phi^i x, \phi^i x')^\alpha .$$

LEMMA 8.1 (Anosov's closing lemma). *There exist $\varepsilon > 0, C(\alpha)$ depending only on ϕ and α such that if $d(\phi^N x, x) < \varepsilon$, then there exists a closed orbit $\ell' = (w, \phi w, \ldots, \phi^{N-1}w), (\phi^N w = w)$ such that*

$$d_\alpha(\ell, \ell') \le C(\alpha) d(x, \phi^N x)^\alpha ,$$

where $\ell = (x, \phi x, \ldots, \phi^{N-1}x)$.

This can be used to prove

LEMMA 8.2. *If $\ell_1, \ldots, \ell_k$ are orbit segments such that (cyclically) the distance d_i between the last point of ℓ_i and the first point of ℓ_{i+1} is less than ε, then there exists a periodic orbit whose consecutive segments $\ell'_1, \ldots, \ell'_k$ satisfy*

$$\sum_{i=1}^{k} d_\alpha(\ell_i, \ell'_i) \le C(\alpha) \max_i d_i^\alpha .$$

Let $\phi z = z$, and let $W^s(z), W^u(z), W(z) = W^s(z) \cap W^u(z)$ be the sets defined in the last section.

LEMMA 8.3. *If $x \in W^u(z), y \in W^s(z)$ and $d(x, y) < \varepsilon$, then there exists $w \in W(z)$ such that*

$$\sum_{i=-\infty}^{0} d(\phi^i x, \phi^i w)^\alpha + \sum_{i=1}^{\infty} d(\phi^i y, \phi^i w)^\alpha \le C(\alpha) d(x, y)^\alpha .$$

LEMMA 8.4. *If $\ell(x)$ is a loop based at z, then $f(\ell(x)), g(\ell(x))$ are conjugate.*

We approximate $\ell(x)$ by the finite orbit $\phi^{-N_n}x, \ldots x, \ldots, \phi^{N_n}x = \ell_n(x)$, where n is large and therefore $d(\phi^{-N_n}x, \phi^{N_n}x)$ is small. This almost closed orbit can then be approximated by a genuinely closed orbit,

$$\ell_n(x) = \phi^{-N_n}w, \ldots, w, \ldots \phi^{N_n}w \quad (\phi^{N_n+1}w = \phi^{-N_n}w),$$

and therefore

$$f\ell_n(w) \equiv f(\phi^{-N_n}w)\cdots f(w)\cdots f(\phi^{N_n}w)$$

and

$$g\ell_n(w) \equiv g(\phi^{-N}w)\cdots g(w)\cdots g(\phi^{N}w)$$

are conjugate, say, $f\ell_n(w) = \lambda_n^{-1}g\ell_n(w)\lambda_n$. This means that $f\ell_n(x)$ and $g\ell_n(x)$ are *almost* conjugate, becoming closer to true conjugacy as $n \longrightarrow \infty$ and w chosen accordingly. Using compactness (and possibly a subsequence of$\{N_n\}$), we conclude that $f\ell(x)$ and $g\ell(x)$ are conjugate.

Now let $\ell(x_1),\ldots,\ell(x_k)$ be k loops based at z. We can approximate these by almost closed orbits $\ell_n(x_1),\ldots,\ell_n(x_k)$ as above. The lemmas can then be used to allow the introduction of an approximating closed orbit $\ell_1(w_1),\ell_2(w_2),\ldots,\ell_k(w_k)$ (these are *contiguous* segments of the closed orbit). As before we have that the sequences $f\ell(w_1)\cdots f\ell(w_k)$ and $g\ell(w_1)\cdots g\ell(w_k)$ are conjugate, from which one concludes that $f\ell_n(x_1)\cdots f\ell_n(x_k)$ and $g\ell_n(x_1)\cdots g\ell_n(x_k)$ are almost conjugate. As long as our limits are taken through our fixed sequence $\{N_n\}$, we conclude as before that

LEMMA 8.5. *If $\ell(x_1),\ldots,\ell(x_k)$ are k loops based at z, then $f\ell(x_1)\cdots f\ell(x_k)$ is conjugate to $g\ell(x_1)\cdots g\ell(x_k)$.*

Now using the compactness of G and the fact that for each $g_1,\ldots,g_m \in G$ there is a subsequence M_n such that $g_i^{M_n} \longrightarrow g_i^{-1}$ $(i = 1,\ldots,m)$, we arrive at the following result.

THEOREM 8.6. *For all $\ell = \ell_1^{\pm1}\ell_2^{\pm1}\cdots\ell_k^{\pm1} \in \Gamma$ we have that $f(\ell)$ and $g(\ell)$ are conjugate.*

The kernel of the map $f : \Gamma \longrightarrow G$ consists of those $\ell = (\ell_1^{\pm1}\ell_2^{\pm1}\cdots\ell_k^{\pm1})$ such that $f(\ell) = 1$, and since $f(\ell), g(\ell)$ are conjugate, this is the case if and only if $g(\ell) = 1$. Hence $\ker(f) \equiv \ker(g)$, and therefore $f(\Gamma), g(\Gamma)$ are isomorphic. Denote the corresponding isomorphism by α. Evidently, α is unambiguously defined by $\alpha(f(\ell)) = g(\ell)$, and it is immediate that α is an isometry and stabilises conjugacy classes. Hence the same is true of its extension, $\alpha : \overline{f(\Gamma)} \longrightarrow \overline{g(\Gamma)}$. When $\overline{f(\Gamma)} = G$, a simple exercise shows that α is surjective, i.e., $\alpha : G \longrightarrow G$ is an automorphism.

9. Conjugate weights theorem

We can now use the results from the previous section to address the questions posed at the beginning of that section.

THEOREM 9.1. *Let $\phi : X \longrightarrow X$ be hyperbolic and let $f, g \in C^\alpha(X, G)$ $(G \subset U(d))$ have conjugate weights for all periodic points. If ϕ_f is ergodic, then there exists an automorphism $\alpha : G \longrightarrow G$ (stabilising conjugacy classes) such that αf and g are cohomologous.*

PROOF. Since $f(\Gamma)$ is dense, α is defined on G. The fact that α is an isometry enables us to conclude that $\alpha : G \longrightarrow G$ is surjective. Clearly

$$\begin{aligned} f_-(x)f(x) &= f(z)^{-1}f(\phi x), \\ g_-(x)g(x) &= g(z)^{-1}g(\phi x) \quad (\text{for } x \in W(z)) \end{aligned}$$

and since α is defined on all of G,

$$\alpha g_-(x)\alpha g(x) = \alpha g(z)^{-1} \cdot \alpha g(\phi x)\,.$$

However, $f(z) = g(z) = \alpha f(z)$ and therefore

$$g(x)^{-1}g_-(x)^{-1}\alpha f_-(x)\alpha f(x) = g(\phi x)^{-1}\alpha f(\phi x).$$

Defining $h(x) = g_-(x)^{-1}\alpha f_-(x)$, we have

$$g(x)^{-1}h(x)\alpha(f(x)) = h(\phi x),$$

i.e., $\alpha(f(x)) = h(x)^{-1}g(x)h(\phi x)$ for $x \in W(z)$. Since $W(z)$ is dense in X (since one can easily see that by irreducibility homoclinic points on Σ are dense and then take the image under π), it suffices to prove that h defined on $W(z)$ is uniformly Hölder continuous.

Let $x, y \in W(z)$, $d(x,y) < \varepsilon$, $\varepsilon > 0$ and let w be the homoclinic point associated with $\{\phi^{-n}x\}$, $n \leq 0$ and $\{\phi^n y\}$, $n \geq 0$. Then

$$|h(x) - h(y)| \leq |h(x) - h(w)| + |h(w) - h(y)|.$$

Evidently

$$\begin{aligned} |h(x) - h(w)| &= |g_-(x)^{-1}\alpha f_-(x) - g_-(w)^{-1}\alpha g_-(w)| \\ &\leq |g_-(x) - g_-(w)| + |f_-(x) - f_-(w)| \end{aligned}$$

since α is an isometry. Furthermore,

$$\begin{aligned} |h(w) - h(y)| &= |g_-(w)^{-1}\alpha f_-(w) - g_-(y)^{-1}\alpha f_-(y)| \\ &= |g(w)g_+(w)\alpha f_+(w)^{-1}\alpha f(w)^{-1} - g(y)g_+(y)\alpha f_+(y)^{-1}\alpha f(y)^{-1}| \\ &\leq |g(w)g_+(w) - g(y)g_+(y)| + |f(w)f_+(w) - f(y)f_+(y)|. \end{aligned}$$

From these estimates it follows that

$$\begin{aligned} |h(x) - h(y)| &\leq \sum_{n=-\infty}^{-1} (|g(\phi^n z) - g(\phi^n y)| + |f(\phi^n x) + f(\phi^n w)|) \\ &+ \sum_{n=0}^{\infty} (|g(\phi^n w) - g(\phi^n y)| + |f(\phi^n w) - f(\phi^n y)|) \\ &\leq \sum_{n=-\infty}^{-1} (|g|_\alpha + |f|_\alpha)\, d(\phi^n x, \phi^n w)^\alpha + \sum_{n=0}^{\infty} (|g|_\alpha + |f|_\alpha) \\ &+ \sum_{n=0}^{\infty} (|g|_\alpha + |f|_\alpha)\, d(\phi^n y, \phi^n w)^\alpha \\ &\leq C(\alpha) d(x,y)^\alpha \end{aligned}$$

and therefore h is uniformly Hölder on $W(z)$. □

COROLLARY 9.2. *Under the above conditions there exists a compact Lie group $G' \supset G$ such that G'/G is finite and f, g are cohomologous with respect to G', i.e., there exists $h \in C^\alpha(x, G')$ such that $f = h^{-1}gh\phi$.*

The proof depends on the facts that

(1) α is an isometry,
(2) the group of automorphisms of a compact simple group modulo inner automorphisms is finite, and
(3) every compact Lie group is the quotient by a finite group of a direct product of a semi-simple Lie group with a torus.

Above we used the ergodic hypothesis to ensure that α was defined everywhere. In principle, this assumption is not needed, or rather it is only a possible topological obstruction which prevents a complete characterisation of the conjugate weights condition. More precisely, the proof in the general case would work if we could assume that an ergodic component had a trivial bundle structure, or equivalently that there was a Hölder cross-section to this component. This is the case, for example, if X is zero-dimensional so that we have

THEOREM 9.3. *If $\phi : X \longrightarrow X$ is hyperbolic, where X is zero-dimensional, and if $f, g \in C^\alpha(X, G)$ $(G \subset U(d))$ satisfy the conjugate weights condition for periodic orbits, then there exist a closed subgroup $H \subset G$, an isometric isomorphism $\alpha : H \longrightarrow \alpha(H)$ which stabilises conjugacy classes, and a function $f' \in C^\alpha(X, H)$ such that f, g are cohomologous to $f', \alpha(f')$, respectively.*

If we replace the conjugate weights condition by the much stronger assumption that f-weights and g-weights of closed orbits are identical, then the analysis is greatly simplified. No discussion of the group Γ is needed and in effect the isomorphism α is the identity.

THEOREM 9.4. *If $f, g \in C^\alpha(X, G)$ and $f(x) \cdots f(\phi^{n-1}x) = g(x) \cdots g(\phi^{n-1}x)$ whenever $\phi^n x = x$, then f and g are cohomologous.*

The proof follows the main lines of the proof of Theorem 9.1.

Our final remarks on the conjugate weights theorem concern shifts of finite type (i.e., the zero-dimensional case). If we suppose our functions f, g are locally constant, there is no harm in supposing that $f(x) = f(x_0, x_1)$, $g(x) = g(x_0, x_1)$ where $\{x_n\} \in X$. One can then show that if $h \in C^\alpha(X, G)$ satisfies

$$f(x) = h(x)^{-1} g(x) h(\phi x),$$

then h is a function of the single coordinate x_0, i.e., $f(x) = f(x_0)$.

The problems we have discussed so far now specialise to the context of graph theory, and Theorems 9.3 and 9.4 become (with much simpler proofs)

THEOREM 9.5. *Let f, g be two edge functions (with values in G) of a finite strongly connected oriented graph. Suppose that there is at most one edge leading from one vertex to another.*

(i) *If the products in G of the edge weights around each closed loop are conjugate, then there exist a closed subgroup H, an isomorphism $\alpha : H \to H$ which stablises conjugacy classes, an H-valued edge function f', and a vertex function h such that $f'(v_0, v_1) = h(v_0)^{-1} f(v_0, v_1) h(v_1)$ and $\alpha(f'(v_0, v_1)) = h(v_0)^{-1} g(v_0, v_1) h(v_1)$.*

(ii) *If the products in G of the edge weights around each closed loop agree, then there exists a vertex function h such that $f(v_0, v_1) = h(v_0)^{-1} g(v_0, v_1) h(v_1)$.*

10. Stable mixing

Our final problem concerns the question as to whether skew products which are 'near' to mixing skew products are themselves mixing. More precisely if $\phi : X \longrightarrow X$ is hyperbolic, m is a (Hölder) equilibrium state, and $f \in C^\alpha(X, G)$ $(G \subset U(d))$, we say that ϕ_f is *stably mixing* if there exists $\varepsilon > 0$ such that ϕ_g is mixing for all g with $\|g - f\| < \varepsilon$. Our aim is to prove that the set of mixing $f \in C^\alpha(X, G)$(i.e., ϕ_f is mixing) contains a dense open set when

(1) G is semi-simple or

(2) X is a connected hyperbolic attractor and G is a torus.

Here we shall consider the semi-simple case. In the next section we deal with the case where G is a torus and X is an attractor. Putting these two cases together, using basic Lie group structure theory, we can claim the following result.

THEOREM 10.1. *If X is a connected hyperbolic attractor and $G \subset U(d)$, then the set of $f \in C^\alpha(X, G)$ for which ϕ_f is mixing contains a dense open set.*

For the rest of this section we assume $G \subset U(d)$ is semi-simple. We shall need the following result due to Field (cf. [**F**]):

LEMMA 10.2. *The set $U = \{(g, h) \in G \times G : \langle g, h \rangle = G\}$ contains a dense open set, where $\langle g, h \rangle$ means the smallest closed subgroup of G which contains g and h.*

Using this, it is straightforward to prove that there exists a dense set of $f \in C^\alpha(X, G)$ for which ϕ_f is mixing. We choose $x, y \in W(z)$ whose orbits are distinct. For any $f \in C^\alpha(X, G)$ we consider the elements $f(\ell(x))$ and $f(\ell(y))$ of G. Together they may or may not generate G. However it is clear from the above lemma that we can make arbitrarily small perturbations of f in neighbourhoods of x and y so that for the resulting function g we have $\langle g\ell(x), g\ell(y) \rangle = G$. By Theorem 2.2 it follows that ϕ_g is mixing.

The following lemma will be used to prove that

$$\{f \in C^\alpha(X, G) : \langle f\ell(x), f\ell(y) \rangle = G\}$$

is open.

LEMMA 10.3. *The map $f \mapsto f\ell(x), C^\alpha(X, G) \longrightarrow G$ is continuous.*

PROOF. Since $f\ell(x) = f_-(x) f(x) f_+(x)$, it will suffice to prove $f \longrightarrow f_-(x)$ (and in the same way $f \longrightarrow f_+(x)$) is continuous. Let $f, g \in C^\alpha(X, G)$ and choose N so that $d(\phi^n x, z) \leq K\lambda^n$ $(0 < \lambda < 1)$ for $|n| \geq N$. Evidently

$$\begin{aligned}
& |f(z)^{-n} f_{-n}(x) - g(z)^{-n} g_{-n}(x)| \\
\leq\ & \left| f(z)^{-N} \left(f(z)^{-(n-N)} f(\phi^{-n}x) \cdots f(\phi^{-N-1}x) \right) f_{-N}(x) - f(z)^{-N} f_{-N}(x) \right| \\
+\ & \left| g(z)^{-N} \left(g(z)^{-(n-N)} g(\phi^{-n}x) \cdots g(\phi^{-N-1}x) \right) g_{-N}(x) - g(z)^{-N} g_{-N}(x) \right| \\
+\ & |f(z)^{-N} f_{-N}(x) - g(z)^{-N} g_{-N}(x)| \\
=\ & \left| f(\phi^{-n}x) \cdots f(\phi^{-(N-1)}x) - f(z)^{-(n-N)} \right| \\
+\ & \left| g(\phi^{-n}x) \cdots g(\phi^{-(N-1)}x) - g(z)^{-(n-N)} \right| \\
+\ & |f(z)^{-N} f_{-N}(x) - g(z)^{-N} g_{-N}(x)| \\
\leq\ & \sum_{i=-n}^{-(N-1)} (|f|_\alpha + |g|_\alpha)\, d(\phi^i x, z)^\alpha + |f(z)^{-N} f_{-N}(x) - g(z)^{-N} g_{-N}(x)| \\
\leq\ & K\,(|f|_\alpha + |g|_\alpha) \frac{\lambda^{\alpha N}}{(1 - \lambda^\alpha)} + |f(z)^{-N} f_{-N}(x) - g(z)^{-N} g_{-N}(x)|,
\end{aligned}$$

which is independent of $n > N$. Hence $|f\ell(x) - g\ell(x)|$ is dominated by the same expression. Given $\varepsilon > 0$, we can choose N so that the first term is less than $\varepsilon/2$, when $\|g - f\| < 1$ (say) and then we can choose δ so that $\|g - f\| < \delta$ implies that the second term is less than $\varepsilon/2$. □

If G is semi-simple, then each f for which $\langle f\ell(x), f\ell(y)\rangle = G$ gives rise to a mixing skew product ϕ_f. Hence

THEOREM 10.4. *When G is semi-simple, the set of $f \in C^\alpha(X,G)$ such that ϕ_f is mixing contains a dense open set.*

PROOF. We need only note that $\{f : \langle f\ell(x), f\ell(y)\rangle \in U\}$, where U is the open set specified in Lemma 10.2, is an open subset of $C^\alpha(X,G)$ since $f \longrightarrow \langle f\ell(x), f\ell(y)\rangle$, $C^\alpha(X,G) \longrightarrow G \times G$ is continuous. □

11. Stability of mixing — the torus case

Let $\phi : X \longrightarrow X$ be a hyperbolic map of a *connected* set, preserving the (Hölder) equilibrium state, m. Here we consider functions $f \in C^\alpha(X,G)$ where G is a k-dimensional torus. In fact we shall take $k = 2$ since proofs are essentially unaltered for the general case $k \geq 2$.

Consider the Bruschlinsky group $H^1(X,\mathbb{Z})$ of continuous maps of X to the circle K modulo null-homotopic maps. Since the latter form a divisible subgroup of $C(X,K)$, we can write $C^\alpha(X,K) = H \times H_0$ where now H_0 is the subgroup of C^α null-homotopic maps. Evidently

$$C^\alpha(X,G) = \bigcup_{\eta,\zeta} H(\eta,\zeta)$$

where

$$H(\eta,\zeta) = \{(f,g) \in C^\alpha(X,G) \times C^\alpha(X,G) : f \text{ is homotopic to } \eta \text{ and } g \text{ is homotopic to } \zeta\}.$$

Here we take $\eta, \zeta \in H$. We fix η,ζ and concentrate on $H(\eta,\zeta)$, which we note is a closed and open set.

For each $F \in C^\alpha(X,K)$ we write $F = \gamma_F \cdot e^{2\pi i r_F}$ where $\gamma_F \in H$, $r_F \in C^\alpha(X,\mathbb{R})$ and for each $\gamma \in H$ we write $\gamma \circ \phi = \phi^*\gamma e^{2\pi i s_\gamma}$ where $s_\gamma \in C^\alpha(X,\mathbb{R})$ and ϕ^* is the induced automorphism on H. Suppose $(f_n, g_n) \in H(\eta,\zeta)$ define non-mixing skew products $\phi_{f_n,g_n} : X \times K \times K \longrightarrow X \times K \times K$ and $(f_n,g_n) \longrightarrow (f,g)$. Then we have non-trivial equations

$$\begin{aligned} F_n \circ \phi / F_n &= e^{2\pi i a_n} \cdot f_n^{k_n} g_n^{l_n} \\ &= e^{2\pi i a_n (k_n u_n + l_n v_n)} \eta^{k_n} \zeta^{l_n} \end{aligned}$$

where $a_n \in \mathbb{R}$, $u_n, v_n \in C^\alpha(X,\mathbb{R})$, $(k_n,l_n) \neq (0,0)$, $f_n = \eta e^{2\pi i u_n}$, $g_n = \zeta e^{2\pi i v_n}$, $f = \eta e^{2\pi i u}$, $g = \zeta e^{2\pi i v}$ and $u_n \longrightarrow u$, $v_n \longrightarrow v$ $(u = r_f,\ v = r_g)$. Moreover,

$$\begin{aligned} F_n \circ \phi / F_n &= (\gamma_{F_n} \circ \phi / \gamma_{F_n}) e^{2\pi i (f_{F_n} \circ \phi - f_{F_n})} \\ &= (\phi^* \gamma_{F_n} / \gamma_{F_n}) e^{2\pi i (s_{\gamma_{F_n}} + f_{F_n} \circ \phi - r_{F_n})}. \end{aligned}$$

Thus $(\phi^*\gamma_{F_n}/\gamma_{F_n}) = \eta^{k_n}\zeta^{l_n}$ and $s_{\gamma_{F_n}} + r_{F_n} \circ \phi - r_{F_n} = a_n + k_n u_n + l_n v_n$ where a_n is a_n' modified by an integer. We note that $s : H \longrightarrow C^\alpha(X,\mathbb{R})/\mathbb{Z}$ is a homomorphism and we let V be the closed subspace of $C^\alpha(X,\mathbb{R})$ spanned by $\{s_\gamma : \gamma \in H\}$ and 1. Hence $k_n u_n + l_n v_n \in V \mod B^\alpha$ where B^α is the *closed* (by an application of Livsic's theorem; cf. [**PP1**]) subspace of $C^\alpha(X,\mathbb{R})$ consisting of coboundaries. When H has finite rank, V is finite dimensional and therefore $V + B^\alpha$ is closed. Clearly u, v are linearly dependent modulo $V + B^\alpha$.

THEOREM 11.1. *If H has finite rank, then $\phi_{f,g}$ is mixing if r_f, r_g are linearly independent modulo $V + B^\alpha$.*

COROLLARY 11.2. *$\{(f,g) \in C^\alpha(X, K \times K)\}$ contains an open dense set when H is of finite rank.*

PROOF. It is not difficult to see that $C^\alpha(X,\mathbb{R})/B^\alpha$ and therefore also $C^\alpha(X,\mathbb{R})/(B^\alpha + V)$ are finite dimensional. It follows that

$$\{(u,v) \in C^\alpha(X, \mathbb{R}\times\mathbb{R}) : u, v \text{ are linearly independent mod } B^\alpha + V\}$$

is open and dense, from which the corollary follows. □

In conclusion it should be noted that Theorem 11.1 applies to the case of a hyperbolic attractor:

PROPOSITION 11.3. *If X is a hyperbolic attractor, then H has finite rank.*

PROOF. Let $\gamma_i, \dots, \gamma_n$ be integrally independent members of H and let $U \supset X$ be a manifold with boundary such that $\phi U \subset$ interior of U and $\bigcap_{n=1}^\infty \phi^n U = X$. Each γ_i can be extended to a circle-valued function γ_i' defined on an open neighbourhood $V \supset X$; so choosing N large enough, we have $\phi^N U \subset V$ so that $\gamma_1', \dots, \gamma_n'$ are integrally independent when restricted to $\phi^N U$. Clearly $\operatorname{rank} H^1(U,\mathbb{Z}) = \operatorname{rank} H^1(\phi^N U, \mathbb{Z}) < \infty$ so that $n \le \operatorname{rank} H^1(U,\mathbb{Z})$, i.e., H has finite rank. □

A somewhat different approach to this and related problems (based on the geometric behaviour near homoclinic orbits) was introduced in [**FMT**]. This has the advantage of giving results for quite general attractors. Moreover, Pugh and Shub have studied stable ergodicity in the context of quite general partially hyperbolic systems [**PS**]. We shall briefly discuss both of these developments in the next section.

12. Related results

12.1. Stable ergodicity and partially hyperbolic systems. In this subsection, we shall consider diffeomorphisms that preserve the normalised Riemannian volume. In recent work Pugh and Shub studied stable ergodicity, in a more general sense, whereby any nearby volume-preserving diffeomorphism on the manifold is required to be ergodic. We shall briefly review the basic results. A very complete survey appears in [**BPSW**].

Let $\phi : M \longrightarrow M$ be a C^∞ diffeomorphism of a compact Riemannian manifold whose metric we denote by d. The diffeomorphism $\phi : M \longrightarrow M$ is said to be *partially hyperbolic* if the tangent bundle TM can be written as a direct sum $TM = E^u \oplus E^0 \oplus E^s$ (continuously split) such that

(1) $D\phi(E^u) = E^u$, $D\phi(E^0) = E^0$, and $D\phi(E^s) = E^s$;
(2) there exist constants $c > 1, \alpha < \beta < 1 < \gamma < \delta$ such that

$$\begin{aligned} \|D\phi^n v\| &\le c\alpha^n \|v\| \quad \text{for } v \in E^s, \\ \frac{1}{c}\delta^n \|v\| &\le \|D\phi^n v\| \quad \text{for } v \in E^u, \\ \frac{1}{c}\beta^n \|v\| &\le \|D\phi^n v\| \le c\gamma^n \|v\| \quad \text{for } v \in E^0 \end{aligned}$$

for $n \ge 0$.

This is a natural generalization of the case of skew products, where ϕ is no longer necessarily an isometry in the direction of the bundle E^0, but instead the distortion associated to $D\phi|E^0$ is dominated by the contraction and expansion in the stable and unstable directions, respectively.

Standard examples of partially hyperbolic diffeomorphisms include not only the skew products we have focused on, but also other familiar examples such as quasi-hyperbolic toral automorphisms, time-one geodesic flows and frame flows.

Recall that we can associate to the bundles E^s, E^u stable and unstable manifolds defined by $W^s(x) = \{y \in M : d(\phi^n x, \phi^n y) \to 0, \text{ as } n \to 0\}$, $W^u(x) = \{y \in M : d(\phi^{-n} x, \phi^{-n} y) \to 0, \text{ as } n \to 0\}$, for each $x \in M$. Following Pugh and Shub, we say that the diffeomorphism ϕ has the *accessibility property* if for any points $x, y \in M$ we can find a path connecting x and y and lying in the union of finitely many pieces of stable and unstable manifolds.

Let us make the following assumptions on ϕ.

(1) $\|D_x f|E_x^0\|.\|(D_x f|E_x^0)^{-1}\|$ is sufficiently close to 1 *(center bunching)*.
(2) The distributions E^0, $E^0 \oplus E^u$ and $E^0 \oplus E^s$ are all integrable and tangent to the associated foliations *(dynamically coherent)*.
(3) Any diffeomorphism sufficiently close to ϕ in the C^2 topology has the accessibility property *(stable accessibility property)*.

Assume that ϕ preserves the normalised volume μ on M. In this more general context, we say that ϕ is *stably ergodic* (in the general sense) if any sufficiently C^2-close volume-preserving diffeomorphism ϕ' to ϕ is also necessarily ergodic. The following remarkable result was shown by Pugh and Shub.

THEOREM 12.1. *Assume that $\phi : M \to M$ is a volume-preserving partially hyperbolic diffeomorphism satisfying hypotheses (1), (2) and (3). Then ϕ is stably ergodic (in the general sense).*

The proof of this theorem is based on a delicate version of the classical Hopf argument. The basic idea is to consider the forward (or backward) Birkhoff averages of continuous functions along orbits, which exist almost everywhere by the pointwise ergodic theorem. Observe that these are equal for points lying on the same stable (or unstable) manifolds. The aim is to show that these averages exist and are equal to a constant value for almost every pair of points $x, y \in M$ by linking them by suitable paths lying in stable and unstable manifolds (using the accessibility property). One can then deduce that the Birkhoff averages are equal for almost every point and thus conclude μ is ergodic. However, the main technical problem is that the continuous foliation corresponding to E^0 is not necessarily absolutely continuous (an essential ingredient in the original Hopf argument). One of the key ideas of Pugh and Shub was to replace absolute continuity by the more general concept of *julienne quasi-conformality* along stable (and unstable) leaves [**BPSW**], which suffices for the Hopf argument and yet holds even for ϕ' in a small C^2 neighbourhood of ϕ.

In practice, there are effective estimates on the approximation required in hypothesis (1). Moreover, one immediately sees that centre bunching holds automatically if $D\phi$ is an isometry on E^0. For example, in the particular case that ϕ is a compact group extension of an Anosov diffeomorphism, we see that it is partially hyperbolic and hypotheses (1) and (2) hold. In this context, Burns and Wilkinson used Theorem 12.1 to prove the following interesting result for skew products [**BW**].

THEOREM 12.2. *Assume ϕ is a skew product with respect to a compact semi-simple Lie group and a volume-preserving Anosov diffeomorphism. Then ϕ is stably ergodic (in the general sense) if and only if ϕ is ergodic.*

Although such results show ergodicity is stable in a much broader sense than we considered earlier, it does require more restrictive hypotheses. Firstly, it requires diffeomorphisms to be close in a stronger sense (i.e., in the C^2 topology rather than the Hölder topology). Secondly, it applies only to the normalized volume, rather than more general equilibrium measures.

12.2. Skew products and homoclinic points. Recently, Field, Melbourne and Török developed another approach to showing stability of mixing for skew products over hyperbolic sets X [**FMT**]. In particular, their approach has the distinct advantage that it works without requiring X to be a connected attractor. Furthermore, they are also able to consider the case of C^r skewing functions.

Let $\phi : M \to M$ be a C^2 diffeomorphism and let G be a compact Lie group. Let $X \subset M$ be a ϕ-invariant hyperbolic set and let m be an equilibrium measure.

THEOREM 12.3. *Let $r > 0$. There is an open and dense set of C^r functions $f : M \to \mathbb{R}$ such that the associated skew product $\phi_f : X \times G \to X \times G$ is ergodic (and mixing).*

We shall briefly explain the main ideas in the proof. Let us assume for definiteness that $G = K$ (the circle) [**FP**]. The starting point is the analogue of the results we described in Section 11. Let us write $f = e^{2\pi i F}$, say. In particular, ϕ_f is ergodic provided $F \notin \overline{V + B}$ where, in the present setting, B is the space of C^r boundaries and V is the space of locally constant functions; and we take the closure in the C^r topology.

A key ingredient in the proof is to show good asymptotic bounds on the average of functions along a sequence of periodic points approximating a homoclinic point. More precisely, consider $F \in C^r(M, \mathbb{R})$. Let p_0 denote a periodic orbit (which, for simplicity, we may assume to be a fixed point). We can also assume, for convenience, that $F(p_0) = 0$. Assume that $x \in W^s(p_0) \cap W^u(p_0) - \{p_0\}$ is an associated (transverse) homoclinic point. In particular, since we have assumed $F(p) = 0$, and F is Lipschitz, the summation of F along the orbit of x, which we can denote by $F^\infty(x) := \sum_{i=-\infty}^{\infty} F(\phi^i x)$, is finite. One can choose a sequence of periodic orbits p_N $(N \geq 1)$ with periods q_N (i.e., $\phi^{q_N}(p_N) = p_N$), say , such that $p_N \to x$ as $N \to +\infty$. In particular, the difference

$$F^\infty(x) - F^{q_N}(p_N)$$

can be made arbitrarily small, by choosing N sufficiently large. If we assume that $F \in \overline{V + B}$, then for N sufficiently large we also know that $F^{q_N}(p_N)$ vanishes. In particular, this implies that $F^\infty(x) = 0$. Using very detailed estimates on the behaviour of orbits, Field, Melbourne and Török show that for generic F (in the C^r topology) one has that $F^\infty(x) \neq 0$. Thus, for such F we deduce that $F \notin \overline{V + B}$ and, in particular, ϕ_f is ergodic.

12.3. Different function spaces. A central idea in these notes has been the regularisation of functions using transfer operators. If we start in the context of hyperbolic maps, this involves using Markov partitions (and Bowen's theorem) to first introduce a two-sided subshift of finite type and then a one-sided version. However, a recent idea of Goetszel and Liverani is suggestive of an alternative approach, which we shall briefly explore [**GL**]. Let $\phi : M \to M$ be a transitive Anosov diffeomorphism on a compact manifold which preserves the normalised Riemannian volume. Consider the operator $U_\phi : L^2(\mu) \to L^2(\mu)$ given by $U_\phi w = |Jac(\phi)| w \circ \phi$.

THEOREM 12.4 ([**GL**]). *There exists an invariant Banach space (of Schwartzian distributions B) for which the operator $U_\phi : B \to B$ is quasi-compact.*

REMARK 12.5. The Banach space B is quite complicated to describe, but its elements are chosen to exploit the hyperbolicity of ϕ. A function in $B \cap L^2(\mu)$ has more regularity than a typical element of $L^2(\mu)$. Let W denote a piece of smooth manifold with the same dimension as, and which lies at a small angle θ, say, to, the unstable manifolds. Consider functions $f : M \to \mathbb{R}$ which are smooth, and consider all derivatives, up to order k, say, when restricted to W. We can then associate linear functionals on $C^l(M)$, for $l \leq k-1$, say, by taking the inner product with these derivatives. Finally, B is the completion of the smooth functions with respect to norm given by the supremum over f in the unit ball of $C^k(W)$ of linear functionals.

The analogue of Theorem 5.5 is the following.

THEOREM 12.6. *The space B decomposes as $B = V_0 + V_1$ where $U_\phi : V_1 \longrightarrow V_1$, $L_f : V_0 \longrightarrow V_0$ and*

(1) $\rho(U_\phi\,|_{V_0}) < 1$,
(2) $\dim(V_1) \leq d$,
(3) *V_1 is spanned by (generalized) eigenvectors corresponding to eigenvalues of modulus* 1.

This suggests the possibility of repeating the proof of Theorem 6.1 to show that any measurable solution to certain coboundary identities must also be in B. More precisely, assume that

$$w \circ \phi = |\mathrm{Jac}(\phi)| w.$$

THEOREM 12.7. *Let $U_\phi w = w$ a.e., where $w \in L^2(\mu, \mathbb{C})$. Then there exists $w_1 \in B$ such that $w = w_1$ a.e.*

PROOF. Let $L^2 = V_1^\perp \oplus V_1$. Then $V_1^\perp \supset V_0$ and $B \cap L^2(\mu) = V_0 \oplus V_1$. Choose $\varepsilon > 0$ and $v \in B \cap L^2(\mu)$ such that $\int |v-w|^2\, d\mu < \varepsilon^2$. Then $v = v_0 + v_1$, $w = w_0 + w_1$ where $v_0 \in V_0$, $v_1 \in V_1$, $w_0 \in V_1^\perp$, $w_1 \in V_1$. Thus

$$\varepsilon^2 \geq \int |v-w|^2\, d\mu = \int |v_0 - w_0|^2\, d\mu + \int |v_1 - w_1|^2\, d\mu$$

and hence $\int |v_0 - w_0|^2\, d\mu \leq (\int |v_0 - w_0|^2\, d\mu)^{1/2} \leq \varepsilon$. Furthermore $U_\phi V_1^\perp \subset V_1^\perp$. To see this, suppose $\langle u, u_1 \rangle = 0$ whenever $U_\phi u_1 = \alpha u_1$ ($|\alpha| = 1$) or equivalently $f u_1 = \alpha u_1 \circ \sigma$. Then $\langle u, \alpha f^{-1} u_1 \circ \sigma \rangle = 0$ which implies that $\langle U_\phi u, \alpha u_1 \rangle = 0$. This shows that $U_\phi u \perp V_1$ when $u \perp V_1$.

Hence $\int |U_\phi^n v_0 - U_\phi^n w_0|\, d\mu \leq \int |v_0 - w_0|\, d\mu \leq \varepsilon$ and since $U_\phi^n v_0 \longrightarrow 0$, we have $\lim_{n\to\infty} \int |U_\phi^n w_0|\, d\mu \leq \varepsilon$, i.e., $w_0 = 0$ a.e. and $w = w_1$ a.e. □

The above identity sometimes occurs quite naturally. For example, if an Anosov diffeomorphism ϕ preserves an absolutely continuous invariant measure, then the density $w \in L^1(\mu)$ satisfies this type of identity. (In the particular case that ϕ preserves the volume, then, of course, we have the trivial solution $w = 1$.) However, in this case, de la Llave, Marco and Moriyan have used Sobolev regularity techniques to show that if ϕ is smooth, then so is the density w.

13. Non-uniformly hyperbolic systems

One can ask if there are Livsic theorems under weaker conditions on ϕ than hyperbolicity. One approach is to consider the broader class of *non-uniformly hyperbolic systems*. Here, the hyperbolicity is typically restricted to a set of full measure, where it is characterized by non-zero Lyapunov exponents. One approach, which has been successful in particular settings, is to replace the subshift of finite type by a *tower*, in the sense of L.-S. Young [**Y**]. This allows the system to be modelled by a countable state subshift. This approach to Livsic theorems was considered by Bruin, Nicol and Holland [**BNH**]. We shall describe an approach which is closer in spirit to the original work of Livsic for hyperbolic systems (cf. also [**Ll**]).

Let $\phi : M \to M$ be a C^∞ diffeomorphism of a compact Riemannian manifold. Assume that

(1) ϕ preserves the normalised volume μ,
(2) the Lyapunov exponents of μ are non-zero.

In the particular case that M is a surface and $h(\mu) > 0$, condition (2) holds automatically.

Let G be a compact Lie group. Given $f \in C^\alpha(M, G)$, consider the coboundary identity

$$f = h^{-1} h \circ \phi \text{ a.e.},$$

where $h : M \to \mathbb{R}$ is measurable. Observe that by Luzin's theorem, for any $\epsilon > 0$ one can find a (compact) set $X \subset M$ with $\mu(M - X) < \epsilon$ and such that $h|X$ is continuous. The following theorem improves this to a Hölder bound.

THEOREM 13.1. *For any $\epsilon > 0$ one can find a set $Y \subset M$ with $\mu(M - Y) < \epsilon$ and $h|Y$ is Hölder continuous.*

When we have non-zero Lyapunov exponents, many of the properties familiar in the hyperbolic theory remain true in a weaker form.

The proof of Theorem 13.1 is based on the following three lemmas. The first is due to Pesin [**Pe**].

LEMMA 13.2. *There exists a family $\Lambda_1 \subset \Lambda_2 \subset \cdots M$ of compact sets such that*

(1) *$\mu(\Lambda_n) \nearrow 1$ as $n \to +\infty$;*
(2) *for each $n \geq 1$, there exists $\delta_n > 0$ such that for $x \in \Lambda_n$ the local stable manifold $W^s_{\text{loc}}(x) = \{y \in M : d(\phi^n x, \phi^n y) \leq \delta_n, n \geq 0\}$ and the local unstable manifold $W^u_{\text{loc}}(x) = \{y \in M : d(\phi^{-n} x, \phi^{-n} y) \leq \delta_n, n \geq 0\}$ are embedded disks of uniform size which meet transversely at x;*
(3) *there is a C^α dependence $\Lambda_n \ni x \mapsto W^s_{\text{loc}}(x)$ and $\Lambda_n \ni x \mapsto W^u_{\text{loc}}(x)$;*
(4) *the laminations $\{W^u_{\text{loc}}(x)\}_{x\in\Lambda_n}$ and $\{W^s_{\text{loc}}(x)\}_{x\in\Lambda_n}$ are both absolutely continuous.*

This lemma naturally leads to non-uniform "local product structure". More precisely, for $n \geq 1$ there exists $\epsilon_n > 0$ such that for $x, y \in \Lambda_n$ with $d(x, y) \leq \epsilon_n$

the local stable and unstable manifolds intersect at a single point $z = z(x,y)$, say, i.e., $W^s_{\mathrm{loc}}(x) \cap W^u_{\mathrm{loc}}(y) = \{z\}$.

Let us assume $\epsilon < \frac{1}{4}$. By part (1) of Lemma 13.2, we can choose n sufficiently large that $\mu(\Lambda_n) \geq 1 - \epsilon$. In particular, if we write $X_0 = X \cap \Lambda_n$, then we have $\mu(X_0) \geq 1 - 2\epsilon > \frac{1}{2}$, say.

LEMMA 13.3.

(1) *There exists $K > 0$ such that for any $x, y \in X_0$ one has $d(x,y) \leq Kd(x,z)$ and $d(x,y) \leq Kd(x,y)$.*
(2) *For a.e. $x \in X_0$, we have that*

$$\lim_{n\to\infty} \frac{1}{n} \sum_{i=1}^{n} \chi_{X_0}(\phi^i x) = \mu(X_0) > \frac{1}{2}.$$

Moreover, the same estimate holds for a.e. $y \in W^s_{\mathrm{loc}}(x)$ and a.e. $z \in W^u_{\mathrm{loc}}(x)$

PROOF. This follows immediately from the Birkhoff ergodic theorem. $\square$

Without loss of generality, we can replace X_0 by its density points (for the measure μ). The following lemma is essentially due to Livsic [**Li**].

LEMMA 13.4.

(1) *For a.e. $x, y \in X_0$ with $d(x,y) < \epsilon_n$, say, one can find x_1, x_2, x_3 such that $x_1 \in W^u_{\mathrm{loc}}(x)$, $x_2 \in W^s_{\mathrm{loc}}(x_1) \cap X_0$ and $x_3 \in W^u_{\mathrm{loc}}(x_2) \cap W^s_{\mathrm{loc}}(y)$.*
(2) *There exists $K > 0$ such that for any $K\ d(x,y) \geq d(x,x_1) + d(x_1,x_2) + d(x_2,x_3) + d(x_3,y)$.*

We can now modify the standard argument of Livsic ([**Li**], p. 1299) to prove Theorem 13.1. Let $x \in X_0 \cap W^s_{\mathrm{loc}}(y)$, say. For any $n \geq 1$ there exist $D > 0$ and $0 < \theta < 1$ such that $d(\phi^n x, \phi^n y) \leq D\theta^n d(x,z)$. For any $n \geq 1$ we can write

$$\begin{aligned}
|h(x) - h(y)| &= |f(x)^{-1} \ldots f(\phi^{n-1}x)^{-1} h(\phi^n x) - f(y)^{-1} \ldots f(\phi^{n-1}y)^{-1} h(\phi^n y)| \\
&\leq |[f(x)^{-1} - f(y)^{-1}] f(\phi x)^{-1} \ldots f(\phi^{n-1}x)^{-1} h(\phi^n x)| \\
&+ |f(y)^{-1}[f(\phi x)^{-1} - f(\phi y)^{-1}] f(\phi^2 x)^{-1} \ldots f(\phi^{n-1}x)^{-1} h(\phi^n x)| \\
&\ldots \\
&+ |f(y)^{-1} \cdots f(\phi^{n-2}y)^{-1}[f(\phi^{n-1}x)^{-1} - f(\phi^{n-1}y)^{-1}] h(\phi^n x)| \\
&+ |f(y)^{-1} \cdots f(\phi^{n-1}y)^{-1}[h(\phi^n x) - h(\phi^n y)]| \\
&\leq Cd(x,y)^\alpha + |h(\phi^n x) - h(\phi^n y)|
\end{aligned}$$

where $C = |f^{-1}|_\alpha D^\alpha/(1-\theta^\alpha)$. By Lemma 13.3 we can find a subsequence n_k such that $\phi^{n_k}(x), \phi^{n_k}(y) \in X_0$. Thus, letting $k \to +\infty$, we see that

$$|h(x) - h(y)| \leq Cd(x,y)^\alpha .$$

Similarly, if $x \in X_0 \cap W^u_{\mathrm{loc}}(y)$, then we can show that $|h(x) - h(y)| \leq Cd(x,y)^\alpha$. Finally, by Lemma 13.4, for $x, y \in X_0$ we can associate x_1, x_2, x_3 and apply the above bounds to write

$$\begin{aligned} & |h(x) - h(y)| \\ \leq \; & |h(x) - h(x_1)| + |h(x_1) - h(x_2)| + |h(x_2) - h(x_3)| + |h(x_3) - h(y)| \\ \leq \; & Cd(x, x_1)^\alpha + Cd(x_1, x_2)^\alpha + Cd(x_2, x_3)^\alpha + d(x_3, y)^\alpha \\ \leq \; & 4CKd(x, y) . \end{aligned}$$

This completes the proof of Theorem 13.1.

14. Some open questions

Despite progress in recent years (or, perhaps, even because of it), there remain a number of interesting open questions and conjectures. In this final section we shall briefly recall a few of these.

QUESTION 14.1. *When can we replace the compact Lie group G by other non-compact groups in the stable ergodicity results?*

In the simplest case $G = \mathbb{R}^d$, Nitica and Pollicott gave conditions (which are open, but far from dense) on the skewing function $f : X \to \mathbb{R}^d$ to make the skew product ϕ_f transitive [**NP**]. This work has been extended in [**FMT**]. On the other hand, a natural candidate as a generalization of ergodicity is *exactness*, which has been studied in the context of $\mathbb{R}^d$ extensions of the one-sided shift $\sigma : \Sigma^+ \to \Sigma^+$ by Guivarc'h and Aaronson and Denker.

Another direction in which one could try to extend the stable ergodicity results is by generalizing the class of base transformations $\phi : M \to M$. For example, an obvious question is the following.

QUESTION 14.2. *If we replace the base transformation $\phi : X \to X$ by a partially hyperbolic diffeomorphism, what are the analogous results on stable ergodicity for skew products?*

There has been work on Livsic type theorems for partially hyperbolic maps by Katok and Kononenko which may have a bearing on this question.

The results of Pugh and Shub depend on various technical hypotheses that may eventually prove to be unnecessary. Indeed, they have formulated the following natural conjecture [**BPSW**].

CONJECTURE 14.3. *On any compact manifold M, ergodicity holds for an open and dense set of C^2 volume-preserving hyperbolic diffeomorphisms.*

There are partial results in this direction in low dimensions.

In [**FMT**], Field, Melbourne and Török show that for a C^2 hyperbolic diffeomorphism ϕ and an equilibrium state the family

$$\{f \in C^r(M, G) : \phi_f \text{ is ergodic}\}$$

is open in the C^2 topology. In [**FMT**] the following question is also posed:

QUESTION 14.4. *Do the corresponding results hold in the case that $f \in C^s$, for $1 \leq s < 2$?*

A closely related concept to stable ergodicity is that of *robust transitivity.* We say that a diffeomorphism $\phi : M \to M$ is robustly transitive if any sufficiently close $\phi' : M \to M$ (in a suitable topology) is also transitive. (Although transitivity is a weaker property than ergodicity, we no longer restrict to volume-preserving diffeomorphisms). This property has been studied by a number of people (most notably by Bonatti and Diaz), but there remain a number of natural open problems.

QUESTION 14.5. *When is a partially hyperbolic diffeomorphism robustly transitive? For example, are time-one geodesic flows and frame flows robustly transitive? When are skew products robustly transitive?*

Finally, we conclude with a very well-known conjecture which has been open for several decades but whose positive solutions would have a strong bearing on the problems of stable ergodicity.

CONJECTURE 14.6. *Let $\phi : M \to M$ be a transitive Anosov diffeomorphism on a compact manifold. The linear action $\phi_* : H_1(M,\mathbb{R}) \to H_1(M,\mathbb{R})$ on the first real homology does not have* 1 *as an eigenvalue.*

An equivalent formulation of this conjecture is that any (Hölder) continuous function $f : M \to \mathbb{R}$ which sums to integer values around closed orbits must necessarily be cohomologous to a constant.

References

[AKS] R. Adler, B. Kitchens, M. Shub, *Stably ergodic skew products,* Discrete Contin. Dynam. Systems **2** (1996), 349–350.

[AW] R. Adler, B. Weiss, *Similarity of automorphisms of the torus,* Memoirs Amer. Math. Soc., **98**, Amer. Math. Soc., Providence, RI, 1970.

[BNH] H. Bruin, M. Nicol, M. Holland, *Livsic regularity for Markov systems,* preprint.

[Bo1] R. Bowen, *Markov partitions for Axiom A diffeomorphisms,* Memoirs Amer. Math. Soc. **92**, Amer. Math. Soc., Providence, 1970.

[Bo2] R. Bowen, *Equilibrium states and the ergodic theory of Anosov diffeomorphisms,* Springer Lecture Notes Math. **470**, Springer, Boston, 1975.

[Bo3] R. Bowen, One-dimensional hyperbolic sets for flows, *J. Differential Eq.* **12** (1972), 173–179.

[BPSW] K. Burns, C. Pugh, M. Shub, A. Wilkinson, *Recent results about stable ergodicity,* to appear.

[Br] M. Brin, *Ergodicity of frame flows,* Ergodic theory and dynamical systems, II (A. Katok, ed.), Progr. Math. 21, Birkhäuser, Boston, 1982, pp. 163–183.

[BW] K. Burns, A. Wilkinson, *Stable ergodicity of skew products,* Ann. Sci. Éc. Norm. Supér. **32** (1999), 859–889.

[D] D. Dolgopyat, *On the mixing properties of compact group extensions of hyperbolic systems,* Israel J. Math. **130** (2002), 157–205.

[F] M. Field, *Generating sets for compact semi-simple Lie groups,* Proc. Amer. Math. Soc. **127** (1999), 3361–3365.

[FMT] M. Field, I. Melboune, A. Török, *Stable ergodicity for smoothcompact Lie group extensions of compact hyperbolic basic sets,* Ergod. Th. Dynam. Sys., to appear.

[FP] M. Field, W. Parry, *Stable ergodicity of skew extensions by compact Lie groups,* Topology **38** (1999), 167–187.

[GL] S. Goetszel, C. Liverani, preprint.

[HPS] M. Hirsch, C. Pugh, M. Shub, *Invariant Manifolds,* Springer Lecture Notes Math. **583**, Springer, Boston, 1977.

[KN] H. Keynes, D. Newton, *Ergodic measures for non-abelian compact group extensions,* Compositio Math. **32** (1976), 1278–1301.

[Li] A. Livsic, *The cohomology of dynamical systems,* Math. USSR – Izvestia **6** (1972), 1278–1301.

[Ll] R. de la Llave, personal communication.

[NP] V. Nitica, M. Pollicott, *Transitivity of Euclidean extensions of Anosov diffeomorphisms,* Ergod. Th. Dynam. Sys., to appear.

[Pa1] W. Parry, *Skew products of shifts with a compact Lie group,* J. London Math. Soc. **56** (1997), 395–404.

[Pa2] W. Parry, *The Livsic periodic point theorem for non-abelian cocycles,* Ergod. Th. Dynam. Sys. **19** (1999), 687–701.

[Pe] Y. Pesin, *Lyapunov characteristic exponents and smooth ergodic theory,* Uspehi Mat. Nauk. **32** (1977), 55–112.

[PP1] W. Parry, M. Pollicott, *Stablility of mixing for toral extensions of hyperbolic systems,* Proc. Steklov Inst. Math. **216** (1997), 350–359.

[PP2] W. Parry, M. Pollicott, *Zeta functions and the closed orbit structure of hyperbolic systems,* Astérisque 187-188, S.M.F, 1990.

[PS] C. Pugh, M. Shub, *Stable ergodicity,* Bull. Amer. Math. Soc. **41** (2004), 1–41.

[R] D. Ruelle, *Thermodynamic Formalism,* Addison-Wesley, London, 1978.

[Si1] Y. Sinai, *Construction of Markov partitions,* Func. Anal. Appl. **2** (1968), 39–52.

[Si2] Y. Sinai, *Gibbs measures in ergodic theory,* Russ. Math. Surveys **27** (1972) 21–69.

[Sm] S. Smale, *Differentiable dynamical systems,* Bull. Amer. Math. Soc. **73** (1967), 747–817.

[W] R. Williams, *Expanding attractors,* Publ. Math. IHES **43** (1974), 169–203.

[Y] L.-S. Young, *Recurrence times and rates of mixing,* Israel J. Math. **110** (1999), 153–188.

Department of Mathematics, University of Warwick, Coventry, CV4 7AL, United Kingdom

Department of Mathematics, University of Warwick, Coventry, CV4 7AL, United Kingdom

E-mail address: mpollic@maths.warwick.ac.uk

Amer. Math. Soc. Transl.
(2) Vol. **217**, 2006

Distribution of Ergodic Sums for Hyperbolic Maps

Mark Pollicott and Richard Sharp

ABSTRACT. In this paper we study statistical properties of hyperbolic maps. In particular, we estimate how sums of functions along orbits are distributed relative to intervals which shrink in size.

Introduction

In this article we shall study statistical properties for the orbits of dynamical systems. Given any measurable map $T : X \to X$ and ergodic probability measure m, we can consider an integrable function $f : X \to \mathbb{R}$ such that $\int f dm = 0$. Let $f^n(x) := f(x) + f(Tx) + \cdots + f(T^{n-1}x)$ denote the sum along the first n points in the orbit of $x \in X$. The Birkhoff Ergodic Theorem implies that $f^n(x)/n \to 0$, as $n \to +\infty$, for almost every $x \in X$ with respect to m. An important problem in ergodic theory is to obtain a more detailed understanding of such ergodic sums $f^n(x)$ and, in particular, the fluctuations from their mean behaviour.

To get interesting results, we need to consider a more restricted class of systems. In particular, we shall study the important class of (mixing) hyperbolic diffeomorphisms and expanding maps $T : X \to X$, where X is a compact subset of a Riemannian manifold M. Let m be a Gibbs state (for a Hölder continuous function $g : X \to \mathbb{R}$) and let $f : X \to \mathbb{R}$ be a Hölder continuous function for which the variance

$$\sigma^2(f) := \lim_{n\to+\infty} \frac{1}{n} \int (f^n)^2 dm$$

is non-zero. In this case, the sums $f^n(x)$ satisfy the stronger Central Limit Theorem [**Ra, CP**], i.e., for any real numbers $a < b$, we have that

$$\lim_{n\to+\infty} m\left\{x \in X : a \leq \frac{f^n(x)}{\sqrt{n}} \leq b\right\} = \frac{1}{\sqrt{2\pi}\sigma} \int_a^b e^{-t^2/2\sigma^2} dt.$$

Moreover, the sums also satisfy the weak invariance principle and the law of the iterated logarithm, both of which are consequences of a more general almost sure invariance principle in [**DP**].

2000 *Mathematics Subject Classification.* Primary 37D20, 37A30, 37D35, 37C30.

Key words and phrases. Hyperbolic systems, ergodicity, thermodynamic formalism, local limit theorems.

The second author was supported by an EPSRC Advanced Research Fellowship.

For non-lattice functions, Lalley showed the following Local Limit Theorem: for any real numbers $a < b$, we have that

$$m\{x \in X : a \leq f^n(x) \leq b\} \sim \frac{(b-a)}{\sqrt{2\pi}\sigma} \frac{1}{\sqrt{n}}, \qquad (0.1)$$

as $n \to +\infty$ [**L**]. (Here we have used the notation $A(n) \sim B(n)$, as $n \to +\infty$, if $\lim_{n\to+\infty} A(n)/B(n) = 1$.) Related results were proved by Guivarc'h and Hardy.

In this paper we strengthen (0.1) to give an asymptotic formula when the interval $[a, b]$ is allowed to shrink, at a suitably slow rate, as n increases. Towards this end, we need to impose modest additional restrictions on the function $f : X \to \mathbb{R}$.

DEFINITION. We say that the function $f : X \to \mathbb{R}$ is *Diophantine* if we can find periodic orbits $T^{n_1}(x_1) = x_1$, $T^{n_2}(x_2) = x_2$ and $T^{n_3}(x_3) = x_3$ such that

$$\alpha = \frac{f^{n_2}(x_2) - f^{n_1}(x_1)}{f^{n_3}(x_3) - f^{n_1}(x_1)}$$

is a Diophantine number (i.e., there exists $C > 0$ and $\gamma > 2$ such that $|\alpha - p/q| \geq C/q^\gamma$, for all $p, q \in \mathbb{N}$).

THEOREM 1. *Let $T : X \to X$ be a C^1 expanding map and let m be the Gibbs state for a Hölder continuous function. Suppose that $f : X \to \mathbb{R}$ is a Hölder continuous function satisfying the Diophantine condition and such that $\int f dm = 0$. Then there exists $\delta > 0$ such that for any $z \in \mathbb{R}$, $a < b$ and sequence $\epsilon_n > 0$ which tends to zero and satisfies $\epsilon_n^{-1} = O(n^\delta)$, we have that*

$$m\{x \in X : \ z + \epsilon_n a \leq f^n(x) \leq z + \epsilon_n b\} \sim \frac{(b-a)}{\sqrt{2\pi}\sigma} \frac{\epsilon_n}{\sqrt{n}},$$

as $n \to +\infty$.

REMARK. It is easy to see that typical functions satisfy the Diophantine condition. Indeed, the Diophantine condition holds for generic functions in various topologies [**PP, FMT**].

Theorem 1 also holds if expanding maps are replaced by one-sided subshifts of finite type. This will be apparent from the proof. We shall also show analogous results for Axiom A diffeomorphisms restricted to a basic set (Theorem 2) and periodic orbits (Theorem 3).

In Section 1 we shall recall some preliminary results. In Section 2, we present the proof of Theorem 1. In Section 3 we extend these results to Axiom A diffeomorphisms. Finally, in Section 4 we will present an analogous result for periodic points.

1. Preliminaries

Let M be a compact connected smooth Riemannian manifold and suppose that $X \subset U \subset M$ with X compact and U open. Let $T : U \to M$ be a C^1 map. Suppose that there exists $\lambda > 1$ such that $||DT_x(v)|| \geq \lambda ||v||$ for all $x \in U$ and all $v \in T_x M$ and that $X = \bigcap_{n\geq 0} T^{-n} U$. We shall then refer to $T : X \to X$ as an *expanding map*. In the special case where $X = U = M$, we shall call T an expanding endomorphism of M; in this case T is topologically conjugate to an expanding endomorphism of an infranilmanifold. In addition, we shall suppose that $T : X \to X$ is topologically mixing.

We recall that two continuous functions $f, f' : X \to \mathbb{R}$ are cohomologous if $f - f' = h \circ T - h$, for some continuous function $h : X \to \mathbb{R}$. By Livsic's Theorem, this is equivalent to the statement that $f^n(x) = (f')^n(x)$ whenever $T^n x = x$ is a periodic point. In particular, the assumption that $\sigma^2 > 0$ is equivalent to the statement that f is not cohomologous to a constant. We say that a function f has integer periods if $\{f^n(x) : T^n x = x, n \geq 1\} \subset \mathbb{Z}$. We recall that f is called a *non-lattice* function if one has the stronger condition that f is not cohomologous to a function of the form $a + b\psi$, where $a, b \in \mathbb{R}$ and ψ is a function with integer periods. If f satisfies the Diophantine condition, then f is a non-lattice function and, in particular, f is not cohomologous to a constant.

We shall write $\mathcal{M}$ for the space of T-invariant probability measures on X. For $\nu \in \mathcal{M}$, we write $h(\nu)$ for the entropy of T with respect to ν. Given a continuous function $g : X \to \mathbb{R}$, we define its pressure $P(g)$ by

$$P(g) = \sup\left\{h(\nu) + \int g d\nu : \nu \in \mathcal{M}\right\}.$$

If g is Hölder continuous, then the above supremum is attained for a unique measure called the equilibrium state for g. If $g - g'$ is cohomologous to a constant, then g and g' have the same equilibrium state.

Given a $k \times k$ matrix A with entries 0 or 1, we define a space

$$\Sigma^+ = \{x = (x_n)_{n=0}^{\infty} : A(x_n, x_{n+1}) = 1 \ \forall n \in \mathbb{Z}^+\}$$

and a shift map $\sigma : \Sigma^+ \to \Sigma^+$ given by $(\sigma x)_n = x_{n+1}$. The pair (Σ^+, σ) is called a (one-sided) shift of finite type. There is a metric on Σ^+ given by $d(x, y) = 2^{-N}$, where $N = \sup\{n : x_i = y_i, i \leq n\}$. The map $\sigma : \Sigma^+ \to \Sigma^+$ is mixing if the matrix A is aperiodic, i.e., there exists $N \geq 1$ such that $A^N(i, j) \geq 1$ for any $1 \leq i, j \leq k$.

An important feature of expanding maps is that they may be modelled by shifts of finite type.

PROPOSITION 1.1. *Let $T : X \to X$ be a (mixing) expanding map. Then there exists a mixing subshift of finite type $\sigma : \Sigma^+ \to \Sigma^+$ and a Hölder continuous map $\pi : \Sigma^+ \to X$ such that*

(i) *$T \circ \pi = \pi \circ \sigma$,*
(ii) *π is surjective, bounded-to-one and one-to-one almost everywhere with respect to any ergodic measure on Σ^+.*

Given $\alpha > 0$, we let $C^\alpha(\Sigma^+)$ be the Banach space of Hölder continuous functions $f : \Sigma^+ \to \mathbb{R}$ with norm $||f|| = |f|_\alpha + |f|_\infty$, where

$$|f|_\alpha = \sup\left\{\frac{|f(x) - f(y)|}{d(x, y)^\alpha} : x, y \in \Sigma^+\right\}$$

and $|f|_\infty$ is the supremum norm. Let $\mathcal{L}_{g+iuf} : C^\alpha(\Sigma^+) \to C^\alpha(\Sigma^+)$ be the transfer operator defined by

$$\mathcal{L}_{g+iuf} w(x) = \sum_{\sigma y = x} e^{g(y)+iuf(y)} w(y).$$

We will say that g is normalized if $\mathcal{L}_g 1 = 1$; by adding a coboundary and a constant, it is always possible to arrange that g is normalized. The following result is standard (see [**PP**] for parts (1) and (2) and [**PS1**] for part (3)).

LEMMA 1.2.

(1) *When* $u = 0$, *the operator* $\mathcal{L}_{g+iuf}$ *has a maximal eigenvalue* $e^{P(g)}$ *and the rest of the spectrum is contained in a disc of strictly smaller radius. In particular, if* g *is normalized, then* $P(g) = 0$ *and* $\mathcal{L}_g^*\mu = \mu$, *where* μ *is the equilibrium state for* g.

(2) *There exists* $a > 0$ *such that, for* $|u| < a$, $\mathcal{L}_{g+iuf}$ *has a simple maximal eigenvalue* $e^{P(g+iuf)}$, *satisfying* $|e^{P(g+iuf)}| \leq e^{P(g)}$, *and the rest of the spectrum is contained in* $\{z : |z| \leq \theta e^{P(g)}\}$, *for some* $0 < \theta < 1$. *In particular,* $u \mapsto e^{P(g+iuf)}$ *is analytic for* $|u| < a$. *Furthermore, we write* $\left.\frac{d^2P(g+iuf)}{du^2}\right|_{u=0} = -\sigma^2$.

(3) *There exists a change of coordinates* $v = v(u)$ *such that for* $|u| < a$, *we can expand* $e^{P(g+iuf)} = e^{P(g)}(1 - v^2 + iQ(v))$, *where* $Q(v)$ *is real valued and satisfies* $Q(v) = O(|v|^3)$. *In particular,* $v'(0) = \sigma/\sqrt{2}$.

The following identity will be important in subsequent calculations.

LEMMA 1.3. *Let* μ *denote the equilibrium state for* g. *If* g *is normalized, then*

$$\int e^{iuf^n(x)}d\mu(x) = \int \mathcal{L}_{g+iuf}^n 1(x)d\mu(x).$$

PROOF. This follows from the identity $\mathcal{L}_g^*\mu = \mu$ by a simple calculation. □

A key point in our proof will be a bound which involves an estimate on iterates of $\mathcal{L}_{g+iuf}$; estimates of the kind we require were developed in [**D**] and [**PS2**].

LEMMA 1.4. *Assume that* f *satisfies the Diophantine condition and that* g *is normalized. Then there exists* $\gamma > 0$, $D > 0$ *and* $C, c > 0$ *such that, for* $|u| \geq a$, *we have that*

$$||\mathcal{L}_{g+iuf}^{2Nm}1||_\infty \leq C\left(1 - \frac{c}{|u|^\gamma}\right)^m, \text{ for } n \geq 1, \tag{1.1}$$

where $N = [D\log|u|]$.

PROOF. Since we are assuming the Diophantine condition, the hypotheses of Proposition 2 in [**PS2**] hold. This gives the inequality (1.1). □

2. Proof of Theorem 1

In this section we will present a proof of Theorem 1 using properties of the transfer operator from Section 1. Let m be the equilibrium state for a Hölder continuous function $g : X \to \mathbb{R}$ and choose $g_0 : \Sigma^+ \to \mathbb{R}$ to be a normalized Hölder continuous function on Σ^+ which is cohomologous to $g \circ \pi$. Let μ denote the equilibrium state for g_0. Given a Hölder continuous function $f : X \to \mathbb{R}$ (with $\int f dm = 0$), we define $f_0 : \Sigma^+ \to \mathbb{R}$ by $f_0 = f \circ \pi$. Since π is Hölder, $f_0 \in C^\alpha(\Sigma^+)$, for some $\alpha > 0$. Then

$$\begin{aligned} &m\{x \in X : \ z + \epsilon_n a \leq f^n(x) \leq z + \epsilon_n b\} \\ &= \mu\left\{x \in \Sigma^+ : \ z + \epsilon_n a \leq f_0^n(x) \leq z + \epsilon_n b\right\}. \end{aligned}$$

Thus, to prove Theorem 1, it suffices to prove the corresponding asymptotic formula for $\mu\{x \in \Sigma^+ : \ z + \epsilon_n a \leq f_0^n(x) \leq z + \epsilon_n b\}$. For the remainder of this section, we shall abuse notation and write f and g for f_0 and g_0.

We shall first prove a modified result, where the interval $[z + \epsilon_n a, z + \epsilon_n b]$ is replaced by a sequence of smooth test functions. Let $\chi : \mathbb{R} \to \mathbb{R}$ be a compactly supported C^k function (where k will be chosen later). We shall write $\chi_n(x) = \chi(\epsilon_n^{-1}(x-z))$ and we note that the Fourier transform satisfies $\widehat{\chi}_n(u) = e^{izu}\epsilon_n\widehat{\chi}(\epsilon_n u)$. Let us define

$$\rho(n) := \int \chi_n(f^n(x))d\mu.$$

PROPOSITION 2.1. *Let γ be as in Lemma 1.4. Then, provided that $\epsilon_n^{-1} = O(n^\delta)$, for some $\delta < 1/\gamma$, we have that*

$$\rho(n) \sim \frac{\int \chi(x)dx}{\sqrt{2\pi}\sigma}\frac{\epsilon_n}{\sqrt{n}}, \quad \textit{as } n \to +\infty.$$

To prove Proposition 2.1, we first use the inverse Fourier transform and Fubini's Theorem to write

$$\begin{aligned}\rho(n) &= \frac{1}{2\pi}\int_{-\infty}^{\infty}\left(\int e^{iuf^n(x)}d\mu(x)\right)\widehat{\chi}_n(u)du \\ &= \frac{1}{2\pi}\int_{-\infty}^{\infty}\left(\int \mathcal{L}_{g+iuf}^n 1(x)d\mu(x)\right)\widehat{\chi}_n(u)du,\end{aligned} \tag{2.1}$$

using Lemma 1.3 for the last equality.

Choose $a > 0$ sufficiently small. Using part (3) of Lemma 1.2, we can change coordinates on $(-a, a)$ to $v = v(u)$ and write $e^{P(g+iuf)} = (1 - v^2 + iQ(v))$, for $|v| < a$, say. If $\mathcal{P}_{g+iuf} : C^\alpha(\Sigma^+) \to C^\alpha(\Sigma^+)$ is the associated one-dimensional eigenprojection, then by perturbation theory $\mathcal{P}_{g+iuf}(1) = 1 + O(|v|)$. Using the formula $\mathcal{L}_{g+iuf}^n 1 = e^{nP(g+iuf)}(1 + O(|v|)) + O(\theta^n)$, we may write

$$\begin{aligned}&\int_{-a}^{a}\left(\int \mathcal{L}_{g+iuf}^n 1(x)d\mu(x)\right)\widehat{\chi}_n(u)du \\ &= \int_{-a}^{a}(1 - v^2 + iQ(v))^n(1 + O(|v|))\,\widehat{\chi}_n(u(v))\frac{du}{dv}dv + O(\theta^n) \\ &= \frac{\epsilon_n\widehat{\chi}(0)\sqrt{2}}{\sigma}\int_{-a}^{a}(1 - v^2 + iQ(v))^n(1 + O(|v|))\,dv + O\left(\frac{\epsilon_n}{n}\right) + O(\theta^n),\end{aligned} \tag{2.2}$$

where the $O(\epsilon_n n^{-1})$ estimate follows from a simple calculation in [**PS1**, p. 409]. Using another easy calculation in [**PS1**, pp. 408–409], we see that the principle term in the last line of (2.2) is asymptotic to $\int_{-a}^{a}(1 - v^2)^n dv$; by making the substitution $w = v^2$, we may estimate this as

$$\begin{aligned}\frac{\epsilon_n\widehat{\chi}(0)\sqrt{2}}{\sigma}\int_{-a}^{a}(1 - v^2)^n dv &= 2\frac{\epsilon_n\widehat{\chi}(0)\sqrt{2}}{\sigma}\int_{0}^{a}(1 - v^2)^n dv \\ &= \frac{\epsilon_n\widehat{\chi}(0)\sqrt{2}}{\sigma}\int_{0}^{a^2}\frac{(1 - w)^n}{w^{1/2}}dw \\ &= \frac{\epsilon_n\widehat{\chi}(0)\sqrt{2}}{\sigma}\int_{0}^{1}\frac{(1 - w)^n}{w^{1/2}}dw + O((1 - a^2)^n) \\ &\sim \sqrt{2\pi}\frac{\widehat{\chi}(0)}{\sigma}\frac{\epsilon_n}{\sqrt{n}},\end{aligned} \tag{2.3}$$

as $n \to +\infty$ (cf. [**St**, p. 236]). Moreover, the term arising from the $O(|v|)$ term in the integrand is of order

$$\int_{-a}^{a} (1-v^2)^n |v| dv = \int_0^{a^2} (1-w)^n dw = O\left(\frac{1}{n}\right).$$

It remains to estimate the integral in (2.1) over $|u| \geq a$. To do this, we shall use the bound on the transfer operators $\mathcal{L}_{g+iuf}$ contained in Lemma 1.4. We shall also use the following simple lemma.

LEMMA 2.2. *If $\chi : \mathbb{R} \to \mathbb{R}$ is C^k and compactly supported, then the Fourier transform $\widehat{\chi}(u)$ satisfies $\widehat{\chi}(u) = O(|u|^{-k})$, as $|u| \to \infty$.*

PROOF. This is a standard application of integration by parts. □

To complete the proof of Proposition 2.1, we can bound

$$\begin{aligned} &\int_{|u|\geq a} \left(\int \mathcal{L}^n_{g+iuf} 1(x) d\mu(x)\right) \widehat{\chi}_n(u) du \\ &= \epsilon_n \int_{|u|\geq a} e^{izu} \left(\int \mathcal{L}^n_{g+iuf} 1(x) d\mu(x)\right) \widehat{\chi}(\epsilon_n u) du \qquad (2.4) \\ &= O\left(\frac{1}{\epsilon_n^{k-1}} \int_a^\infty \left(1 - \frac{c}{u^\gamma}\right)^{n/2[D \log |u|]} u^{-k} du\right). \end{aligned}$$

We need to show that this quantity tends to zero more quickly than $\epsilon_n n^{-1/2}$. To see this, we shall split the integral in (2.4) into two parts:

$$\begin{aligned} &\int_a^\infty \left(1 - \frac{c}{u^\gamma}\right)^{n/2[D \log |u|]} u^{-k} du \\ &= \int_a^{n^{\delta'}} \left(1 - \frac{c}{u^\gamma}\right)^{n/2[D \log |u|]} u^{-k} du + \int_{n^{\delta'}}^\infty \left(1 - \frac{c}{u^\gamma}\right)^{n/2[D \log |u|]} u^{-k} du, \end{aligned}$$

where we choose $\delta < \delta' < 1/\gamma$. The first integral may be bounded by

$$\int_a^{n^{\delta'}} \left(1 - \frac{c}{u^\gamma}\right)^{n/2[D \log |u|]} u^{-k} du = O\left(n^{\delta'} \left(1 - \frac{c}{n^{\delta'\gamma}}\right)^{n/2D\delta' \log n}\right)$$

and, since $\delta'\gamma < 1$, this tends to zero faster than the reciprocal of any polynomial. The second integral may be bounded by

$$\int_{n^{\delta'}}^\infty \left(1 - \frac{c}{u^\gamma}\right)^n u^{-k} du = O(n^{(1-k)\delta'}).$$

Combining these estimates, we see that

$$\int_{|u|\geq a} \left(\int \mathcal{L}^n_{g+iuf} 1(x) d\mu(x)\right) \widehat{\chi}_n(u) du = O(\epsilon_n^{-(k-1)} n^{(1-k)\delta'}) = O(n^{(k-1)(\delta-\delta')}).$$

We obtain the required bound by choosing k sufficiently large that $(k-1)(\delta-\delta') < -\delta - 1/2$.

Finally, Theorem 1 may be deduced from Proposition 2.1 by a simple approximation argument. More precisely, given $\eta > 0$, we can choose smooth functions

$\chi^- \leq \chi_{[a,b]} \leq \chi^+$, where $\chi_{[a,b]}$ denotes the indicator function of the interval $[a, b]$, such that $b - a - \eta \leq \int \chi^-(x)dx \leq \int \chi^+(x)dx \leq b - a + \eta$. Then

$$\limsup_{n\to+\infty} \frac{n^{1/2}}{\epsilon_n} \mu \left\{x \in \Sigma^+ : \ z + a\epsilon_n \leq f^n(x) \leq z + \epsilon_n b\right\}$$
$$\leq \limsup_{n\to+\infty} \frac{n^{1/2}}{\epsilon_n} \int \chi_n^+(f^n(x))d\mu \leq \frac{b-a+\eta}{\sqrt{2\pi}\sigma}$$

and

$$\liminf_{n\to+\infty} \frac{n^{1/2}}{\epsilon_n} \mu \left\{x \in \Sigma^+ : \ z + a\epsilon_n \leq f^n(x) \leq z + \epsilon_n b\right\}$$
$$\geq \liminf_{n\to+\infty} \frac{n^{1/2}}{\epsilon_n} \int \chi_n^-(f^n(x))d\mu \geq \frac{b-a-\eta}{\sqrt{2\pi}\sigma}.$$

Since $\eta > 0$ is arbitrary, this gives the result.

3. Axiom A diffeomorphisms

In this section we shall show how the results of Theorem 1 can be extended to invertible systems. This requires some technical details which we shall describe in this section.

Let $T : M \to M$ be a C^1 diffeomorphism. We call an T-invariant set X a basic set if

(i) we have a DT-invariant splitting $T_X M = E^s \oplus E^u$ such that $\exists C > 0, 0 < \lambda < 1$, such that $||DT^n|E^s|| \leq C\lambda^n$ and $||DT^{-n}|E^u|| \leq C\lambda^n$;
(ii) $\exists$ open set $U \supset X$ such that $X = \bigcap_{n=-\infty}^{\infty} T^{-n}U$;
(iii) $T : X \to X$ is transitive; and
(iv) the periodic orbits for $T|X$ are dense in X.

We say that T satisfies Axiom A if the non-wandering set Ω is hyperbolic. In particular, Ω is a finite union of hyperbolic fixed points and basic sets.

The analogue of Theorem 1 for Axiom A diffeomorphisms is the following.

THEOREM 2. *Let $T : X \to X$ be an Axiom A diffeomorphism restricted to a non-trivial basic set. Suppose that $T : X \to X$ is mixing and let m be the Gibbs state for a Hölder continuous function. Suppose that $f : X \to \mathbb{R}$ is a Hölder continuous function satisfying the Diophantine condition and such that $\int f dm = 0$. Then there exists $\delta > 0$ such that for any $z \in \mathbb{R}$, $a < b$ and sequence $\epsilon_n > 0$ which tends to zero and satisfies $\epsilon_n^{-1} = O(n^\delta)$, we have that*

$$m\left\{x \in \Lambda : \ z + \epsilon_n a \leq f^n(x) \leq z + \epsilon_n b\right\} \sim \frac{(b-a)}{\sqrt{2\pi}\sigma} \frac{\epsilon_n}{\sqrt{n}},$$

as $n \to +\infty$.

We begin by introducing the two-sided version of subshifts of finite type.

3.1. Symbolic dynamics. As in Section 1, given a $k \times k$ matrix A with entries 0 or 1, we define a space

$$\Sigma = \{x = (x_n)_{n=-\infty}^{\infty} : A(x_n, x_{n+1}) = 1 \ \forall n \in \mathbb{Z}\}$$

and a shift map $\sigma : \Sigma \to \Sigma$ given by $(\sigma x)_n = x_{n+1}$. The pair (Σ, σ) is called a (two-sided) shift of finite type. There is a metric on Σ given by $d(x, y) = 2^{-k}$, where $k = \sup\{n : x_i = y_i, |i| \leq n\}$.

The following result reduces this to the study of the subshift $\sigma : \Sigma \to \Sigma$.

PROPOSITION 3.1. *Let $T : X \to X$ be an Axiom A diffeomorphism restricted to a non-trivial basic set and suppose that T is mixing. Then there exists a mixing subshift of finite type $\sigma : \Sigma \to \Sigma$ and a Hölder continuous map $\pi : \Sigma \to X$ such that*

(i) $T \circ \pi = \pi \circ \sigma$;
(ii) *π is surjective, bounded to-one and one-to-one almost everywhere with respect to any ergodic measure on Σ.*

Given $\alpha > 0$, we let $C^\alpha(\Sigma)$ be the Banach space of Hölder continuous functions $f : \Sigma \to \mathbb{R}$ with norm $||f|| = |f|_\alpha + ||f||_\infty$, where

$$|f|_\alpha = \sup\left\{\frac{|f(x) - f(y)|}{d(x,y)^\alpha} : x, y \in \Sigma, x \neq y\right\}$$

and $||f||_\infty$ is the supremum norm. Observe that for any $0 < \beta < \alpha$ we have that $|f|_\beta \leq |f|_\alpha$.

Suppose that m is the Gibbs state for the Hölder continuous function $g : X \to \mathbb{R}$ and let μ be the Gibbs state for $g \circ \pi : \Sigma \to \mathbb{R}$. As in Section 2, we have

$$\begin{aligned} & m\{x \in X : \ z + \epsilon_n a \leq f^n(x) \leq z + \epsilon_n b\} \\ & = \mu\{x \in \Sigma : \ z + \epsilon_n a \leq (f \circ \pi)(x) \leq z + \epsilon_n b\}, \end{aligned}$$

so it suffices to consider functions on Σ. We shall suppose that α is chosen so that $f \circ \pi, g \circ \pi \in C^\alpha(\Sigma)$. Once again, we shall abuse notation and write f and g instead of $f \circ \pi$ and $g \circ \pi$.

In order to obtain the asymptotics in Theorem 2, we need to relate f to functions defined on the corresponding one-sided shift Σ^+; then we can apply the analysis of Section 2. It is well known that it is possible to find a Hölder continuous function defined on Σ^+ which is cohomologous to f; however,

$$\mu\{x : z + \epsilon_n a \leq f(x) \leq z + \epsilon_n b\}$$

is not invariant under this change. It is therefore necessary to employ a slightly more sophisticated approach involving approximations.

3.2. Introducing functions $\widetilde{f}_k$ on Σ^+. The first step in the proof is to approximate $f \in C^\alpha(\Sigma)$ by functions that only go finitely far into the "past" (i.e., we choose $f_k : \Sigma \to \mathbb{R}$ depending only on the co-ordinates $x_{-k}, x_{-k+1}, x_{-k+2}, \ldots$) sufficiently close to f, in a suitable sense. In particular, we want to let $k = k(n) = \eta \log n$, where $\eta = (1+\gamma)(\alpha \log 2)^{-1} > 0$, where γ will be specified later. We then choose $f_k(x) = \inf\{f(y) : y_i = x_i, \ i \geq -k\}$.

The following relates f_k to f and reduces Theorem 2 to proving the corresponding result for f_k, for some sufficiently small $\delta > 0$ (independent of k).

LEMMA 3.2. *Assume that $\gamma > 1 > \delta$ and let $\epsilon_n^{-1} = o(n^\delta)$. Then*

$$\begin{aligned} & \mu\left\{x \in \Sigma : z + \epsilon_n a + n^{-\delta} \leq f_k^n(x) \leq z + \epsilon_n b - n^{-\delta}\right\} \\ & \leq \mu\{x \in \Sigma : z + \epsilon_n a \leq f^n(x) \leq z + \epsilon_n b\} \\ & \leq \mu\left\{x \in \Sigma : z + \epsilon_n a - n^{-\delta} \leq f_k^n(x) \leq z + \epsilon_n b + n^{-\delta}\right\}, \end{aligned} \tag{3.1}$$

for all sufficiently large n.

PROOF. This is similar to the approach in [**L**]. Observe that $2^{-\alpha k} = n^{-(1+\gamma)} = o(n^{-1}\epsilon_n)$. Then $||f_k^n - f^n||_\infty \le n|f|_\alpha 2^{-\alpha k} = o(n^{-\delta})$, provided that $\delta < \gamma$. In particular, we can compare

$$\begin{aligned}
&\mu\left\{x \in \Sigma : \ z + \epsilon_n a - n|f|_\alpha 2^{-\alpha k} \le f_k^n(x) \le z + \epsilon_n b + n|f|_\alpha 2^{-\alpha k}\right\} \\
&\le \mu\left\{x \in \Sigma : \ z + \epsilon_n a \le f^n(x) \le z + \epsilon_n b\right\} \qquad (3.2)\\
&\le \mu\left\{x \in \Sigma : \ z + \epsilon_n a - n|f|_\alpha 2^{-\alpha k} \le f_k^n(x) \le z + \epsilon_n b + n|f|_\alpha 2^{-\alpha k}\right\}.
\end{aligned}$$

This implies the required result. □

Thus we see that establishing asymptotic results for f^n suffices to show the required results for f_k^n, where $k = k(n)$ is as defined earlier. The basic idea is to show that γ sufficiently large gives rise to a suitable δ since the first and last terms in (3.2) are asymptotic (with an expression involving constants in terms of f, and independent of n).

In order to introduce transfer operators, we first want to shift each truncated function f_k into a function depending on the co-ordinates $x_0, x_1, x_2, \ldots$. More precisely, we shall write $\widetilde{f}_k := f_k \circ \sigma^k \in C^\alpha(\Sigma^+)$, for $k \ge 0$. Since μ is σ-invariant, we can write

$$\begin{aligned}
&\mu\left\{x \in \Sigma : \ z + \epsilon_n a \le f_k^n(x) \le z + \epsilon_n b\right\} \\
&= \mu\left\{x \in \Sigma^+ : \ z + \epsilon_n a \le \widetilde{f}_k^n(x) \le z + \epsilon_n b\right\},
\end{aligned}$$

for each $k \ge 0$.

As in Section 2, we need to introduce a sufficiently differentiable test function $\chi : \mathbb{R} \to \mathbb{R}$. By replacing f by $\widetilde{f}_k$ in (2.1), we see that we want to estimate

$$\rho_k(n) = \frac{1}{2\pi}\int_{-\infty}^{\infty}\left(\int \mathcal{L}^n_{g+iu\widetilde{f}_k} 1(x)d\mu(x)\right)\widehat{\chi}_n(u)du. \qquad (3.3)$$

For future reference, we shall write $w_k := f_k + f_k \circ \sigma + \ldots + f_k \circ \sigma^{k-1}$. In particular, we have the trivial relation $\widetilde{f}_k = f_k + w_k \circ \sigma - w_k \in C^\alpha(\Sigma^+) \subset C^\beta(\Sigma^+)$, provided $\beta < \alpha$.

3.3. Introducing functions $\overline{f}_k$ on Σ^+. Although the shifted functions $\widetilde{f}_k$ give the above expression for $\rho_k(n)$, we have little a priori control over how the corresponding transfer operators $\mathcal{L}_{g+iu\widetilde{f}_k}$ behave as $k \to +\infty$. To address this problem, it is convenient to introduce a second sequence of better behaved functions $\overline{f}_k$, such that $\overline{f}_k$ is cohomologous to f_k, for each $k \ge 1$.

The properties of these new functions are described in the following simple lemma.

LEMMA 3.3. *Let $\beta < \alpha$. There exist $\overline{f}, \overline{f}_k \in C^{\alpha/2}(\Sigma^+)$, for $k \ge 1$, and $u_{f_k} \in C^{\alpha/2}(\Sigma)$ and $C, C_0 > 0$ such that $\widetilde{f}_k = \overline{f}_k + u_{f_k} \circ \sigma - u_{f_k}$, where*

(1) $||\overline{f} - \overline{f}_k||_\infty \le C|f|_\alpha 2^{-\alpha k}$, *for* $k \ge 1$;
(2) $|\overline{f} - \overline{f}_k|_\beta \le C|f|_\beta 2^{-(\alpha-\beta)k}$, *for* $k \ge 1$; *and*
(3) $|u_f|_{\beta/2} \le C_0|f|_\alpha$.

PROOF. Following [**Si**] and [**P**], we may define a linear operator $\tau : C^\alpha(\Sigma) \to C^{\alpha/2}(\Sigma^+)$ such that $\tau : f \mapsto \overline{f} = f + u_f \circ \sigma - u_f$. More precisely, one can write

$f(x) = \sum_{n=0}^{\infty} \mathfrak{f}_n(x)$ where $\mathfrak{f}_n(x) = \mathfrak{f}_n(x_{-n}, \ldots, x_0, \ldots, x_n)$ and $||\mathfrak{f}_n||_\infty \leq |f|_\alpha 2^{-\alpha n}$. Let

$$u_f(x) = \sum_{n=0}^{\infty} \sum_{m=0}^{n-1} \mathfrak{f}_n \circ \sigma^m.$$

Then one sees this is well defined in $C^0(\Sigma^+)$ and

$$||u_f||_\infty \leq ||f||_\infty \sum_{n=0}^{\infty} n 2^{-\alpha n} \leq \frac{||f||_\infty}{(1-2^{-\alpha})^2} < +\infty.$$

To check the coboundary identity, we see that

$$u_f \circ \sigma - u_f = \sum_{n=0}^{\infty} \mathfrak{f}_n \circ \sigma^n - \sum_{n=0}^{\infty} \mathfrak{f}_n = \overline{f} - f.$$

Parts (1) and (2) then come from the corresponding properties of f and $\mathfrak{f}_k$:

(i) $||f - \mathfrak{f}_k||_\infty \leq |f|_\alpha 2^{-\alpha k}$, for $k \geq 1$; and
(ii) $|f - \mathfrak{f}_k|_\beta \leq |f|_\beta 2^{-(\alpha \ \beta)k}$, for $k \geq 1$.

These results can be found in [**Ru1**].

To bound the norm of $u_f \in C^{\alpha/2}(\Sigma)$, assume that $x_i = y_i$, $i = -N, \ldots, N$. Then

$$\begin{aligned} |u_f(x) - u_f(y)| &= \sum_{n=0}^{\infty} \sum_{m=0}^{n-1} |\mathfrak{f}_n(x_{-n+m}, \ldots, x_{n+m}) - \mathfrak{f}_n(y_{-n+m}, \ldots, y_{n+m})| \\ &\leq 2|f|_\alpha \sum_{n=[(N-1)/2]}^{\infty} \sum_{m=N-n}^{n} 2^{-\alpha(N-m)} \\ &\leq 2|f|_\alpha \left(\sum_{n=[(N-1)/2]}^{\infty} \frac{2^{-\alpha n/2}}{1-2^{-\alpha/2}} \right) \\ &\leq \left(\frac{2|f|_\alpha}{\left(1-2^{-\alpha/2}\right)^2} \right) 2^{-\alpha N/2}. \end{aligned}$$

Thus we deduce that $|u_f|_{\alpha/2} \leq C_0 |f|_\alpha$, where $C_0 = 2(1-2^{-\alpha/2})^{-2}$. Since $|u_f|_{\beta/2} \leq |u_f|_{\alpha/2}$, this shows (3). □

Later, we shall want to choose a conveniently small value of $\beta > 0$.

3.4. Comparing transfer operators for $\widetilde{f}_k$ and $\overline{f}_k$. We now have two sequences of functions $\overline{f}_k, \widetilde{f}_k \in C^{\beta/2}(\Sigma^+)$, $k \geq 1$, each cohomologous to f_k, and we need to understand the coboundary that relates them. This is the purpose of the next lemma.

LEMMA 3.4. *Let us denote $v_k := w_k + u_{f_k}$. Then $\widetilde{f}_k - \overline{f}_k = v_k \circ \sigma - v_k$ and $v_k \in C^{\beta/2}(\Sigma^+)$.*

PROOF. Fix $k \geq 1$. Then, a priori, we only know that $v_k \in C^{\beta/2}(\Sigma)$. However, since $\overline{f}_k - \widetilde{f}_k \in C^{\beta/2}(\Sigma^+)$, we can apply Livsic's Theorem for periodic orbits [**PP**, p. 45] to $\sigma : \Sigma^+ \to \Sigma^+$ to deduce that $v_k \circ \sigma - v_k = g_k \circ \sigma - g_k$, for some $g_k \in C^{\beta/2}(\Sigma^+)$. Thus $(v_k - g_k) \circ \sigma = (v_k - g_k)$, and by ergodicity (with respect to any Gibbs measure, say) we see that $v_k = g_k + K \in C^{\beta/2}(\Sigma^+)$, for some constant K. This completes the proof. □

We have now constructed two sequences of pairwise cohomologous functions $\widetilde{f}_k$ and $\overline{f}_k$, $k \geq 1$. The former are best suited for studying $\rho_k(n)$ in (3.3), but the latter are best suited for taking limits. The following is useful to relating the spectrum of the associated transfer operators.

LEMMA 3.5. *The transfer operators* $\mathcal{L}_{g+iu\widetilde{f}_k}$ *and* $\mathcal{L}_{g+iu\overline{f}_k}$, $k \geq 1$, *are conjugate. In particular,*

$$\mathcal{L}_{g+iu\widetilde{f}_k} = \Delta(e^{iuv_k})\mathcal{L}_{g+iu\overline{f}_k}\Delta(e^{-iuv_k}),$$

where $\Delta(e^{iuv_k}) : C^{\beta/2}(\Sigma^+) \to C^{\beta/2}(\Sigma^+)$ *denotes multiplication by the function* $e^{iuv_k} \in C^{\beta/2}(\Sigma^+)$.

PROOF. This follows immediately from the definition of the transfer operator and the identity $\widetilde{f}_k - \overline{f}_k = v_k \circ \sigma - v_k$ in Lemma 3.4. □

Thus, in order to compare bounds on the (common) spectra of the two transfer operators, we need to understand better the conjugating operator. This is achieved using the following lemma.

LEMMA 3.6. *There exists* $C_1 > 0$ *such that, for any* $h \in C^{\beta/2}(\Sigma^+)$, *we have the bounds* $||\Delta(e^{iuv_k})h||_\infty \leq ||h||_\infty$ *and* $|\Delta(e^{iuv_k})h|_{\beta/2} \leq C_1 2^{k\beta/2}|u||f|_\alpha|h|_\infty + |h|_{\beta/2}$.

PROOF. The first inequality is obvious.

Using that $|h_1h_2|_{\beta/2} \leq |h_1|_{\beta/2}|h_2|_\infty + |h_2|_{\beta/2}|h_1|_\infty$, we can bound

$$\begin{aligned} |e^{iuv_k}h|_{\beta/2} &\leq |e^{iuv_k}|_{\beta/2}|h|_\infty + |e^{iuv_k}|_\infty |h|_{\beta/2} \\ &= |e^{iuv_k}|_{\beta/2}|h|_\infty + |h|_{\beta/2}. \end{aligned} \tag{3.4}$$

We therefore need to bound

$$|e^{iuv_k}|_{\beta/2} \leq |e^{iuu_{f_k}}|_{\beta/2} + |e^{iuw_k}|_{\beta/2} \leq |u| \left(|u_{f_k}|_{\beta/2} + |w_k|_{\beta/2}\right), \tag{3.5}$$

say. Observe here that whereas we can interpret $v_k \in C^{\beta/2}(\Sigma^+)$ (by Lemma 3.4), we still have to view $u_{f_k}, w_k \in C^{\beta/2}(\Sigma)$. However, this does not affect the bound on v_k.

Firstly, we can bound $|u_{f_k}|_{\beta/2} \leq C_0|f|_\alpha$ using Lemma 3.3. Secondly, from the definition of $|\cdot|_{\beta/2}$ we see that $|f_k \circ \sigma^i|_{\beta/2} = 2^{i\beta/2}|f_k|_{\beta/2}$, for $k \geq 1$. Thus, by the triangle inequality, we can also bound for each $k \geq 1$

$$\begin{aligned} |w_k|_{\beta/2} &= |f_k + f_k \circ \sigma + \ldots + f_k \circ \sigma^{k-1}|_{\beta/2} \\ &\leq |f_k|_{\beta/2} + |f_k \circ \sigma|_{\beta/2} + \ldots + |f_k \circ \sigma^{k-1}|_{\beta/2} \\ &\leq |f_k|_{\beta/2} \left(1 + \sum_{j=0}^{k-1} 2^{j\beta/2}\right) \\ &\leq \frac{2^{k\beta/2}}{2^{\beta/2} - 1}|f|_{\beta/2}, \end{aligned} \tag{3.6}$$

and thus by comparing (3.5) and (3.6), we get

$$|e^{iuv_k}|_{\beta/2} \leq \left(C_0 + \frac{2^{k\beta/2}}{2^{\beta/2} - 1}\right)|u||f|_{\beta/2},$$

and since $|f|_{\beta/2} \leq |f|_\alpha$, the result follows from substituting into (3.4). □

3.5. Perturbation theory for u small. We return to the problem of relating the spectra of the two families of operators. This is easiest for the maximal eigenvalue. Since we now know that the functions $\overline{f}_k$ and $\widetilde{f}_k$ differ by a coboundary, it is immediate that the operators $\mathcal{L}_{g+iu\overline{f}_k}, \mathcal{L}_{g+iu\widetilde{f}_k} : C^{\beta/2}(\Sigma^+) \to C^{\beta/2}(\Sigma^+)$ have exactly the same maximal eigenvalues $e^{P(g+iu\overline{f}_k)} = e^{P(g+iu\widetilde{f}_k)}$ (provided $|u| < a$, where a is sufficiently small to make the "complex presssure" well defined [**PP**]).

By standard perturbation theory, both the maximal eigenvalue

$$C^{\beta/2}(\Sigma^+) \ni g_1 \mapsto e^{P(g_1)} \in \mathbb{R}$$

and the associated eigenprojection

$$C^{\beta/2}(\Sigma^+) \ni g_1 \mapsto \mathcal{P}_{g_1} \in B\left(C^{\beta/2}(\Sigma^+), C^{\beta/2}(\Sigma^+)\right)$$

(taking values in the space of bounded linear operators) are analytic [**PP**]. Moreover, we have the following.

LEMMA 3.7. *There exist $\epsilon > 0$ and $C > 0$ such that, for $||g_1||, ||g_2|| \leq \epsilon$, we have that*

(i) $||e^{P(g_1)} - e^{P(g_2)}|| \leq C||g_1 - g_2||$, *and*
(ii) $||\mathcal{P}_{g_1} - \mathcal{P}_{g_2}|| \leq C||g_1 - g_2||$,

where $||\cdot|| = |\cdot|_\infty + ||\cdot||_\infty$.

For $|u| < a$, we can apply the bound in Lemma 3.7 for the maximal eigenvalue to bound

$$\begin{aligned}|e^{nP(g+iu\overline{f}_k)} - e^{nP(g+iu\overline{f})}| &= O\left(n||\overline{f} - \overline{f}_k||\right) \\ &= O\left(n|\overline{f}|_\beta \left(\frac{1}{2}\right)^{(\alpha-\beta)k}\right) \\ &= O\left(n^{1-(1-\frac{\beta}{\alpha})(1+\gamma)}|\overline{f}|_\beta\right),\end{aligned}$$

using Lemma 3.3 and the definition of $k = k(n)$. Thus, since the pressure is unchanged by adding coboundaries, we can estimate

$$\begin{aligned}e^{nP(g+iu\widetilde{f}_k)} &= e^{nP(g+iu\overline{f})} + \left(e^{nP(g+iu\overline{f}_k)} - e^{nP(g+iu\overline{f})}\right) \\ &= e^{nP(g+iu\overline{f})} + O\left(|\overline{f}|_\beta n^{1-(1-\frac{\beta}{\alpha})(1+\gamma)}\right) \\ &= e^{nP(g+iu\overline{f})} + O\left(\frac{1}{n}\right),\end{aligned} \tag{3.7}$$

provided $\gamma > 1$ is chosen sufficiently large and β is chosen sufficiently small. (For definiteness, if we assume $\beta < \alpha/2$, then it would suffice that $\gamma > 3$.) As in Section 2, we can use the Morse Lemma to change coordinates in a neigbourhood of the origin to $v = v(u)$ and write $e^{P(g+iu\overline{f})} = (1 - v^2 + iQ(v))$. In addition, using Lemma 3.7 (ii), we can estimate

$$\begin{aligned}\mathcal{P}_{g+iu\widetilde{f}_k}1 &= 1 + O\left(\min\left\{1, |u||\widetilde{f}_k|_{\beta/2}\right\}\right) \\ &= 1 + O\left(\min\left\{1, |u|n^{\frac{\beta}{2\alpha}(1+\gamma)}\right\}\right),\end{aligned} \tag{3.8}$$

where we have used $|\widetilde{f}_k|_{\beta/2} \leq 2^{\beta k/2}|f_k|_{\beta/2} = O(2^{\beta k/2}) = O(n^{\frac{\beta}{2\alpha}(1+\gamma)})$, by definition of k.

We recall the following standard result on the spectral gap for the transfer operators $\mathcal{L}_{g+iu\overline{f}_k}$.

LEMMA 3.8. *We can choose $0<\theta<1$ and $C>0$ such that*

$$||\mathcal{L}^n_{g+iu\overline{f}_k} - e^{nP(g+iu\overline{f}_k)}\mathcal{P}^n_{g+iu\overline{f}_k}|| \le C\theta^n, \ \textit{for } n \ge 1,$$

whenever $|u|\le a$ and $k \ge 1$.

We can use Lemmas 3.6 and 3.8 to get the slightly weaker bounds for $\mathcal{L}^n_{g+iu\widetilde{f}_k}$:

$$\begin{aligned}
&||(\mathcal{L}^n_{g+iu\widetilde{f}_k} - e^{nP(g+iu\overline{f}_k)}\mathcal{P}_{g+iu\widetilde{f}_k})1||_\infty \\
&= ||\Delta(e^{iuv_k})\left(\mathcal{L}^n_{g+iu\overline{f}_k} - e^{nP(g+iu\overline{f}_k)}\mathcal{P}_{g+iu\overline{f}_k}\right)(e^{-iuv_k})||_\infty \\
&= ||\Delta(e^{iuv_k})||_\infty||\mathcal{L}^n_{g+iu\overline{f}_k} - e^{nP(g+iu\overline{f}_k)}\mathcal{P}_{g+iu\overline{f}_k}||||\Delta(e^{-iuv_k})|| \\
&= O\left(\theta^n 2^{\beta/2k}\right) = O\left(\theta^n n^{\frac{\beta}{2\alpha}(1+\gamma)}\right),
\end{aligned}$$

where $||\Delta(e^{-iuv_k})|| = O\left(2^{\beta k/2}\right)$, by Lemma 3.6. Thus we can write

$$\mathcal{L}^n_{g+iu\widetilde{f}_k}1 = \left(e^{nP(g+iu\overline{f})} + O\left(\frac{1}{n}\right)\right)\left(1+O\left(|v|n^{\frac{\beta}{\alpha}(1+\gamma)}\right)\right) + O\left(\theta^n n^{\frac{\beta}{\alpha}(1+\gamma)}\right). \tag{3.9}$$

3.6. The integral for $|u|$ small. Using (3.9), we can now estimate the part of the integral (3.3) with $|u|<a$ by

$$\begin{aligned}
&\frac{1}{2\pi}\int_{-a}^{a}\left(\int \mathcal{L}^n_{g+iu\widetilde{f}_k}1(x)d\mu(x)\right)\widehat{\chi}_n(u)du \\
&= \int_{-a}^{a}(1-v^2+iQ(v))^n\left(1+O\left(|v|n^{\frac{\beta}{2\alpha}(1+\gamma)}\right)\right)\widehat{\chi}_n(u(v))\frac{du}{dv}dv \\
&\qquad + O\left(\theta^n n^{\frac{\beta}{2\alpha}(1+\gamma)}\right) \qquad (3.10)\\
&= \frac{\epsilon_n\widehat{\chi}(0)\sqrt{2}}{\sigma}\int_{-a}^{a}(1-v^2+iQ(v))^n\left(1+O\left(|v|n^{\frac{\beta}{2\alpha}(1+\gamma)}\right)\right)dv \\
&\qquad + O\left(\frac{\epsilon_n}{n}\right) + O\left(\theta^n n^{\frac{\beta}{2\alpha}(1+\gamma)}\right).
\end{aligned}$$

We can bound the error term coming from the integral by

$$n^{\frac{\beta}{\alpha}(1+\gamma)}\int_0^a v(1-v^2)^n dv = O\left(n^{\frac{\beta}{\alpha}(1+\gamma)-1}\right).$$

In particular, we can make all of the error terms of order $O\left(n^{-(\delta+\frac{1}{2})}\right)$ by choosing $\beta=\beta(\gamma)>0$ sufficiently small. As in Section 2, we see that the remaining principle asymptotic is $\widehat{\chi}(0)(\sqrt{2\pi}\sigma)^{-1}\epsilon_n n^{-1/2}$.

3.7. The integral for $|u|$ large. Finally, we want to bound

$$\int_{|u|\ge a}\left(\int \mathcal{L}_{g+iu\widetilde{f}_k}1d\mu\right)\widehat{\chi}_n(u)du \le \int_{|u|\ge a}||\mathcal{L}_{g+iu\widetilde{f}_k}1||_\infty|\widehat{\chi}_n(u)|du. \tag{3.11}$$

Observe that for $|u| \geq a$ we can relate the operators associated to $\widetilde{f}_k$ and $\overline{f}_k$ by

$$\begin{aligned} ||\mathcal{L}^n_{g+iu\widetilde{f}_k}1||_\infty &\leq ||\Delta(e^{iuv_k})||_\infty ||\mathcal{L}^n_{g+iu\overline{f}_k}|| ||\Delta(e^{iuv_k})|| \\ &\leq \left(||\mathcal{L}^n_{g+iu\overline{f}} - \mathcal{L}^n_{g+iu\overline{f}_k}|| + ||\mathcal{L}^n_{g+iu\overline{f}}||\right) ||\Delta(e^{iuv_k})||. \end{aligned} \tag{3.12}$$

Assuming χ is C^κ, say, the results in Section 2 applied to $\overline{f}$, which also satisfies the Diophantine Condition, show that we can bound

$$\begin{aligned} \int_{|u|\geq a} ||\mathcal{L}_{g+iu\overline{f}}1|| |\widehat{\chi}_n(u)| du &= O\left(n^{-(1-\kappa)(\delta'-\delta)} 2^{\beta k/2}\right) \\ &= O\left(n^{-(1-\kappa)(\delta'-\delta)} n^{(1+\gamma)\beta/(2\alpha)}\right), \end{aligned} \tag{3.13}$$

where $\delta < \delta' < \frac{1}{\gamma}$, say. This contribution is of order $O(n^{-(\delta+\frac{1}{2})})$ if $\kappa > 0$ is sufficiently large and then $\beta = \beta(\kappa, \gamma) > 0$ is chosen sufficiently small.

To bound the difference between (3.11) and (3.13), we shall use the trivial identity

$$\mathcal{L}^n_{g+iu\overline{f}} - \mathcal{L}^n_{g+iu\overline{f}_k} = \sum_{j=1}^{n-1} \mathcal{L}^j_{g+iu\overline{f}} \left(\mathcal{L}_{g+iu\overline{f}} - \mathcal{L}_{g+iu\overline{f}_k}\right) \mathcal{L}^{(n-j)}_{g+iu\overline{f}_k}. \tag{3.14}$$

By Lemma 3.7, we can bound

$$||\mathcal{L}_{g+iu\overline{f}} - \mathcal{L}_{g+iu\overline{f}_k}|| \leq C|u| ||\overline{f} - \overline{f}_k||, \tag{3.15}$$

provided $|u|$ is sufficiently small. The following bound will also be quite useful.

LEMMA 3.9. *There exists $C > 0$ such that*

$$||\mathcal{L}^m_{g+iu\overline{f}}|| \leq C|u| \text{ and } ||\mathcal{L}^m_{g+iu\overline{f}_k}|| \leq C|u|, \text{ for } m \geq 1,$$

for all $k \geq 1$.

Let $\rho > 0$ be a value to be specified later. Using (3.14), we can bound

$$\begin{aligned} &\int_a^{n^\rho} ||\mathcal{L}^n_{g+iu\overline{f}} - \mathcal{L}^n_{g+iu\overline{f}_k}|| ||\widehat{\chi}_n(u)||_\infty du \\ &\leq \int_a^{n^\rho} \left(\sum_{j=1}^{n-1} ||\mathcal{L}^j_{g+iu\overline{f}}||_\infty ||\mathcal{L}_{g+iu\overline{f}} - \mathcal{L}_{g+iu\overline{f}_k}|| ||\mathcal{L}^{(n-j)}_{g+iu\overline{f}_k}|| ||e^{iuv_k}||\right) ||\widehat{\chi}_n(u)||_\infty du. \end{aligned} \tag{3.16}$$

Furthermore, we can use Lemmas 3.3, 3.6 and 3.9 to obtain a bound for (3.16) of order

$$\begin{aligned} &O\left(\frac{2^{k\beta/2}}{\epsilon_n^{\kappa-1}}(n-1)\int_a^{n^\rho} |u|^2 ||\overline{f} - \overline{f}_k|| u^{-\kappa} du\right) \\ &= O\left(2^{k\beta/2} n^{(\kappa-1)\delta} n^{1+\rho} \left(\frac{1}{2}\right)^{k(\alpha-\beta/2)}\right) \\ &= O\left(n^{(1+\gamma)\beta/(2\alpha)} n^{(\kappa-1)\delta} n^{1+\rho} n^{-(1+\gamma)(1-\beta/(2\alpha))}\right). \end{aligned} \tag{3.17}$$

The remaining contribution to the integral can be bounded as

$$\begin{aligned}\int_{n^\rho}^{\infty} ||\mathcal{L}^n_{g+iu\bar{f}} - \mathcal{L}^n_{g+iu\bar{f}_k}|| \, |\widehat{\chi}_n(u)|du &\leq \int_{n^\rho}^{\infty} 2C|u| \, |\widehat{\chi}_n(u)|du \\ &= O\left(\frac{1}{\epsilon_n^{\kappa-1}} \int_{n^\rho}^{\infty} \frac{1}{u^{\kappa-1}} du\right) \\ &= O\left(\frac{n^{\delta(\kappa-1)}}{n^{\rho(\kappa-2)}}\right).\end{aligned} \tag{3.18}$$

We can first choose $\rho = \rho(\kappa)$ sufficiently large that the contribution of (3.18) is $O(n^{-(\delta+\frac{1}{2})})$. If we then assume that $\gamma = \gamma(\kappa, \rho)$ is sufficiently large, the contribution from (3.17) is of the same size. (If $\beta < \alpha/2$, then γ can be chosen independently of β. This is necessary, since earlier in the proof we need to choose $\beta = \beta(\gamma)$ small.) This finally completes the proof of Theorem 2.

4. Periodic orbits

In this final section, we shall sketch the proof of a version of Theorems 1 and 2 for sums over periodic points. In this case the result holds for hyperbolic diffeomorphisms, as well as expanding maps. We shall suppose that $T : X \to X$ is either the restriction of an Axiom A diffeomorphism to a non-trivial basic set or an expanding map. As above, we additionally assume that T is topologically mixing.

Let f be a non-lattice function. Suppose that there exists a Gibbs state m such that $\int f dm = 0$. Without loss of generality, we may choose the measure m to be the Gibbs state for a function $g = \xi f$, for some unique $\xi \in \mathbb{R}$ [**PS1**, Lemma 5]. With this choice, the supremum

$$\beta = \sup \left\{ h(\nu) \, : \, \nu \in \mathcal{M} \text{ such that } \int f d\nu = 0 \right\}$$

is attained at $\nu = m$, i.e., $\beta = h(m)$.

We recall the following asymptotic formula proved in [**PS1**].

PROPOSITION 4.1. *Let f be a non-lattice function. Suppose that there exists a Gibbs state m such that $\int f dm = 0$. Then, for any real numbers $a < b$, we have that*

$$\# \{x \in \mathrm{Fix}(T^n) \, : \, a \leq f^n(x) \leq b\} \sim \frac{1}{\sqrt{2\pi}\sigma} \left(\int_a^b e^{-\xi t} dt\right) \frac{e^{\beta n}}{\sqrt{n}}, \tag{4.1}$$

as $n \to +\infty$, where ξ is defined as above.

We shall strengthen (4.1) to give an asymptotic formula when the interval $[a, b]$ is allowed to shrink at a subexponential rate.

THEOREM 3. *Suppose that $f : X \to \mathbb{R}$ is a Hölder continuous function satisfying the Diophantine condition and that $\int f dm = 0$, where m is the Gibbs state for ξf. Then there exists $\delta > 0$ such that for any $z \in \mathbb{R}$, $a < b$ and sequence $\epsilon_n > 0$ which tends to zero and satisfies $\epsilon_n^{-1} = O(n^\delta)$, we have that*

$$\# \{x \in \mathrm{Fix}(T^n) \, : \, z + a\epsilon_n \leq f^n(x) \leq z + \epsilon_n b\} \sim \frac{1}{\sqrt{2\pi}\sigma} \left(\int_a^b e^{-\xi t} dt\right) \frac{\epsilon_n e^{\beta n}}{\sqrt{n}},$$

as $n \to +\infty$.

The proof follows the same lines as that of Theorem 1. The diffeomorphism $T : X \to X$ is modelled by a two-sided subshift of finite type $\sigma : \Sigma \to \Sigma$, where

$$\Sigma = \{x = (x_n)_{n=-\infty}^{\infty} : A(x_n, x_{n+1}) = 1 \ \forall n \in \mathbb{Z}\}.$$

The correspondence between periodic points for $T : X \to X$ and $\sigma : \Sigma \to \Sigma$ is not one-to-one but this discrepancy does not affect the asymptotics.

In order to use the transfer operator analysis from the earlier sections, one may pass from $\sigma : \Sigma \to \Sigma$ to the one-sided subshift $\sigma : \Sigma^+ \to \Sigma^+$. To do this, we apply the following standard lemma [**Si**] (cf. Lemma 3.2).

LEMMA 4.2. *Let $f : \Sigma \to \mathbb{R}$ be a Hölder continuous function. Then there exists a Hölder continuous function $f' : \Sigma \to \mathbb{R}$ which is cohomologous to f and has the property that $f'(x) = f'(y)$ if $x_n = y_n$ for $n \geq 0$. In particular, we may regard f' as a function $f' : \Sigma^+ \to \mathbb{R}$.*

For periodic orbits, Lemma 1.3 is replaced by the following result, which can be easily deduced from results in [**Ru2**].

LEMMA 4.3. *There exists $0 < \theta < 1$ such that, for any $x_0 \in \Sigma^+$,*

$$\sum_{\sigma^n x = x} e^{(\xi+iu)f^n(x)} = (\mathcal{L}^n_{(\xi+iu)f}1)(x_0)(1 + O(\max\{1, |u|\}n\theta^n)).$$

To proceed, we can use the identity

$$\begin{aligned} \rho(n) &= \frac{1}{2\pi}\int_{-\infty}^{\infty}\left(\sum_{\sigma^n x = x} e^{iuf^n(x)}\right)\widehat{\chi}_n(u)du \\ &= \frac{1}{2\pi}\int_{-\infty}^{\infty}\Big((\mathcal{L}^n_{(\xi+iu)f}1)(x_0)(1 + O(\max\{1, |u|\}n\theta^n))\Big)\widehat{\chi}_n(u)du. \end{aligned} \tag{4.2}$$

The proof now follows the same lines as in Theorem 1. Using the bounds on the transfer operators from Section 2, the same arguments give the estimate

$$\frac{1}{2\pi}\int_{-\infty}^{\infty}\Big((\mathcal{L}^n_{(\xi+iu)f}1)(x_0)\Big)\widehat{\chi}_n(u)du \sim \frac{\int \chi(x)dx}{\sqrt{2\pi}\sigma}\frac{\epsilon_n}{\sqrt{n}}, \quad \text{as } n \to +\infty.$$

The contribution of the second term on the right-hand side of (4.2) is dominated by this principal term because of the θ^n factor.

Finally, the proof of Theorem 3 is completed by an approximation argument, as at the end of Section 2.

REMARK 4.4. The method of proof of the results in this paper should also lend itself to proving uniform "local limit theorems" for shrinking intervals (cf. [**Ro**]).

References

[CP] Z. Coelho, W. Parry, *Central limit asymptotics for shifts of finite type,* Israel J. Math. **69** (1990), 235–249.

[DP] M. Denker, W. Philipp, *Approximation by Brownian motion for Gibbs measures and flows under a function,* Ergodic Theory Dynam. Systems **4** (1984), 541–552.

[D] D. Dolgopyat, *Prevalence of rapid mixing in hyperbolic flows,* Ergodic Theory Dynam. Systems **18** (1998), 1097–1114.

[FMT] M. Field, I. Melbourne, A. Török, *Stability of rapid mixing for hyperbolic flows,* preprint, 2004.

[L] S. Lalley, *Ruelle's Perron-Frobenius theorem and the central limit theorem for additive functionals of one-dimensional Gibbs states,* Adaptive statistical procedures and related topics (Upton, N.Y., 1985), IMS Lecture Notes Monogr. Ser., vol. 8, Inst. Math. Statist., Hayward, CA, 1986, pp. 428–446.

[P] W. Parry, *Bowen's equidistribution theory and the Dirichlet density theorem,* Ergodic Theory Dynam. Systems **4** (1984), 117–134.

[PP] W. Parry, M. Pollicott, *Zeta functions and the periodic orbit structure of hyperbolic dynamics,* Astérisque **187–188** (1990), 1–268.

[PS1] M. Pollicott, R. Sharp, *Rates of recurrence for* $\mathbb{Z}^q$ *and* $\mathbb{R}^q$ *extensions of subshifts of finite type,* J. London Math. Soc. **49** (1994), 401–418.

[PS2] M. Pollicott, R. Sharp, *Error terms for closed orbits of hyperbolic flows,* Ergodic Theory Dynam. Syst. **21** (2001), 545–562.

[Ra] M. Ratner, *The central limit theorem for geodesic flows on* n*-dimensional manifolds of negative curvature,* Israel J. Math. **16** (1973), 181–197.

[Ro] J. Rousseau-Egele, *Un théorème de la limite locale pour une classe de transformations dilatantes et monotones par morceaux,* Ann. Probab. **11** (1983), 772–788.

[Ru1] D. Ruelle, *Thermodynamic Formalism,* Encyclopedia of Mathematics and its Applications, vol. 5, Addison-Wesley, Reading, MA, 1978.

[Ru2] D. Ruelle, *An extension of the theory of Fredholm determinants,* Inst. Hautes Études Sci. Publ. Math. **72** (1990), 175–193.

[Si] Ya. Sinai, *Gibbs measures in ergodic theory,* Uspehi Mat. Nauk **27** (1972), 21–64.

[St] G. Stephenson, *Mathematical methods for science students,* Longman, London, 1961.

Department of Mathematics, University of Warwick, Coventry, CV4 7AL, United Kingdom
E-mail address: mpollic@maths.warwick.ac.uk

Department of Mathematics, University of Manchester, Oxford Road, Manchester, M13 9PL, United Kingdom
E-mail address: sharp@maths.man.ac.uk

Amer. Math. Soc. Transl.
(2) Vol. **217**, 2006

Transverse Properties of Dynamical Systems

Jean Renault

ABSTRACT. Roughly speaking, two dynamical systems are transversally equivalent if they have the same space of orbits; a property is transverse if it is preserved under transverse equivalence. Various notions of transverse equivalence have been defined: among them, similarity of measured groupoids, Morita equivalence of locally compact groupoids, stable orbit equivalence of measure equivalence relations. After reviewing some of these notions, we go on to discussing the Morita equivalence of groupoids and give examples and applications, in particular in connection with operator algebras.

1. Introduction

The dynamical systems of the present paper are pairs $(X, \mathcal{G})$ consisting of a space X and a family $\mathcal{G}$, usually a group or a semi-group, of transformations of X. They also include foliations $(X, \mathcal{F})$. The orbits of a dynamical system (or the leaves of a foliation) form a partition of the space X. Roughly speaking, a transverse property of the dynamical system is a property of its orbit space $X/\mathcal{G}$ (or its leaf space $X/\mathcal{F}$ in the case of a foliation). The kind of question we may ask is, for example, "Are there transverse properties which distinguish two irrational rotations of the circle?" The answer is not obvious because their orbit spaces are singular and not amenable to classical analysis. Therefore, we need a precise definition saying when two orbit spaces are the same (one also says that the dynamical systems are "orbit equivalent"). Of course, it will depend on the way we view the orbit space, for example as a measured space or as a topological space.

Groupoids provide a convenient framework for giving such a definition. There are groupoids (usually one) naturally attached to the dynamical systems we consider here. After introducing a suitable notion of equivalence of groupoids, a transverse property of a dynamical system can be defined as a property of the equivalence class of its groupoid. This idea appears in the work of G. Mackey (cf. [**Ma**]) in the framework of ergodic theory (for a different purpose, namely the study of unitary group representations). In this context, he considers measured groupoids, and the notion of equivalence he uses is similarity of measured groupoids; he defines

2000 *Mathematics Subject Classification.* Primary 37D35; Secondary 46L85.
Key words and phrases. Locally compact groupoids, C*-algebras, Morita equivalence, amenability, groupoid cohomology, Brauer group.

a virtual group as a similarity class of measured groupoids. This notion has been studied extensively by A. Ramsay (e.g. in [**Ra**]).

Specialized to countable standard measured equivalence relations, as defined by Feldman and Moore in [**FM**], similarity agrees with Kakutani equivalence. Here X is a standard Borel space, R is an equivalence relation on X which has countable classes and a Borel graph (also denoted by R), and $[\mu]$ is the class of a σ-finite measure on X quasi-invariant under R. The action of a countable group of Borel automorphisms of X preserving the measure class $[\mu]$ provides such an equivalence relation (in fact, they can all be obtained that way). First, two countable standard measured equivalence relations $(X_i, R_i, [\mu_i]), i = 1, 2$, are said to be *isomorphic* if the measured groupoids $(R_i, [\mu_i])$ are isomorphic; this amounts to saying that there exists a Borel isomorphism $\varphi : X_1' \to X_2'$, where X_i' is a conull set in X_i, sending μ_1 onto a measure equivalent to μ_2 and such that for all $x \in X_1'$, $\varphi(R_1(x)) = R_2(\varphi(x))$. Two countable standard measured equivalence relations $(X_i, R_i, [\mu_i]), i = 1, 2$, are said to be *similar* if there exists a countable standard measured equivalence relations $(Y, S, [\mu])$ and Borel subsets $Y_i, i = 1, 2$, of Y having positive measure and meeting almost every orbit of S such that $(X_i, R_i, [\mu_i])$ are, respectively, isomorphic to $(Y_i, S_{|Y_i}, [\mu|Y_i])$. The most studied case is when there is a probability measure μ on X invariant under the equivalence relation R (when the relation arises from the action of a countable group, this agrees with the usual definition of an invariant measure). One says then that (X, R, μ) is a probability measure preserving (p.m.p.) equivalence relation. An isomorphism of two ergodic p.m.p. equivalence relations $(X_i, R_i, \mu_i), i = 1, 2$, automatically preserves the invariant probability measures. In the non-ergodic case, one may want to add this condition. This notion of isomorphism of p.m.p. equivalence relations is crucial in the classification of p.m.p. actions of countable groups (in this context, one often says that the actions are *orbit equivalent* when their orbit equivalence relations are isomorphic), starting with the earlier work by Murray and von Neumann [**MN**] and followed by H. Dye [**D1, D2**], A. Vershik, R. Belinskaya and others. The recent survey [**G**] by D. Gaboriau gives a clear account of the present theory of orbit equivalence of group actions.

Let us consider now topological dynamical systems. We have in mind for example a locally compact group acting continuously on a locally compact space. Then there is a topological groupoid, called its semi-direct product, attached to such a dynamical system. There is a convenient notion of equivalence of topological groupoids, introduced in [**Re2**], which replaces Mackey's similarity of measured groupoids. Since it is prompted by the notion of Morita equivalence of C*-algebras introduced by M. Rieffel in [**Ri**] and gives an equivalence of categories, it is often called Morita equivalence of groupoids. It is the object of this communication.

The theory of foliations and the study of orbifolds teach us that, due to the possible presence of holonomy, it is more natural to introduce a groupoid than an equivalence relation. I should point out that, even when applied to a free action of a group, where the semi-direct product groupoid is an equivalence relation, the notions of isomorphism or equivalence of groupoids discussed in this talk are not the notions of topological orbit equivalence and Kakutani equivalence used by Giordano, Putnam and Skau in [**GPS**]. Indeed, let $(X_i, R_i), i = 1, 2$, be equivalence relations on topological spaces. Orbit equivalence in the sense of [**GPS**] is given by a homeomorphism $\varphi : X_1 \to X_2$ such that for all $x \in X_1$, $\varphi(R_1(x)) = R_2(\varphi(x))$. In the context discussed here, the equivalence relations R_i are topological groupoids, and isomorphism requires that the induced bijection $\varphi \times \varphi : R_1 \to R_2$ respects the topology, which is a stronger requirement. It should also be said that a conjugation of two dynamical systems gives an isomorphism of the corresponding groupoids but not conversely. On the other hand, the notion of equivalence of groupoids used here and its derived notion of transverse property agree with that used in the theory of foliations (see [**H2**] and [**Ml**, Section 2.7]).

The next section recalls the definition of Morita equivalence of C*-algebras and topological groupoids. One of the most basic transverse invariants is K-theory, which is briefly discussed in Section 3. Topological amenability, which is a transverse property, and its applications are reviewed in Section 4. Section 5 briefly discusses cohomology of topological groupoids and gives a construction of the Dixmier-Douady class of an elementary equivariant C*-bundle. This is interesting for our purpose since it makes good use of both Morita equivalence of C*-algebras and equivalence of groupoids. The definitions and results of Sections 2–4 are well known. Section 5 relies on [**T2, TXL**] and[**KMRW**]. Although I have no explicit reference for it, the construction of the Dixmier-Douady class presented here will look familiar to the specialists. It is based on a talk which I gave at a conference on groupoids and stacks held in Luminy, June 28–July 2, 2004. I thank the participants of both conferences for stimulating discussions and in particular J.-L. Tu and X. Tang.

2. Morita equivalence of C*-algebras and groupoids

2.1. C*-algebras. Let us recall the now classical notion of Morita equivalence of C*-algebras, introduced by M. Rieffel in [**Ri**]. It is relevant here for several reasons: it will appear a few times and it has prompted the definition of equivalence of groupoids given below.

DEFINITION 2.1 ([**Ri**, Definition 2.1]). Let B be a C*-algebra. A right *B-C*-module* is a complex linear space E endowed with

(i) a structure of right-B-module
$$(x, b) \in E \times B \mapsto xb \in E,$$
(ii) a B-valued "inner product"
$$(x, y) \in E \times E \mapsto \langle x, y\rangle \in B$$
satisfying the following:
(a) it is B-linear in the second variable y,
(b) $\langle y, x\rangle^* = \langle x, y\rangle$,
(c) $\langle x, x\rangle$ is a positive element of B,
(d) $\|x\| = \|\langle x, x\rangle\|^{1/2}$ is a complete norm on E.

We say that a B-C*-module E is *full* if the linear span of the range of the inner product is dense in B.

One recognizes the definition of a Hilbert space when $B = \mathbf{C}$. The notions of bounded and compact linear operators can be defined in this new setting. Given two C*-modules E and F, one defines $\mathcal{L}_B(E, F)$ as the space of bounded B-linear operators $T : E \to F$ which admit an adjoint and $\mathcal{K}_B(E, F)$ as the norm-closure of the linear span of the rank-one operators $\theta_{x,y}$, where $x \in E, y \in E$ and for $z \in E$,
$$\theta_{x,y}(z) = x\langle y, z\rangle.$$
For $F = E$, one gets the C*-algebra $\mathcal{L}_B(E)$ and its closed ideal $\mathcal{K}_B(E)$.

DEFINITION 2.2 ([**Ri**, Definition 4.19]). Let A and B be C*-algebras. An (A, B)-*C*-correspondence* (or a correspondence from A to B) is a right B-C*-module E together with a $*$-homomorphism $\pi : A \to \mathcal{L}_B(E)$.

We shall usually view an (A, B)-C*-correspondence as an (A, B)-bimodule. A $*$-homomorphism $\pi : A \to B$ defines an (A, B)-C*-correspondence, by considering $E = B$ as a right B-C*-module (then $\mathcal{L}_B(E) = B$). It is useful to view C*-correspondences as generalized $*$-homomorphisms. There is a composition of correspondences extending composition of $*$-homomorphisms: given a C*-correspondence E from A to B and a C*-correspondence F from B to C, one can construct the C*-correspondence $E \otimes_B F$ from A to C. It is the C-C*-module obtained by separation and completion of the ordinary tensor product $E \otimes F$ with respect to the inner product

$$\langle x \otimes y, x' \otimes y' \rangle = \langle \langle x, x' \rangle_B, y' \rangle_C, \qquad x, x' \in E, y, y' \in F;$$

the left A action is given by $a(x \otimes y) = ax \otimes y$ for $a \in A$. Associativity holds up to isomorphism. More precisely, given C*-algebras A, B, C, D and C*-correspondences E from A to B, F from B to C and G from C to D, there is a canonical isomorphism $(E \otimes_B F) \otimes_C G \to E \otimes_B (F \otimes_C G)$.

We say that an (A, A)-correspondence E is an identity if it is isomorphic to the correspondence A (defined by the identity map) and that an (A, B)-correspondence E is invertible if there exists a (B, A)-correspondence F such that $E \otimes_B F$ and $F \otimes_A F$ are identity correspondences. One has the following easy characterization of invertible correspondences.

PROPOSITION 2.3 ([**Ri**, Section 6]). *Let A and B be C*-algebras and let E be a linear space. Then the following conditions are equivalent:*

(i) *E is invertible (A, B)-C*-correspondence,*
(ii) *E is a full right B-C*-module and there is a $*$-isomorphism $\pi : A \to \mathcal{K}_B(E)$,*
(iii) *E is a full right B-C*-module and a full left A-C*-module such that for a in A, x, y, z in E and b in B,*
  (a) $(ax)b = a(xb)$,
  (b) ${}_A\langle x, y \rangle z = x \langle y, z \rangle_B$,
  (c) $\langle ax, ax \rangle_B \leq \|a\|^2 \langle x, x \rangle_B$,
  (d) ${}_A\langle xb, xb \rangle \leq \|b\|^2 \, {}_A\langle x, x \rangle$.

DEFINITION 2.4 ([**Ri**, Definition 6.10]). An (A, B)-bimodule as in the above proposition is called an (A, B)-*Morita equivalence*. Two C*-algebras A and B are said to be *Morita equivalent* if there exists an (A, B)-Morita equivalence.

We leave it to the reader to check that A^n is an $(M_n(A), A)$-Morita equivalence (where $M_n(A)$ is the C*-algebra of $n \times n$ matrices over A). Another related basic example is the following: let C be a C*-algebra and let e, f be two projections in C such that $CeC = CfC$. Then eCf is a (eCe, fCf)-Morita equivalence. One should note the analogy with the definition of Kakutani equivalence recalled in the introduction. Let E be an (A, B)-Morita equivalence. We define an inverse (B, A)-Morita equivalence $\tilde{E}$ by a bijection $x \in E \mapsto \tilde{x} \in \tilde{E}$ and the following operations:

(i) $\tilde{x}+\tilde{y}=\widetilde{(x+y)}$, $\widetilde{(\lambda x)}=\overline{\lambda}\tilde{x}$ for $x,y\in X$ and $\lambda\in\mathbf{C}$,

(ii) $b^*\tilde{x}a^*=\widetilde{axb}$ for $(a,x,b)\in A\times E\times B$,

(iii) $\langle\tilde{x},\tilde{y}\rangle_A={}_A\langle x,y\rangle$ and ${}_B\langle\tilde{x},\tilde{y}\rangle=\langle x,y\rangle_B$.

Then $E\otimes_B\tilde{E}$ [resp. $\tilde{E}\otimes_A E$] is isomorphic to the identity correspondence A [resp. B] via the inner product.

2.2. Groupoids. Let us review some of the definitions. More information for this section can be found in [**Re1, Re2, MRW**]. The most concise way to define a groupoid is to say that it is a small category with inverses. More concretely, we have a set of units $G^{(0)}$, with elements denoted by $x,y,\ldots$, a set of arrows G, with elements denoted by $\gamma,\gamma',\ldots$, the range and the source maps $r,s:G\to G^{(0)}$, the product map from $G^{(2)}=\{(\gamma,\gamma')\in G\times G:s(\gamma)=r(\gamma')\}$ to G and the inverse map $\gamma\in G\mapsto\gamma^{-1}\in G$. We view $G^{(0)}$ as a subset of G so that we have $r(\gamma)=\gamma\gamma^{-1}$ and $s(\gamma)=\gamma^{-1}\gamma$. Let us give a few examples, arising mostly from dynamical systems.

Groups. This is the case when $G^{(0)}$ has a single element.

Equivalence relations. Suppose that R is an equivalence relation on the set X. Then, $G^{(0)}=X$, $G=\{(x,y)\in X\times X:xRy\}$ is the graph of the equivalence relation (from now on, we shall simply write $G=R$); the range and source maps are, respectively, the first and the second projections, the product is given by $(x,y)(y,z)=(x,z)$ and the inverse by $(x,y)^{-1}=(y,x)$. The identification of $G^{(0)}$ as a subset of G is given by the diagonal map $x\mapsto(x,x)$.

Group actions. This is the classical case of a dynamical system. A group Γ acts on a space X (for convenience here on the right). We denote by $(x,t)\in X\times\Gamma\mapsto xt\in X$ the action map. Then we define $G^{(0)}=X$ and the groupoid of the action, also called the semi-direct product,

$$G=X\rtimes\Gamma=\{(x,t,y)\in X\times\Gamma\times X:xt=y\}.$$

The range and source maps are, respectively, given by $r(x,t,y)=x$ and $s(x,t,y)=y$. The product map is given by $(x,t,y)(y,t',z)=(x,tt',z)$ and the inverse by $(x,t,y)^{-1}=(y,t^{-1},x)$. Note that G is a subgroupoid of the product groupoid $\Gamma\times(X\times X)$ where $X\times X$ is the graph of the trivial equivalence relation. The image of the projection of G into $X\times X$ is the graph of the orbit equivalence relation. The action is free if and only if this map is one-to-one.

Endomorphisms. Suppose that we have a map T of X into itself, not necessarily invertible. We define $G^{(0)}=X$ and the groupoid of the endomorphism

$$G=\{(x,m-n,y)\in X\times\mathbf{Z}\times X:m,n\in\mathbf{N},T^mx=T^ny\}.$$

The maps and operations are the same as above. When T is invertible, G is the groupoid of the action of $\mathbf{Z}$ given by T. The following subgroupoid of G is also relevant:

$$R=\{(x,y)\in X\times\mathbf{Z}\times X:\exists m\in\mathbf{N},T^mx=T^my\}.$$

These groupoids have been studied in particular by A. Vershik, who also introduced a generalization of them, with the notion of polymorphisms ([**V2**]; see also [**ArR**]).

Foliations. We introduce briefly this important example, which goes back to C. Ehresmann. Let $\mathcal{F}$ be a foliation, i.e., a partition into leaves, of the space X.

A path γ with endpoints $s(\gamma)$ and $r(\gamma)$ traced on a leaf defines a holonomy; we identify two paths defining the same holonomy and denote by $[\gamma]$ the holonomy classes $[\gamma]$. Then we define $G^{(0)} = X$ and the holonomy groupoid

$$G = \{(x, [\gamma], y) : x, y \in X, r(\gamma) = x, s(\gamma) = y\}.$$

The maps and operations are the same as above, with $[\gamma][\gamma'] = [\gamma\gamma']$, where $\gamma\gamma'$ is the concatenated path. One also considers transverse holonomy groupoids: they are reductions $G_{|T}$ to a full (i.e., meeting every leaf) transversal $T \subset X$.

Groupoid actions can be defined as functors. More concretely, a left G-space, where G is a groupoid, consists of a space Z, a projection map $r : Z \to G^{(0)}$ (assumed to be surjective) and an action map from $G * Z$ to Z, where $G * Z$ is the set of composable pairs (γ, z) with $s(\gamma) = r(z)$, satisfying the usual associativity rules; the image of (γ, z) is denoted by γz. A right G-space is defined similarly. One says that the action is *free* if the equality $\gamma z = z$ implies that γ is the unit $r(z)$.

Two basic examples of actions of a groupoid G are, on one hand, its action on its unit space $G^{(0)}$ given by $\gamma y = x$ where $x = r(\gamma)$ and $y = s(\gamma)$ and, on the other hand, its action on itself by left multiplication. This latter action is always free.

The definitions of a *topological groupoid* and of a continuous action are straightforward: with the above notation, G, $G^{(0)}$ and Z are topological spaces and all the maps we have introduced are continuous. There is a slight difficulty here because one may have to consider non-Hausdorff groupoids, for example the holonomy groupoid of the Reeb foliation. For the sake of simplicity, we assume that these topological spaces are Hausdorff. There is a further assumption that we make: the range, source and projection maps are assumed to be open. We then say the topological G-space Z is *proper* if the map from $G * Z$ (endowed with the topology induced by the product topology of $G \times Z$) to $Z \times Z$ which sends (γ, z) to $(\gamma z, z)$ is proper. For example the left action of G on itself is always proper.

The examples given above provide topological groupoids. For example, the groupoid of a continuous group action, endowed with the topology induced by the product topology of $X \times \Gamma \times X$, is a topological groupoid. The holonomy groupoid of a smooth foliation $\mathcal{F}$ of dimension p on a manifold of dimension $p + q$ has a natural structure of manifold of dimension $2p + q$, which turns it into a topological groupoid. The groupoid of an endomorphism which is a local homeomorphism has also a natural topology (see for example [**ArR**]).

Equivalence of topological groupoids can be presented in a fashion analogous to Morita equivalence of C*-algebras.

DEFINITION 2.5. Let G and H be topological groupoids. A (G, H)-*correspondence* (also called a correspondence from G to H) is a right free and proper H-space Z endowed with a left action of G such that its projection map $r : Z \to G^{(0)}$ gives a homeomorphism of Z/H onto $G^{(0)}$.

A groupoid homomorphism $\phi : G \to H$ defines a (G, H)-correspondence

$$Z = G^{(0)} * H = \{(x, \eta) \in G^{(0)} \times H : \phi^{(0)}(x) = r(\gamma)\},$$

where H acts on the right by$(x, \eta)\eta' = (x, \eta\eta')$, and G acts on the left by $\gamma(s(\gamma), \eta) = (r(\gamma), \phi(\gamma)\eta)$. Correspondences appear under various names in the literature; they are called morphisms of quotient spaces in [**HS**], generalized morphisms in [**L1**] and

Hilsum–Skandalis maps in [**Mr**]. There is a composition of correspondences extending the composition of groupoid homomorphisms: given a (G, H)-correspondence X and an (H, K)-correspondence Y, one can construct the (G, K)-correspondence $X \circ Y = X * Y/H$, where $X * Y$ is the subspace of $X \times Y$ consisting of pairs (x, y) such that $s(x) = r(y)$ and H acts by $(x, y)\eta = (x\eta, \eta^{-1}y)$. We say that a (G, G)-correspondence Z is an identity if it is isomorphic to the correspondence G (defined by the identity map) and that a (G, H)-correspondence X is invertible if there exists a (H, G)-correspondence Y such that $X \circ Y$ and $Y \circ X$ are identity correspondences. One has the following easy characterization of invertible correspondences.

PROPOSITION 2.6. *Let G and H be topological groupoids and let Z be a topological space endowed with commuting left G- and right H-actions. Then the following conditions are equivalent:*

(i) *Z is an invertible (G, H)-correspondence.*
(ii) *The actions are free and proper; the projection map $r : Z \to G^{(0)}$ gives a homeomorphism of Z/H onto $G^{(0)}$ and projection map $s : Z \to H^{(0)}$ gives a homeomorphism of $G\backslash Z$ onto $H^{(0)}$.*

DEFINITION 2.7 ([**Re2**, Section 3], [**MRW**, Definition 2.1]). A (G, H)-space as in the above proposition is called a (G, H)-*equivalence*. Two topological groupoids G and H are said to be *equivalent* if there exists a (G, H)-equivalence.

Let Z be a free and proper right H-space. The graph $Z * Z$ of the equivalence relation defined by the projection map $s : Z \to H^{(0)}$ is a topological groupoid and so is its quotient $Z * Z/H$ by the diagonal action of H. This groupoid has an obvious left action on Z which turns Z into a $(Z * Z/H, H)$-equivalence. Conversely, if Z is a (G, H)-equivalence, then G is isomorphic to $Z * Z/H$.

EXAMPLE 2.8. Let $p : Z \to X$ be a surjective continuous and open map, where Z, X are topological spaces. Then, $Z * Z$ (fibred product above X), with the topological groupoid structure inherited from $Z \times Z$ is equivalent to X via Z. In particular, if $\mathcal{U} = \{U_i\}_{i \in I}$ is an open cover of X, if $Z = X_{\mathcal{U}}$ is the disjoint union of the U_i's and if $p : Z \to X$ is the obvious map, we get the groupoid associated with the open cover:

$$G_{\mathcal{U}} = \{(i, x, j) : x \in U_{ij}\}$$

which is equivalent to X.

EXAMPLE 2.9. Let H be a closed subgroup of a topological group G. Then, the right action of H on G by right multiplication is free and proper. Moreover, the semi-direct groupoid $G \ltimes (G/H)$ acts on G on the left (the projection map $r : G \to G/H$ is the quotient map). This makes G into a $(G \ltimes (G/H), H)$-equivalence. This equivalence between H and $G \ltimes (G/H)$ is at the heart of the theory of induced representations.

EXAMPLE 2.10. More generally, let P and Γ be two closed subgroups of a topological group G. Then, the semi-direct product groupoids $\Gamma \ltimes (G/P)$ and $(\Gamma\backslash G) \rtimes P$ are equivalent via the equivalence G.

EXAMPLE 2.11. Let T and T' be transversals of a foliation $(X, \mathcal{F})$ which meet each leaf. Then the holonomy groupoid $G = G(X, \mathcal{F})$ and its reductions to the transversals $G_{|T}$ and $G_{|T'}$ are all equivalent. For example $G^T_{T'} = r^{-1}(T) \cap s^{-1}(T')$ is a $(G_{|T}, G_{|T'})$ equivalence. Let us give as an elementary example the Kronecker foliation of irrational slope α on the 2-torus $\mathbf{T}^2$. The reduction of its holonomy groupoid to a closed transversal given by a line of rational slope is the groupoid of an irrational rotation of angle $2\pi\beta$, where β is in the orbit of α under the action of $PSL(2,\mathbf{Z})$; moreover, any β in this orbit can be obtained. Therefore, two irrational rotations of respective angles $2\pi\beta$ and $2\pi\beta'$, where β and β' are in the same orbit under $PSL(2,\mathbf{Z})$, have equivalent groupoids.

The classifying space of a topological groupoid G, defined as the quotient $BG = EG/G$ of a universal G-space EG (say within the category of free and proper G-spaces) provides some of the most basic transverse invariants (see [**H2**, Section 3]). In the example of the irrational rotation, this classifying space is, up to homotopy equivalence, the 2-torus itself. In general, it is not so accessible.

3. C*-algebras and analytic K-theory

As pointed out earlier, our point of view owes very much to the theory of operator algebras. The analogy of the presentation of Morita equivalence for C*-algebras and for topological groupoids can be given a mathematical form via a suitable form [**Re1**] of the Murray-von Neumann "group-measure construction" [**MN**]. We have to impose additional assumptions on our topological groupoids: we require that they are second countable and locally compact and that they have a Haar system.

Let us explain this notion. Suppose that G is a locally compact groupoid and that Z is a locally compact left G-space, with the projection map $r : Z \to G^{(0)}$. An *r-system* is a family $(\alpha^x), x \in G^{(0)}$, where α^x is a Radon measure on $Z^x = r^{-1}(x)$. It is said to be continuous if for all f in the space $C_c(Z)$ of continuous complex-valued functions on Z with compact support, the function $x \mapsto \alpha^x(f)$ is continuous. It is said to be invariant if for all $\gamma \in G$, the image of $\alpha^{s(\gamma)}$ by $\gamma : Z^{s(\gamma)} \to Z^{r(\gamma)}$ is $\alpha^{r(\gamma)}$. A *Haar system* for the locally compact groupoid G is a continuous and invariant r-system $(\lambda^x), x \in G^{(0)}$ for the left G-space G. We implicitly assume that all λ^x are non-zero.

Contrary to the case of locally compact groups, neither existence nor uniqueness hold in general. Existence of a Haar system is a topological assumption on G. For example, if the range map $r : G \to G^{(0)}$ has countable fibers, G has a Haar system if and only if r is a local homeomorphism. Then, the counting measures form a Haar system and one says that G is an *étale* groupoid. This is a large and interesting class of groupoids including groupoids of discrete group actions, groupoids of endomorphisms provided they are themselves local homeomorphisms and transverse holonomy groupoids. The groupoid $X \rtimes \Gamma$ of the continuous action of a locally compact groupoid Γ on a locally compact space has a Haar system of the form $\lambda^x = \delta_x \times \lambda$, where λ is a (left) Haar measure for Γ and δ_x is the point mass at x. The holonomy groupoid of a foliation also has Haar systems, with measures in the Lebesgue class.

Suppose that G is a locally compact groupoid with Haar system λ. Then $C_c(G)$ can be made into a $*$-algebra via the formulas:

$$f * g(\gamma) = \int f(\gamma\gamma')g(\gamma'^{-1})d\lambda^{s(\gamma)}(\gamma'),$$

$$f^*(\gamma) = \overline{f(\gamma^{-1})}.$$

Then one defines the full C*-algebra $C^*(G,\lambda)$ as its universal C*-completion. One also defines the reduced C*-algebra $C^*_{red}(G,\lambda)$ by introducing the regular representation (it acts no longer on a Hilbert space, as in the case of a group, but on the $C_0(G^{(0)})$-C*-module $L^2(G,\lambda)$). When $G = X \rtimes \Gamma$ comes from a group action, $C^*(G,\lambda)$ and $C^*_{red}(G,\lambda)$ are, respectively, the classical crossed-products $C^*(X,\Gamma)$ and $C^*_{red}(X,\Gamma)$.

THEOREM 3.1 ([**MRW**, Theorem 2.8]). *Let (G,λ) and (H,λ') be second countable locally compact groupoids with Haar systems. Then, for any (G,H)-equivalence Z, $C_c(Z)$ can naturally be completed into a $(C^*(G,\lambda), C^*(H,\lambda'))$-Morita equivalence.*

The same result (with the same proof) holds with the reduced C*-algebras. More generally, if Z is a free and proper right H-space which has an s-system α (here, $s : Z \to H^{(0)}$ denotes the projection map), one can complete $C_c(Z)$ into a right $C^*(H,\lambda')$-C*-module $C^*(Z,\alpha)$. If moreover, Z is a (G,H)-correspondence, $C^*(Z,\alpha)$ is also a right $C^*(G,\lambda)$-module and this makes it into a correspondence from $C^*(G,\lambda)$ to $C^*(H,\lambda')$. With suitable choices of systems of measures, this construction is functorial, in the sense that it respects composition up to isomorphism.

COROLLARY 3.2. *Let G be a second countable locally compact groupoid with Haar system λ. Then for $i = 0,1$ the analytic K-theory groups $K_i(C^*(G,\lambda))$ and $K_i(C^*_{red}(G,\lambda))$ are transverse invariants.*

This results from the theorem and the fact that an (A,B)-Morita equivalence of C*-algebras gives an isomorphism of their K-theory groups. These groups are called the analytic K-theory groups of the groupoid G. An early landmark was the computation by M. Pimsner and D. Voiculescu of these groups for the irrational rotation. As abelian groups ($K_0 = \mathbf{Z}^2, K_1 = \mathbf{Z}^2$), they do not distinguish two irrational rotations. However, K_0 has a natural order and this ordered group is also invariant under Morita equivalence; hence it is also a transverse invariant. For the irrational rotation of angle $2\pi\alpha$, this ordered group is $\mathbf{Z} + \alpha\mathbf{Z}$, with the order inherited from $\mathbf{R}$. Our earlier remark shows that this ordered group is a complete transverse invariant. At this point, let us briefly mention the Baum-Connes or assembly map. These authors have constructed a natural group homomorphism

$$\mu : K^*_{top}(G) \to K_*(C^*_{red}(G,\lambda))$$

where the left-hand side group is the so-called geometrical K-theory group; it involves the universal proper G-space (see [**C**] or [**T1**] for a precise definition). This map is compatible with groupoid equivalence. It has been shown to be an isomorphism in many cases.

4. Amenability

This notion, which plays an important role in the harmonic analysis of locally compact groups, can also be defined for dynamical systems. Here we shall only consider a definition of amenability for locally compact groupoids (topological amenability) and refer the reader to [**AnR**] for further information and references. Amenability of group actions also appears in V. Kaimanovich's talk.

The nicest kind of amenability is the following:

DEFINITION 4.1. We say that a locally compact groupoid G is *properly amenable* if it admits a Haar system $(m^x), x \in G^{(0)}$ consisting of probability measures m^x.

When G already admits a Haar system, this condition is equivalent to the properness of G, in the sense that $G^{(0)}$ is a proper G-space. In the case of a group, this amounts to saying that the group is compact and, in the case of a group action, that the action is proper. We look for a weakening of this condition.

DEFINITION 4.2. We say that a locally compact groupoid G is *topologically amenable* if it admits a *continuous approximate invariant mean*, i.e., a net $(m_i = \{m_i^x\}_{x\in G^{(0)}})$ of continuous systems of probability measures m_i^x on G^x which is approximately invariant in the sense that the function $\gamma \mapsto \|\gamma m_i^{s(\gamma)} - m_i^{r(\gamma)}\|_1$ tends to zero uniformly on the compact subsets of G.

In the case of the semi-direct product groupoid $G = X \rtimes \Gamma$ of a continuous action of a locally compact group Γ on a locally compact space X, we identify G^x with Γ and view m_i^x as a probability measure on Γ; the approximate invariance condition becomes: $(x,t) \mapsto \|tm_i^{xt} - m_i^x\|_1$ tends to zero uniformly on the compact subsets of $X \times \Gamma$. Amenability of the action of Γ on X is a synonym for amenability of the groupoid $G = X \rtimes \Gamma$. Any action of an amenable group is amenable. Non-amenable groups admit interesting amenable actions. For example, boundary actions of hyperbolic groups are amenable. Note that these actions cannot preserve any probability measure. An important property of topological amenability is its invariance under topological equivalence of groupoids.

THEOREM 4.3 ([**AnR**, 2.2.17]). *Topological amenability of locally compact groupoids is a transverse property.*

When G is an amenable locally compact groupoid with Haar system, its full and its reduced C*-algebras coincide [**AnR**, 6.1.8] and this C*-algebra is nuclear [**AnR**, 6.2.14]. Moreover, the Baum-Connes map is an isomorphism ([**T1**]). As an application of this notion, one deduces that groups which admit an amenable action on a compact space (e.g. boundary actions) satisfy the Novikov conjecture [**HR**]. Amenability for measured groupoids had been introduced earlier by R. Zimmer [**Z1, Z2**]; among other equivalent definitions, there is one analogous to Definition 4.2 [**Re1**, II.3.4]. It is invariant under similarity.

As is well known, there are striking results concerning amenable measured equivalence relations. One of them is that, for countable ergodic standard measured equivalence relations, amenability is equivalent to hyperfiniteness [**CFW**]. Another, mentioned in the introduction, is the classification, up to isomorphism, of amenable countable ergodic standard measured equivalence relations. The status of amenable locally compact groupoids, even when restricted to minimal equivalence relations, is very different. T. Giordano, I. Putnam and C. Skau have initiated a program of classification of minimal actions of $\mathbf{Z}^d$ (and more general amenable groups) on the Cantor space, completed for $d = 1$ [**GPS**] and with partial results for $d = 2$ (C. Skau) which uses, as mentioned before, a weaker notion of orbit equivalence.

In some sense, AF equivalence relations [**Re4**] can be viewed as a topological version of hyperfinite equivalence relations. These are the relations which can be obtained as tail equivalence relations of a Bratteli diagram. W. Krieger has shown in [**Kr**] that they are classified up to isomorphism by their dimension range, therefore, up to equivalence, by their dimension group, which is the ordered group K_0 which appeared in the previous section. Thus, its dimension group is a complete transverse invariant for an AF equivalence relations. The analogous statement for AF algebras had been known earlier. Another class of nuclear C*-algebras has a good classification theory, namely the simple purely infinite ones are classified by their KK-theory (Kirchberg-Phillips). The C*-algebras associated with boundary actions of discrete groups enter into this class [**A1**] and are classified up to Morita equivalence by their K-theory. However, to my knowledge, the classification of boundary actions up to groupoid equivalence has not been completed. A step in that direction is the computation by C. Anantharaman-Delaroche in [**A2**] of these K-theory groups for the action of a Fuchsian group on $P^1(\mathbf{R})$, which shows that non-conjugate actions can give isomorphic K-theory groups.

5. Cohomology and the Dixmier-Douady class

Since topological groupoids generalize both topological spaces and topological groups, it is worthwhile to define a cohomology theory which unifies cohomology of spaces and group cohomology. Grothendieck's equivariant sheaf cohomology, originally defined for discrete group actions, has been extended to étale topological groupoids by A. Haefliger [**H1**] and by A. Kumjian [**Ku**]. As shown recently by J.-L. Tu in [**T2**], sheaf and Čech cohomology can be defined for arbitrary topological groupoids as a particular case of topological simplicial spaces. This definition subsumes the previous definitions.

Given a topological groupoid G, let us define a G-bundle $\mathcal{A}$ as a topological G-space with projection map $r : \mathcal{A} \to G^{(0)}$ such that its fibres $\mathcal{A}^x = r^{-1}(x)$ have an additional structure which is compatible with the topology of $\mathcal{A}$ and the action of G. For instance, we can speak of a G-bundle of groups or a G-bundle of Banach spaces. When the projection map is a local homeomorphism, one speaks of a G-sheaf instead of a G-bundle and of a stalk instead of a fibre. Given a topological groupoid G and a G-sheaf $\mathcal{A}$ of abelian groups, we denote by $H^n(G, \mathcal{A}), n \in \mathbf{N}$, these cohomology groups (we do not here distinguish, as Tu does, sheaf and Čech cohomology). These are transverse invariants:

THEOREM 5.1 ([**T2**, Section 8]). *Let G be a topological groupoid and let $\mathcal{A}$ be a G-sheaf $\mathcal{A}$ of abelian groups. Then, the cohomology groups $H^n(G, \mathcal{A})$, $n \in \mathbf{N}$, are invariant under Morita equivalence of topological groupoids.*

For $n = 0$ and $n = 1$, these groups have a simple geometrical description: $H^0(G, \mathcal{A})$ is the group of (global) continuous G-equivariant sections of the projection map. In order to define $H^1(G, \mathcal{A})$, let us introduce the notion of $(G, \mathcal{A})$-affine space, which is quite similar to the notion of $(G, \mathcal{A})$-correspondence given in Definition 2.5. The difference is that we require that the maps $r, s : Z \to G^{(0)}$ coincide and we replace the commutation of the actions $\gamma(za) = (\gamma z)a$ by the equivariance condition $\gamma(za) = (\gamma z)(\gamma a)$ for $(\gamma, z, a) \in G * Z * A$. Then $H^1(G, \mathcal{A})$ is the group of isomorphism classes of $(G, \mathcal{A})$-affine spaces.

Just as for groups, the cohomology group $H^2(G, \mathcal{A})$ is related to extensions of G. Here are some definitions, but we refer the reader to [**KMRW, T2**] for a more precise account. Let G be a topological groupoid and let A be a G-bundle of abelian groups. An *extension* of a topological groupoid G by A is a topological groupoid E such that $E^{(0)} = G^{(0)}$, together with groupoid homomorphisms $i : A \to E$ (where A is viewed as a groupoid) and $p : E \to G$ such that $i^{(0)}$ and $p^{(0)}$ are the identical map, i is injective, p is surjective, for $\sigma \in E$, $p(\sigma)$ is a unit if and only if σ is in the image of A and $\sigma i(a)\sigma^{-1} = i(p(\sigma).a)$, where $\gamma.a$ is the image of $(\gamma, a) \in G * A$ by the action map $G * A \to A$.

In order to integrate equivalence of groupoids, we also consider the extensions of the equivalent groupoid $G_Z = Z * Z/G$ by the induced group bundle $A_Z = Z * A/G$, where Z is a right free and proper G-space. We say that the extensions E_i of G_{Z_i}, where $i = 1, 2$, are equivalent if there is an equivalence Z of the groupoids E_i, $i = 1, 2$, such that the actions of A_{Z_i} agree and $A\backslash Z$ is isomorphic to $Z_1 * Z_2/G$ as (G_{Z_1}, G_{Z_2})-equivalence . Equipped with the Baer sum of extensions, the equivalence classes of such extensions form a group $Ext(G, A)$ isomorphic to $H^2(G, \mathcal{A})$ ([**T2**, Proposition 5.6]), where $\mathcal{A}$ is the sheaf of germs of continuous sections of A.

The case when A is the constant bundle $S = G^{(0)} \times \mathbf{S}^1$, where $\mathbf{S}^1$ is the circle group and the action of G is trivial, has received special attention. The corresponding extensions are called *twists* over G [**Ku**]. They appear in particular in the theory of projective unitary representations of topological groups and in the Dixmier–Douady classification of certain continuous fields of elementary C*-algebras ([**DD**]; see also [**B**, Chapter 4]). A general framework is given in [**KMRW**].

We present it here briefly and under a slightly different form. This theory concerns G-bundles of elementary C*-algebras. The notion of Morita equivalence of C*-algebras which we have reviewed in Section 2 is easily adapted to G-bundles of C*-algebras (some authors, e.g. [**L2**], use the related notion of G-C*-algebra): given a topological space X and two bundles of C*-algebras $A \to X$ and $B \to X$, an X-Morita equivalence is a bundle of Banach spaces $E \to X$ and a structure of (A_x, B_x)-Morita equivalence on the fibre E_x for all $x \in X$ such that all the structure maps are continuous. If $X = G^{(0)}$ and A, B are G-bundles of C*-algebras, we shall say that E is a G-Morita equivalence if it is a G-bundle of Banach spaces and the action of G preserves the structure maps. Details can be found in [**Re3**]. By definition, an *elementary* C*-algebra A is a C*-algebra isomorphic to the C*-algebra $\mathcal{K}(H)$ of compact operators on some Hilbert space H. Equivalently, it is a C*-algebra which is Morita equivalent to the C*-algebra $\mathbf{C}$.

DEFINITION 5.2. A bundle of C*-algebras $A \to X$ is *elementary* if it is X-Morita equivalent to the constant bundle $X \times \mathbf{C}$. Equivalently, there exists a bundle of Hilbert spaces $H \to X$ such that the bundles $A \to X$ and $\mathcal{K}(H) \to X$ are isomorphic. A bundle of C*-algebras $A \to X$ is *locally elementary* if each point $x \in X$ has a neighborhood U such that the reduced bundle $A_{|U} \to V$ is elementary.

The classical Dixmier–Douady invariant classifies locally elementary C*-bundles over X up to X-Morita equivalence. It is an element $\delta_X(A) \in H^2(X, \mathcal{S})$, where $\mathcal{S}$ is the sheaf of germs of continuous functions from X to the circle, which can be described as follows. Let $A \to X$ be a locally elementary C*-bundle. By definition, there exist an open cover $\mathcal{U} = \{U_i\}_{i \in I}$ of X and for all i a U_i-Morita equivalence H_i from $A_i = A_{|U_i}$ to $U_i \times \mathbf{C}$. By choosing if necessary a finer cover, there exists for all i a section ξ_i of $H_i \to U_i$ such that $\|\xi_i(x)\| = 1$ for all $x \in U_i$. It defines a section p_i of $A_i \to U_i$ such that for all $x \in U_i$, $p_i(x) =_{A_i} \langle \xi(x), \xi(x) \rangle$ is a one-dimensional projection of $A(x)$. For all (i, j) such that $U_{ij} = U_i \cap U_j$ is non-empty and $x \in U_{ij}$, we define $L_{ij}(x) = p_i(x)A(x)p_j(x)$. Then L_{ij} is a Hermitian line bundle over U_{ij}. Given i, j, k such that $U_{ijk} = U_i \cap U_j \cap U_k$ is nonempty and $x \in U_{ijk}$, the product in $A(x)$ gives a product $L_{ij}(x) \otimes L_{jk}(x) \to L_{ik}(x)$. For $x \in U_{ij}$, the involution in $A(x)$ gives an involution $L_{ij}(x) \to L_{ji}(x)$. For $v \in L_{ij}(x)$, we have $vv^* = \|v\|^2 p_i(x)$ and $v^*v = \|v\|^2 p_j(x)$. We define

$$E = \{(i, x, j, v) : x \in U_{ij}, v \in L_{ij}(x), \|v\| = 1\}.$$

Endowed with the multiplication $(i,x,j,v)(j,x,k,w) = (i,x,k,vw)$ and the inverse map $(i,x,j,v)^{-1} = (j,x,i,v^*)$, it is an extension of the groupoid associated with the open cover

$$G_{\mathcal{U}} = \{(i,x,j) : x \in U_{ij}\}$$

by the constant bundle $X \times \mathbf{S}^1$. Moreover, these groupoids are topological. One can check that the class of this twist E in $Ext(X, X \times \mathbf{S}^1)$ does not depend on our choices. Viewed as an element of $H^2(X, \mathcal{S})$, it is the Dixmier–Douady invariant $\delta_X(A)$. The main result of [**DD**] is that $\delta_X(A)$ classifies the C*-bundle $A \to X$ up to X-Morita equivalence.

Let us consider next the locally elementary C*-bundles over X which are endowed with a G-action, where G is a given topological groupoid with unit space $G^{(0)} = X$. Let $A \to X$ be such a bundle. There exist an open cover $\mathcal{U} = \{U_i\}_{i \in I}$ of X and sections p_i over U_i consisting of one-dimensional projections as above. We now introduce the groupoid

$$G_{\mathcal{U}} = \{(i,\gamma,j) : \gamma \in G, r(\gamma) \in U_i, s(\gamma) \in U_j\}$$

which is equivalent to G. For $(i,\gamma,j) \in G_{\mathcal{U}}$, we define the Hermitian line

$$L(i,\gamma,j) = p_i(r(\gamma))A(r(\gamma))(\gamma p_j(s(\gamma))).$$

We have a product $L(i,\gamma,j) \otimes L(j,\gamma',k) \to L(i,\gamma\gamma',k)$ given by $a \otimes b \mapsto a(\gamma b)$ and an involution $L(i,\gamma,j) \to L(j,\gamma^{-1},i)$ given by $a \mapsto \gamma^{-1}(a^*)$, which make

$$E = \{(i,\gamma,j,v) : (i,\gamma,j) \in G_{\mathcal{U}}, v \in L(i,\gamma,j), \|v\| = 1\}$$

into a twist over $G_{\mathcal{U}}$ defining the Dixmier–Douady invariant $\delta_G(A) \in H^2(G, \mathcal{S})$.

THEOREM 5.3 ([**KMRW**, Theorem 10.1]). *Let G be a topological groupoid. Two locally elementary G-C*-bundles are G-Morita equivalent if and only if they have the same Dixmier–Douady invariant.*

In fact, the result of [**KMRW**] is more precise. It is first shown there that the G-Morita equivalence classes of locally elementary G-C*-bundles, with tensor product of C*-bundles as composition law, form a group; by analogy with a classical notion in algebra, this group is called the Brauer group of G and is denoted by $Br(G)$. Then, it is shown that the above Dixmier–Douady map is an isomorphism from $Br(G)$ onto $H^2(G, \mathcal{S})$. The notion of K-orientation of a G-real vector bundle used in [**HS**] to construct elements of Kasparov KK-groups is related to the Dixmier–Douady invariant: a G-Euclidean vector bundle of even dimension (the general case follows by standard constructions) is K-oriented if and only if the Dixmier–Douady invariant $\delta_G(A) \in H^2(G, \mathcal{S})$ of the associated complex Clifford algebra bundle A vanishes. These cohomology groups have been mostly studied for étale groupoids and in particular in the context of foliations [**H1**] and for groupoids of endomorphisms [**DKM**]. As a corollary of Theorem 5.1, these results can be applied to those groupoids which are equivalent to étale groupoids. There are also partial results on $Br(G)$ when G is the semi-direct product $X \rtimes \Gamma$ of the action of a group Γ on a space X relating it to the group cohomology of Γ and the Čech cohomology of X (e.g. [**EN**]).

References

[A1] C. Anantharaman-Delaroche, *Purely infinite C^*-algebras arising from dynamical systems*, Bull. Soc. Math. France **125** (1997), 199–225.

[A2] C. Anantharaman-Delaroche, *C^*-algèbres de Cuntz-Krieger et groupes fuchsiens*, Proceedings of the 16th OT Conference (Timisoara, 1996), the Theta Foundation, 1997.

[AnR] C. Anantharaman-Delaroche, J. Renault, *Amenable groupoids*, Monographie de l'Enseignement Mathématique, no. **36**, Genève, 2000.

[ArR] V. Arzumanian, J. Renault, *Examples of pseudogroups and their C^*-algebras*, Operator Algebras and Quantum Field Theory, S. Doplicher, R. Longo, J. E. Roberts and L. Zsido, editors, International Press, 1997, pp. 93–104.

[AV] V. Arzumanian, A. Vershik, *Star-algebras associated with endomorphisms*, Operator algebras and group representations, Proc. Int. Conf., vol. 1, Pitman, Boston, 1984, pp. 17–27.

[B] J.-L. Brylinski, *Loop spaces, Characteristic classes and Geometric Quantization*, Progress in Mathematics **107**, Birkhäuser, Boston, 1993.

[C] A. Connes, *Noncommutative geometry*, Academic Press, 1994.

[CFW] A. Connes, J. Feldman, B. Weiss, *An amenable equivalence relation is generated by a single transformation,* Ergodic Theory and Dynamical Systems **1** (1981), 431–450.

[D1] H. Dye, *On groups of measure preserving transformations. I*, Amer. J. of Math. **81** (1959), 119–159.

[D2] H. Dye, *On groups of measure preserving transformations. II*, Amer. J. of Math. **85** (1963), 551–576.

[DD] J. Dixmier, A. Douady, *Champs continus d'espaces Hilbertiens et C^*-algèbres*, Bull. Soc. Math. France **91** (1963), 227–284.

[DKM] V. Deaconu, A. Kumjian, P. Muhly, *Cohomology of topological graphs and Cuntz-Pimsner algebras*, J. Operator Theory **46** (2001), 251–264.

[E] G. Elliott, *On the classification of inductive limits of sequences of semisimple finite-dimensional algebras*, J. Algebra **38** (1976), 29–44.

[EN] S. Echterhoff, R. Nest, *The structure of the Brauer group and crossed products of $C_0(X)$-linear group actions on $C_0(X, K)$*, Trans. Amer. Math. Soc. **353** (2001), 3685–3712.

[FM] J. Feldman, C. Moore, *Ergodic equivalence relations, cohomologies, von Neumann algebras, I and II*, Trans. Amer. Math. Soc. **234** (1977), 289–359.

[G] D. Gaboriau, *On orbit equivalence of measure preserving actions*, Rigidity in dynamics and geometry (Cambridge, 2000), Springer, Berlin, 2002, pp. 167–186.

[GPS] T. Giordano, I. Putnam, C. Skau, *Topological orbit equivalence and C^*-crossed products*, J. Reine Angew. Math. **469** (1995), 51–111.

[H1] A. Haefliger, *Differentiable cohomology*, CIME Lectures, Varenna, 1976, (1979), 19–70.

[H2] A. Haefliger, *Groupoides d'holonomie et classifiants*, Structure transverse des feuilletages, Toulouse 1982, Astérisque **116** (1984), 70–97.

[HR] N. Higson, J. Roe, *Amenable group actions and the Novikov conjecture*, J. Reine Angew. Math. **519** (2000), 143–153.

[HS] M. Hilsum, G. Skandalis, *Morphismes K-orientés d'espaces de feuilles et fonctorialité en théorie de Kasparov*, Ann. Sci. E.N.S. **20** (1987), 325–390.

[KMRW] A. Kumjian, P. Muhly, J. Renault, D. Williams, *The Brauer group of a locally compact groupoid*, Amer. J. of Math. **120** (1998), 901–954.

[Kr] W. Krieger, *On a dimension for a class of homeomorphism groups*, Math. Ann. **252** (1980), 87–95.

[Ku] A. Kumjian, *On equivariant sheaf cohomology and elementary C^*-bundles*, J. Operator Theory **20**(1988), 207–240.

[L1] P.-Y. Le Gall, *Théorie de Kasparov équivariante et groupoïdes I*, K-Theory **16** (1999), 361–390.

[L2] P.-Y. Le Gall, *Groupoid C^*-algebras and operator K-theory*, Groupoids in Analysis, Geometry, and Physics, Boulder, 1999, Contemporary Mathematics (AMS), **282** (2001), 137–146.

[Ma] G. Mackey, *Ergodic theory and virtual groups*, Math. Ann. **166** (1966), 187–207.

[Mk] I. Moerdijk, *Orbifolds as Groupoids, an Introduction*, K-Theory **18**(1999), 235–253.

[Ml] P. Molino, *Riemannian foliations*, Progress in Mathematics **73**, Birkhäuser, Boston, 1988.

[MN] F. Murray, J. von Neumann, *On rings of operators*, Ann. of Math. **37** (1936), 116–229.

[Mr] J. Mrcun, *Functoriality of the bimodule associated to a Hilsum–Skandalis map*, K-Theory **18** (1999), 235–253.

[MRW] P. Muhly, J. Renault, D. Williams, *Equivalence and isomorphism for groupoid C^*-algebras*, J. Operator Theory **17** (1987), 3–22.

[Ra] A. Ramsay, *Virtual groups and group actions*, Adv. in Math. **6** (1971), 253–322.

[Re1] J. Renault, *A groupoid approach to C^*-algebras*, Lecture Notes in Mathematics, vol. **793**, Springer-Verlag, Berlin, Heidelberg, New York, 1980.

[Re2] J. Renault, *C^*-algebras of groupoids and foliations*, Operator algebras and applications, Kingston, 1980, Proc. Sympos. Pure Math., vol. **38** (1982), part I, pp. 339–350.

[Re3] J. Renault, *Représentations des produits croisés d'algèbres de groupoïdes*, J. Operator Theory **18** (1987), 67–97.

[Re4] J. Renault, *AF equivalence relations and their cocycles*, Proceedings of the OAMP Conference (Constanta, 2001), the Theta Foundation.

[Ri] M. Rieffel, *Induced representations of C^*-algebras*, Adv. in Math. **13** (1974), 176–257.

[T1] J.-L. Tu, *La conjecture de Baum-Connes pour les feuilletages moyennables*, K-theory **17** (1999), 215–264.

[T2] J.-L. Tu, *Groupoid cohomology and extensions*, Trans. Amer. Math. Soc., to appear, `arXiv:math.OA/0404257`.

[TXL] J.-L. Tu, Ping Xu, C. Laurent, *Twisted K-theory of Differentiable Stacks*, Ann. Sci. E.N.S. **37** (2004), 841–910.

[V1] A. Vershik, *Non measurable partitions, trajectory theory, operator algebras*, Soviet. Math. Dokl. **199** (1971).

[V2] A. Vershik, *Many-valued measure-preserving mappings (polymorphisms)and Markovian operators*, translated from Zapiski Nauchnnyk Seminarov Leningradskogo Otdeleniya Matematicheskogo Instituta im. V. A. Steklova Acad. Nauk SSSR **72** (1977), 26-61.

[Z1] R.J. Zimmer, *Amenable ergodic group actions and an application to Poisson boundaries of random walks,* J. Funct. Anal. **27** (1978), 350–372.

[Z2] R.J. Zimmer, *Hyperfinite factors and amenable ergodic actions,* Invent. Math. **41** (1977), 23–31.

Département de Mathématiques, Université d'Orléans, 45067 Orléans, France
E-mail address: `renault@labomath.univ-orleans.fr`

Amer. Math. Soc. Transl.
(2) Vol. **217**, 2006

Minimal One-sided Markov Shifts and Their Cofiltrations

Ben-Zion Rubshtein

ABSTRACT. A classification of one-sided Markov shifts T_G and their cofiltrations $\xi(T_G) = \left\{T_G^{-n}\varepsilon\right\}_{n=0}^{\infty}$ with respect to a measure-preserving isomorphism is given in the terms of the corresponding stochastic graphs G. Any stochastic graph G determines a (unique up to isomorphism) "minimal" stochastic graph $M = M(G)$ majorized by G. Two cofiltrations $\xi(T_{G_1})$ and $\xi(T_{G_2})$ are called isomorphic (resp., finitely isomorphic) if there exists an isomorphism Φ such that $\Phi\left(T_{G_1}^{-n}\varepsilon\right) = T_{G_2}^{-n}\varepsilon$ for all n (resp., for every n there exists an isomorphism Φ_n such that $\Phi_n\left(T_{G_1}^{-k}\varepsilon\right) = T_{G_2}^{-k}\varepsilon$ for all $k \leq n$). Under the assumption of irreducibility and positive recurrence the following conditions are equivalent:

(i) The minimal stochastic graphs $M(G_1)$ and $M(G_2)$ are isomorphic.
(ii) The minimal one-sided Markov shifts $T_{M(G_1)}$ and $T_{M(G_2)}$ are isomorphic.
(iii) The minimal cofiltrations $\xi(T_{M(G_1)})$ and $\xi(T_{M(G_2)})$ are isomorphic.
(iv) The minimal cofiltrations $\xi(T_{M(G_1)})$ and $\xi(T_{M(G_2)})$ are finitely isomorphic.
(v) The cofiltrations $\xi(T_{G_1})$ and $\xi(T_{G_2})$ are finitely isomorphic.

Introduction

In this paper we consider the classification problem for one-sided Markov shifts with respect to measure-preserving isomorphism. Let G be a finite or countable stochastic graph, i.e., a directed graph whose edges $g \in G$ are equipped with positive weights $p(g)$. The weights $p(g)$ determine transition probabilities of a Markov chain on the discrete state space G. The corresponding one-sided Markov shift T_G acts on the backward G-path space $\left(X_G^-, m_G^-\right)$, where $X_G^- = G^{-\mathbb{N}}$, and m_G^- is a Markov measure corresponding to the stationary Markov chain $\ldots, Z_{-3}, Z_{-2}, Z_{-1}$ induced by the graph G on X_G^-. We deal only with irreducible positively recurrent Markov chains, so that the shift T_G is an ergodic endomorphism of the probability Lebesguespace $\left(X_G^-, m_G^-\right)$.

2000 *Mathematics Subject Classification.* Primary 37A35; Secondary 28D05, 60J10.
Key words and phrases. Isomorphism of one-sided Markov shifts, cofiltration, stochastic graph.

The problem under consideration is:

- When, for two given stochastic graphs G_1 and G_2, does there exist an isomorphism $\Phi : X_{G_1}^- \to X_{G_2}^-$ such that $m_{G_2}^- = m_{G_1}^- \circ \Phi^{-1}$ and $\Phi \circ T_{G_1} = T_{G_2} \circ \Phi$?

It is obvious that any (weight-preserving) graph isomorphism $\phi : G_1 \to G_2$ generates such an isomorphism $\Phi = \Phi_\phi$, but non-isomorphic graphs can generate the same shift T_G.

J. Ashley, B. Marcus and S. Tuncel [**AMT**] solved the classification problem for one-sided Markov shifts T_G corresponding to *finite* Markov chains, i.e., in the case when the graph G is finite. In particular, they prove that for any (finite) stochastic graph G there exists a unique (up to isomorphism) *minimal* stochastic graph $M = M(G)$, which is characterized by the following two properties:

- The graph M is *majorized* by G, i.e., there exists a homomorphism $\phi : G \to M$.
- The graph M is *minimal*, i.e., any homomorphism $\phi : M \to H$ of M onto a stochastic graph H is an isomorphism.

It should be emphasized that the argument in [**AMT**] uses a special kind of stochastic graph homomorphisms, which are assumed to be weight preserving and deterministic (right resolving) (cf. Definition 2.1).

It also follows from the results of [**AMT**] that two one-sided Markov shifts T_{M_1} and T_{M_2} corresponding to finite minimal stochastic graphs M_1 and M_2 are isomorphic iff the graphs M_1 and M_2 are isomorphic.

In our recent paper [**R7**], we solved the classification problem for a wide class of one-sided Markov shifts, called finitely Bernoulli or ρ-uniform. This class consists of all one-sided Markov shifts T_G such that the corresponding minimal shifts $T_{M(G)}$ are isomorphic to one-sided Bernoulli shifts. We show in [**R7**] that any ergodic ρ-uniform Markov shift T can be represented in a canonical form $T = T_G$ by means of a canonical (uniquely determined by T) stochastic graph G. In the canonical form, two such shifts T_{G_1} and T_{G_2} are isomorphic if and only if their canonical stochastic graphs G_1 and G_2 are isomorphic.

We use in [**R7**] as well as in our previous work [**R6**] a new approach to classification of one-sided Markov shifts T_G, based on the study of cofiltrations $\xi(T_G)$ generated by the shifts. Recall that the cofiltration $\xi(T)$ generated by an endomorphism T of a Lebesgue space (X, m) is the decreasing sequence $\{\xi_n\}_{n=0}^{\infty}$ of the measurable partitions $\xi_n = T^{-n}\varepsilon_X$ of the space X onto inverse images $T^{-n}x$. If two endomorphisms T_1 and T_2 are isomorphic, i.e., there exists an isomorphism Φ such that $\Phi \circ T_1 = T_2 \circ \Phi$, then $\Phi\left(T_1^{-n}\varepsilon_X\right) = T_2^{-n}\varepsilon_X$ for all n. This means that the cofiltrations $\xi(T_1)$ and $\xi(T_2)$ are isomorphic as soon as the endomorphisms T_1 and T_2 are isomorphic. The converse is not true in general.

If the cofiltrations $\xi(T_1)$ and $\xi(T_2)$ are isomorphic, then they are also finitely isomorphic, i.e., for every n there exists an isomorphism Φ_n such that $\Phi_n\left(T_1^{-k}\varepsilon_X\right) = T_2^{-k}\varepsilon_X$ for all $0 \le k \le n$. The converse is not true in general.

The isomorphism problem for endomorphisms is decomposed into three parts:

(1) When are two cofiltrations $\xi(T_1)$ and $\xi(T_2)$ finitely isomorphic?
(2) When are two finitely isomorphic cofiltrations $\xi(T_1)$ and $\xi(T_2)$ isomorphic?

(3) When are the endomorphisms T_1 and T_2 isomorphic provided that $\xi(T_1)$ and $\xi(T_2)$ are isomorphic?

All three parts of the problem, especially (2) and (3), are quite nontrivial. Various classes of cofiltrations were considered by A. M. Vershik [**Ve1**]–[**Ve4**], V. G. Vinokurov [**Vi**], A. M. Stepin [**S**] and by author [**R1**]–[**R7**]. New remarkable progress in the theory is due to J. Feldman, D. J. Rudolph, D. Heicklen and C. Hoffman; see [**FR, HH, HHR, H, HR**].

Returning to the case when $T = T_G$ is a one-sided Markov shift, we claim that the results of [**R7**] can be carried over to general one-sided Markov shifts by using the corresponding minimal stochastic graphs. Keeping in mind the aim, we solve here the first (and the simplest) part of the problem. Namely, we prove the following.

- For any irreducible stochastic graph G there exists a unique up to isomorphism minimal stochastic graph $M = M(G)$ majorized by G (Theorem 3.4).
- Let T_{G_1} and T_{G_2} be two ergodic one-sided Markov shifts. Then the corresponding cofiltrations $\xi(T_{G_1})$ and $\xi(T_{G_2})$ are finitely isomorphic iff the minimal graphs $M(G_1)$ and $M(G_2)$ are isomorphic (Theorem 4.8).

A one-sided Markov shift T_G is called *minimal* if it is isomorphic to a one-sided Markov shift T_M corresponding to a minimal stochastic graph M.

We show (Theorem 4.9) that classification of minimal one-sided Markov shifts and their cofiltrations is reduced to classification of corresponding minimal stochastic graphs. Namely, we prove the following.

- For any two minimal irreducible stochastic graphs M_1 and M_2, the following conditions are equivalent:
  - (i) The stochastic graphs M_1 and M_2 are isomorphic.
  - (ii) The one-sided Markov shifts T_{M_1} and T_{M_2} are isomorphic.
  - (iii) The cofiltrations $\xi(T_{M_1})$ and $\xi(T_{M_2})$ are isomorphic.
  - (iv) The cofiltrations $\xi(T_{M_1})$ and $\xi(T_{M_2})$ are finitely isomorphic.

Note that we use here just the same definitions of stochastic graph and of minimal graph (Definitions 2.1 and 3.3) as in [**AMT**], but with no assumption that the graphs under consideration are finite.

On the other hand, our approach to constructing and studying minimal graphs is quite new and different from the method used in [**AMT**]. Namely, we apply the *backward cofiltrations* $\xi_G^- = \xi(T_G)$ and also *forward filtrations* η_G^+ corresponding to stochastic graphs G.

In contrast to ξ_G^- the filtration η_G^+ is an increasing sequence of measurable partitions defined on the forward G-path measure space (X_G^+, m_G^+), where $X_G^+ = G^{\mathbb{N}}$, and m_G^+ is a Markov measure corresponding to the stationary Markov chain $Z_1, Z_2, Z_3, \ldots$ induced by the graph G on X_G^+; see Sections 2.3 and 2.4. We use the filtrations η_G^+ to prove the existence and uniqueness of minimal graphs (Theorem 3.4).

On the other hand, the main Theorems 4.8 and 4.9 are based on the following result.

- Let $M(G_1)$ and $M(G_2)$ be the minimal graphs corresponding to irreducible stochastic graphs G_1 and G_2. Then the following conditions are equivalent:
  (i) The minimal graphs $M(G_1)$ and $M(G_2)$ are isomorphic.
  (ii) The minimal filtrations $\eta^+_{M(G_1)}$ and $\eta^+_{M(G_2)}$ are isomorphic.
  (iii) The cofiltrations $\xi^-_{G_1}$ and $\xi^-_{G_2}$ are finitely isomorphic.

Namely, (i) $\iff$ (ii) by Proposition 4.1, and (ii) $\iff$ (iii) by Proposition 4.7.

In contrast to cofiltrations $\xi^-_{G_1}$ and $\xi^-_{G_2}$, the filtrations $\eta^+_{G_1}$ and $\eta^+_{G_2}$ are isomorphic iff they are finitely isomorphic.

1. Filtrations and cofiltrations

1.1. Measurable partitions and homomorphisms. Let $(X, \mathcal{F}, m)$ be a probability Lebesgue space, and let ζ be a measurable partition of X. Elements of ζ are considered as Lebesgue spaces $\left(C, m^C\right)$, $C \in \zeta$ with a canonical system of conditional measures $\left\{m^C, C \in \zeta\right\}$. We denote by $m^C(A)$ the conditional measures $m^C(A \cap C)$ of a set $A \in \mathcal{F}$ in the element C of ζ. Let $C_\zeta(x)$ denote the element of ζ containing a point $x \in X$. Then $m^{C_\zeta(x)}(A)$ is the conditional probability of $A \in \mathcal{F}$ with respect to the σ-algebra $\mathcal{F}(\zeta)$ of all ζ-measurable sets.

The trivial partition of X is denoted by $\nu = \nu_X$, and $\varepsilon = \varepsilon_X$ denotes the partition of X consisting of separate points.

Recall that two measurable partitions ζ and θ are *conditionally independent* with respect to a third measurable partition ω if

$$m^{C_\omega(x)}(A \cap B) = m^{C_\omega(x)}(A)\, m^{C_\omega(x)}(B), \quad A \in \mathcal{F}(\zeta), B \in \mathcal{F}(\theta)$$

for a.e. $x \in X$. We shall write $\zeta \perp \theta \pmod{\omega}$ in this case.

DEFINITION 1.1. Let $\mathcal{Z} = \{\zeta_i\}_{i \in I}$ be a finite or countable family of measurable partitions on (X, m). A measurable partition θ *factorizes* the family $\mathcal{Z}$ if

$$\zeta_i \perp \theta \pmod{\zeta_i \wedge \theta}, \quad i \in I .$$

i.e., ζ_i and θ are conditionally independent with respect to $\zeta_i \wedge \theta$ for all $i \in I$.

If θ factorizes a family $\mathcal{Z} = \{\zeta_i\}_{i \in I}$, we can consider the measurable factor partitions $\zeta_i/\theta := (\zeta_i \wedge \theta)/\theta$, $i \in I$, on the factor space $(X/\theta, m/\theta)$ and the corresponding family $\mathcal{Z}/\theta = \{\zeta_i/\theta, i \in I\}$ on $(X/\theta, m/\theta)$. It should be emphasized that we use here and throughout such factor partitions ζ/θ, while $\zeta_i \not\leq \theta$ in general.

Two finite or countable families $\mathcal{Z} = \{\zeta_i\}_{i \in I}$ and $\mathcal{Z}' = \{\zeta'_i\}_{i \in I}$ of measurable partitions on Lebesgue spaces (X, m) and (X', m'), respectively, are called *isomorphic* if there exists a measure space isomorphism $\Phi : (X, m) \to (X', m')$ such that $\Phi^{-1}\zeta'_i = \zeta_i$ for all $i \in I$.

The families $\mathcal{Z}$ and $\mathcal{Z}'$ are called *finitely isomorphic* if for every finite subset $J \subseteq I$ there exists an isomorphism $\Phi_J : (X, m) \to (X', m')$ such that $\Phi_J^{-1}\zeta'_i = \zeta_i$ for all $i \in J$.

Two infinite isomorphic families are obviously finitely isomorphic, but the converse is not true in general.

We shall use in this paper the following "special kind" of homomorphisms for families of measurable partitions.

DEFINITION 1.2. Let $\mathcal{Z} = \{\zeta_i\}_{i\in I}$ and $\mathcal{Z}' = \{\zeta'_i\}_{i\in I}$ be two families of measurable partitions on Lebesgue spaces (X, m) and (X', m'), respectively. A map $\Phi : X \to X'$ is called a *homomorphism* of $\mathcal{Z}$ onto $\mathcal{Z}'$ if the following three conditions hold:

(i) The map $\Phi : (X, m) \to (X', m')$ is a measure space homomorphism.
(ii) The partition $\Phi^{-1}\varepsilon_{X'}$ factorizes the family $\mathcal{Z}$.
(iii) $\Phi^{-1}\zeta'_i = \zeta_i \wedge \Phi^{-1}\varepsilon_{X'}$ for all $i \in I$.

For any such homomorphism we shall write $\Phi : (X, m, \mathcal{Z}) \to (X', m', \mathcal{Z}')$ and denote the set of all these homomorphisms by $\mathcal{H}om\,(\mathcal{Z}, \mathcal{Z}')$.

1.2. Projective systems and infinite filtrations. Let $\eta = \{\eta_n\}_{n=0}^{\infty}$ be an *infinite filtration*, i.e., an infinite increasing sequence of measurable partitions on a probability Lebesgue space (X, m). Homomorphisms of such filtrations are introduced in accordance with Definition 1.2. Namely, two infinite filtrations $\eta = \{\eta_n\}_{n=0}^{\infty}$ and $\eta' = \{\eta'_n\}_{n=0}^{\infty}$ on (X, m) and (X', m'), respectively, are *isomorphic* if there is an isomorphism $\Phi : (X, m) \to (X', m')$ such that $\Phi^{-1}\eta'_n = \eta_n$ for all $n \in \mathbb{Z}^+$. The filtrations η and η' are *finitely isomorphic* if for every n there is an isomorphism $\Phi_n : (X, m) \to (X', m')$ such that $\Phi_n^{-1}\eta'_k = \eta_k$ for all $0 \le k \le n$.

Let (X_n, m_n) , $n \in \mathbb{Z}^+$, be a sequence of purely atomic probability Lebesgue spaces. A system of homomorphisms

$$\pi_{kn} : (X_n, m_n) \to (X_k, m_k)\,, \quad k \le n\,, \quad k, n \in \mathbb{Z}^+, \tag{1.1}$$

is called a *projective system* if

$$\pi_{kk} = id_{X_k}\,, \quad \pi_{kl} \circ \pi_{ln} = \pi_{kn}\,, \quad k \le l \le n\,, \quad k, l, n \in \mathbb{Z}^+\,. \tag{1.2}$$

A Lebesgue space (X, m) is called a *projective limit*

$$(X, m) \;=\; \operatorname*{proj\,lim}_{n\to\infty} (X_n, m_n) \tag{1.3}$$

of (X_n, m_n) with respect to the projective system (1.1) if there exists a system of homomorphisms

$$\pi_n : (X, m) \to (X_n, m_n)\,, \quad n \in \mathbb{Z}^+, \tag{1.4}$$

such that

$$\pi_{kn} \circ \pi_n = \pi_k\,, \quad k \le n\,, \quad k, n \in \mathbb{Z}^+, \tag{1.5}$$

and $\eta_n := \pi_n^{-1}\varepsilon_{X_n} \nearrow \varepsilon_X$. Thus, the partitions η_n form a filtration $\eta = \{\eta_n\}_{n=0}^{\infty}$ on (X, m) such that $\bigvee_{n=0}^{\infty} \eta_n = \varepsilon_X$.

We say that a system of homomorphisms (1.4) is *compatible* with a projective system (1.1) if the conditions (1.5) hold.

It is known that for any projective system (1.1) there exists a projective limit (1.3) and it is unique in the following sense: Let (X, m) and (X', m') be two projective limits of (X_n, m_n) with respect to a projective system (1.1), and let $\{\pi_n\}$ and $\{\pi'_n\}$ be the corresponding projections. Then there exists an isomorphism $\Phi : (X, m) \to (X', m')$ such that $\pi'_n \circ \Phi = \pi_n$ for all n.

Any projective system (1.1) generates a filtration $\eta = \{\eta_n\}_{n=0}^{\infty}$ on its projective limit (X, m), where $\eta_n := \pi_n^{-1}\varepsilon_{X_n} \nearrow \varepsilon_X$.

Conversely, let $\eta = \{\eta_n\}_{n=0}^{\infty}$ be a filtration on a Lebesgue space (X, m) such that the factor spaces $(X/\eta_n, m/\eta_n)$ are purely atomic and $\eta_n \nearrow \varepsilon_X$. Then (X, m)

is represented as a projective limit of the factor spaces $(X/\eta_n, m/\eta_n)$ with respect to the projective system

$$(1.6) \qquad \pi_{kn} : (X/\eta_n, m/\eta_n) \to (X/\eta_k, m/\eta_k), \quad 0 \le k \le n,$$

with the natural projections

$$\pi_n : (X, m) \to (X/\eta_n, m/\eta_n), \quad n \ge 0.$$

Let $\eta = \{\eta_n\}_{n=0}^{\infty}$ and $\eta' = \{\eta'_n\}_{n=0}^{\infty}$ be two such filtrations on (X, m) and (X', m'), respectively. Let also $\{\pi_{kn}, k \le n\}_{k,n \in \mathbb{Z}^+}$ and $\{\pi'_{kn}, k \le n\}_{k,n \in \mathbb{Z}^+}$ be the projective systems (1.6), corresponding to η and η', respectively. Then the following two conditions are equivalent:

- The filtrations η and η' are isomorphic, i.e., there exists an isomorphism $\Phi : (X, m) \to (X', m')$ such that $\Phi^{-1}\eta'_n = \eta_n$ for all n.
- The projective systems $\{\pi_{kn}, k \le n\}$ and $\{\pi'_{kn}, k \le n\}$ are isomorphic, i.e., there exist isomorphisms

$$\phi_n : (X/\eta_n, m/\eta_n) \to (X/\eta'_n, m/\eta'_n), \quad n \in \mathbb{Z}^+,$$

such that $\pi'_{kn} \circ \phi_n = \phi_k \circ \pi_{kn}$ for all $k \le n$.

PROPOSITION 1.3. *Let $\eta = \{\eta_n\}_{n=0}^{\infty}$ and $\eta' = \{\eta'_n\}_{n=0}^{\infty}$ be two filtrations on (X, m) and (X', m'), respectively, such that $\eta_n \nearrow \varepsilon_X$ and $\eta'_n \nearrow \varepsilon_{X'}$, and all factor spaces $(X/\eta_n, m/\eta_n)$ and $(X'/\eta'_n, m'/\eta'_n)$ are purely atomic. Then the filtrations η and η' are isomorphic iff they are finitely isomorphic.*

PROOF. Let $\{\pi_{kn}, k \le n\}_{k,n \in \mathbb{Z}^+}$ and $\{\pi'_{kn}, k \le n\}_{k,n \in \mathbb{Z}^+}$ be the projective systems (1.6) corresponding to η and η', respectively.

For every $n \in \mathbb{Z}^+$ we denote by $\mathcal{I}_n$ the set of all isomorphisms $\phi_n : (X_n, m_n) \to (X'_n, m'_n)$, where $(X_n, m_n) = (X/\eta_n, m/\eta_n)$ and $(X'_n, m'_n) = (X'/\eta'_n, m'/\eta'_n)$.

Suppose η and η' are finitely isomorphic. Then (X_n, m_n) and (X'_n, m'_n) are isomorphic, and $\mathcal{I}_n \ne \emptyset$ for all n. The set $\mathcal{I}_n$ is equipped with the metric

$$d(\phi_n, \phi'_n) = \sum_{x_n \in X_n} \delta(\phi_n(x_n), \phi'_n(x_n)) \, m_n(x_n),$$

where $\delta(a, b) = 1$ if $a \ne b$, and $\delta(a, b) = 0$ if $a = b$. It is easy to check that $\mathcal{I}_n$ is compact for every $n \in \mathbb{Z}^+$.

We set $\mathcal{I} := \prod_{n=0}^{\infty} \mathcal{I}_n$, and for all $n \in \mathbb{Z}^+$

$$\mathcal{I}^{(n)} := \{\phi = \{\phi_n\}_{n=0}^{\infty} \in \mathcal{I} : \pi'_{kl} \circ \phi_l = \phi_k \circ \pi_{kl}, \ 0 \le k \le l \le n\}.$$

Then $\mathcal{I}$ is compact, and $\mathcal{I}^{(n)}$ is a closed subset of $\mathcal{I}$.

On the other hand, the subsets $\mathcal{I}^{(n)}$ are not empty, since the filtrations η and η' are finitely isomorphic. Therefore, $\mathcal{I}^{(\infty)} := \bigcap_{n=0}^{\infty} \mathcal{I}_n \ne \emptyset$.

We have

$$\phi = \{\phi_n\}_{n=0}^{\infty} \in \mathcal{I}^{(\infty)} \implies \pi'_{kl} \circ \phi_l = \phi_k \circ \pi_{kl}, \quad 0 \le k \le l < \infty.$$

Hence, the projective systems $\{\pi_{kn}, k \le n\}_{k,n \in \mathbb{Z}^+}$ and $\{\pi'_{kn}, k \le n\}_{k,n \in \mathbb{Z}^+}$, corresponding to the filtrations η and η', are isomorphic.

Thus the filtrations η and η' are isomorphic as soon as they are finitely isomorphic. The converse is obvious. □

In this proof we use that all the spaces (X_n, m_n) are purely atomic, and only such kind of filtrations is treated in this paper. It can be shown, however, that Proposition 1.3 holds without this restriction.

1.3. Finite minimal filtrations. Let $\eta = \{\eta_k\}_{k=0}^n$ be a *finite filtration*, i.e., an increasing sequence of measurable partitions on a probability Lebesgue space (X, m). Given two finite filtrations $\eta = \{\eta_k\}_{k=0}^n$ and $\eta' = \{\eta'_k\}_{k=0}^n$, we denote by $\mathcal{H}om\,(\eta, \eta')$ the set of all filtration homomorphisms of η onto η', which is introduced in accordance with Definition 1.2.

A filtration $\eta = \{\eta_k\}_{k=0}^n$ on a non-atomic space (X, m) is called *simple* if

(1) the factor space $\left(\bar{X}_n, \bar{m}_n\right) := (X/\eta_n, m/\eta_n)$ is purely atomic,
(2) every element $\left(C, m^C\right)$ of η_n is a non-atomic space.

For every $C \in \eta_0$ we consider the restriction $\eta_{k,C} := \eta_k|_C$ of η_k to $\left(C, m^C\right)$ and denote by $\eta_C := \{\eta_{k,C}\}_{k=0}^n$ the restriction of the filtration η to C.

Let us consider the partition $\omega(\eta) \leq \eta_0$ defined by

$$C \overset{\omega(\eta)}{\sim} C' \iff \eta_C \simeq \eta_{C'} \quad , \quad C, C' \in \eta_0 \, , \tag{1.7}$$

where "$\simeq$" means a filtration isomorphism. A filtration $\eta = \{\eta_k\}_{k=0}^n$ is called *minimal* if all the filtrations η_C , $C \in \eta_0$, are pairwise non-isomorphic.

If η is simple, there exists a partition θ_n such that

(i) $\theta_n \vee \eta_0 = \eta_n$, $\theta_n \wedge \eta_0 = \omega(\eta)$,
(ii) $\theta_n \perp \eta_k \pmod{\theta_n \wedge \eta_k}$, $0 \leq k \leq n$.

This means that the partition θ factorizes the filtration η in the sense of Definition 1.1 and it "gathers" all isomorphic filtrations $\{\eta_C \, , \, C \in \eta_0\}$.

Given such a *minimal factorizing* θ_n, we have a filtration $\theta = \{\theta_k\}_{k=0}^n$, where $\theta_k = \theta_n \wedge \eta_k$. The factor partitions $\hat{\eta}_k = \theta_k/\theta_n$ and the factor filtration $\hat{\eta} = \eta/\theta_n = \{\hat{\eta}_k\}_{k=0}^n$ are well defined on the purely atomic factor space $(\hat{X}, \hat{m}) = (X/\theta_n, m/\theta_n)$. The natural projection

$$\hat{\pi} \, : \, (X, m, \xi) \to (\hat{X}, \hat{m}, \hat{\eta})$$

is a filtration homomorphism. Since $\theta_n \wedge \eta_0 = \omega(\eta)$, the filtrations θ and $\hat{\eta}$ are *minimal*, i.e., all the filtrations θ_C , $C \in \theta_0$, and all the filtrations $\hat{\eta}_{\hat{C}}$, $\hat{C} \in \hat{\eta}_n$, are pairwise non-isomorphic, $\omega(\theta) = \theta_0$ and $\omega(\hat{\eta}) = \hat{\eta}_0$.

The filtration $\hat{\eta}$ is determined by η uniquely up to an isomorphism, and it will be called the *minimal filtration majorized by* $\boldsymbol{\eta}$. Since $\hat{\eta}_n = \varepsilon_{\hat{X}}$, the minimality property of $\hat{\eta}$ implies that any filtration homomorphism of $(\hat{X}, \hat{m}, \hat{\eta})$ is a filtration isomorphism.

All elements (C, m_C) of η_n, and hence elements (C, m_C) of θ_n, are non-atomic spaces. Therefore, they are isomorphic to a fixed non-atomic Lebesgue space (Y, m_Y), and there exists a decomposition

$$(X, m, \theta) \simeq \left(\hat{X} \times Y, \hat{m} \otimes m_Y, \hat{\eta} \times \nu_Y\right),$$

where $\theta_k \simeq \hat{\eta}_k \times \nu_Y$ for $0 \leq k \leq n$, and $\omega(\eta) = \theta_0 \simeq \varepsilon_{\hat{X}} \times \nu_Y$.

A filtration η is minimal iff $\eta_0 = \theta_0 = \omega(\eta)$; hence any minimal η has the form $\hat{\eta} \times \nu_Y$ in the above representation.

Since the representation of η is determined uniquely up to an isomorphism, we have

PROPOSITION 1.4. *Let $\eta = \{\eta_k\}_{k=0}^n$ and $\eta' = \{\eta'_k\}_{k=0}^n$ be two finite simple filtrations, and let $\hat{\eta} = \{\hat{\eta}_k\}_{k=0}^n$ and $\hat{\eta}' = \{\hat{\eta}'_k\}_{k=0}^n$ be their minimal filtrations.*

(i) *If η and η' are isomorphic, then $\hat{\eta}$ and $\hat{\eta}'$ are isomorphic.*

(ii) *If η and η' are minimal, then they are isomorphic iff $\hat{\eta}$ and $\hat{\eta}'$ are isomorphic.*

Note that two finite simple filtrations η and η' with isomorphic $\hat{\eta}$ and $\hat{\eta}'$ may be non-isomorphic, since the corresponding factor partitions $\omega(\eta)/\eta_0$ and $\omega(\eta')/\eta_0'$ may be non-isomorphic.

1.4. Infinite minimal filtrations. Let $\eta = \{\eta_n\}_{n=0}^{\infty}$ be an infinite filtration. We denote by $\eta^{(n)} = \{\eta_k\}_{k=0}^{n}$, $n \in \mathbb{Z}^+$, the corresponding finite filtrations. For every $C \in \eta_0$ we consider the restriction $\eta_{n,C} := \eta_n|_C$ of η_n to (C, m^C) and denote by $\eta_C := \{\eta_{n,C}\}_{n=0}^{\infty}$ the restriction of the filtration η to C.

Consider the partition $\omega(\eta) \le \eta_0$ defined by

$$(1.8) \qquad C \overset{\omega(\eta)}{\sim} C' \iff \eta_C \simeq \eta_{C'}, \quad C, C' \in \eta_0,$$

and set $\omega_n(\eta) := \omega(\eta^{(n)})$, $n \in \mathbb{Z}^+$. Then $\omega_n(\eta) \nearrow \omega(\eta)$ by Proposition 1.3.

A filtration $\eta = \{\eta_k\}_{k=0}^{n}$ on a non-atomic space (X, m) is called *simple* if

(1) $\eta_n \nearrow \varepsilon_X$,
(2) every finite filtration $\eta^{(n)}$ is simple,
(3) the factor space $(X/\omega(\eta), m/\omega(\eta))$ is purely atomic.

A simple filtration η is called *minimal* if $\omega(\eta) = \eta_0$, i.e., all the filtrations η_C, $C \in \eta_0$, are pairwise non-isomorphic. Notice that the finite filtrations $\eta^{(n)}$ need not be minimal even if η is minimal.

Given an infinite simple filtration η and $n \in \mathbb{Z}^+$, we can consider a minimal finite filtration $\hat{\eta}^{(n)} = \left\{\hat{\eta}_k^{(n)}\right\}_{k=0}^{n}$ majorized by $\eta^{(n)} = \{\eta_k\}_{k=0}^{n}$. The filtration is defined on a purely atomic space $\left(\hat{X}_n, \hat{m}_n\right)$ by means of a corresponding finite filtration $\theta^{(n)} = \{\theta_k\}_{k=0}^{n}$ on (X, m), where $\left(\hat{X}_n, \hat{m}_n\right) \simeq (X/\theta_n, m/\theta_n)$.

PROPOSITION 1.5. *Let $\eta = \{\eta_n\}_{n=0}^{\infty}$ and $\eta' = \{\eta_n'\}_{n=0}^{\infty}$ be minimal simple filtrations on the spaces (X, m) and (X', m'), respectively. Let $\hat{\eta}^{(n)} = \left\{\hat{\eta}_k^{(n)}\right\}_{k=0}^{n}$ and $\hat{\eta}'^{(n)} = \left\{\hat{\eta}_k'^{(n)}\right\}_{k=0}^{n}$ be minimal filtrations majorized by $\eta^{(n)} = \{\eta_k\}_{k=0}^{n}$ and $\eta'^{(n)} = \{\eta_k'\}_{k=0}^{n}$, respectively. Then the following two conditions are equivalent:*

(i) *η and η' are isomorphic.*
(ii) *$\hat{\eta}^{(n)}$ and $\hat{\eta}'^{(n)}$ are isomorphic for every n.*

PROOF. (i) $\Longrightarrow$ (ii) by Proposition 1.4. However, it is not applicable to the implication (ii) $\Longrightarrow$ (i), since the finite filtrations $\eta^{(n)}$ and $\eta'^{(n)}$ need not be minimal.

Let $\eta = \{\eta_n\}_{n=0}^{\infty}$ be a minimal simple filtration on (X, m). It is not hard to construct (by induction) a filtration $\theta = \{\theta_n\}_{n=0}^{\infty}$ on X such that θ_n is a minimal factorizing partition for $\eta^{(n)} = \{\eta_k\}_{k=0}^{n}$, i.e.,

(i) $\theta_n \vee \eta_0 = \eta_n$, $\theta_n \wedge \eta_0 = \omega_n(\eta)$,
(ii) $\theta_n \perp \eta_k \pmod{\theta_n \wedge \eta_k}$, $0 \le k \le n$,

and in addition,

(iii) $\theta_k \le \theta_n \wedge \eta_k$, $0 \le k \le n$,

for all $n \in \mathbb{Z}^+$.

We can set $\theta_k^{(n)} := \theta_n \wedge \eta_k$ and $\hat{\eta}_k^{(n)} := \theta_k^{(n)}/\theta_n$ for all $0 \le k \le n < \infty$. Then $\theta^{(n)} := \left\{\theta_k^{(n)}\right\}_{k=0}^n$ is a minimal filtration, $\hat{\eta}^{(n)} = \left\{\hat{\eta}_k^{(n)}\right\}_{k=0}^n$ is a minimal filtration majorized by $\eta^{(n)}$, and the natural projection

$$\pi_n = \pi_{\theta_n} : \left(X, m, \eta^{(n)}\right) \to \left(\hat{X}_n, \hat{m}_n, \hat{\eta}^{(n)}\right) = \left(X/\theta_n, m/\theta_n, \eta^{(n)}/\theta_n\right)$$

is a filtration homomorphism. The projections $\{\pi_n , n \in \mathbb{Z}^+\}$ induce a projective system $\hat{\pi}_{kn} : \left(\hat{X}_n, \hat{m}_n\right) \to \left(\hat{X}_k, \hat{m}_k\right)$, $0 \le k \le n$, and hence the projective limit $\left(\hat{X}, \hat{m}\right) = \operatorname{proj\,lim}_{n\to\infty} \left(\hat{X}_n, \hat{m}_n\right)$ with the corresponding projections $\hat{\pi}_n : \left(\hat{X}, \hat{m}\right) \to \left(\hat{X}_n, \hat{m}_n\right)$.

Notice now that $\theta_n \nearrow \varepsilon_X$. Indeed, $\theta_n \ge \theta_n \wedge \eta_0 = \omega_n(\eta)$, whence

$$\theta^* := \bigvee_{n=0}^{\infty} \theta_n \ge \bigvee_{n=0}^{\infty} \omega_n(\eta) = \omega(\eta) = \eta_0 , \tag{1.9}$$

on account of minimality of the filtration η. On the other hand, $\theta_n \vee \eta_0 = \eta_n$, whence $\theta^* \vee \eta_0 = \bigvee_{n=0}^{\infty} \eta_n = \varepsilon_X$. This implies $\theta^* = \varepsilon_X$ by (1.9).

Since $\theta_n \nearrow \varepsilon_X$, the projection $\pi : (X, m) \to \left(\hat{X}_n, \hat{m}_n\right)$ defined by the system $\{\pi_n, \hat{\pi}_n , n \in \mathbb{Z}^+\}$ is in fact an isomorphism. Thus we may assume that

$$(X, m) = \operatorname*{proj\,lim}_{n\to\infty} \left(\hat{X}_n, \hat{m}_n\right)$$

and $\pi_n = \hat{\pi}_n$ for all n.

The constructed projective system

$$\hat{\pi}_{kn} : \left(\hat{X}_n, \hat{m}_n\right) \to \left(\hat{X}_k, \hat{m}_k\right), \ 0 \le k \le n < \infty,$$

has the following property:

(a) $\hat{\pi}_{kn}^{-1}\, \hat{\eta}_l^{(k)} = \hat{\eta}_l^{(n)}$ for all $0 \le l \le k \le n < \infty$.

Let $\hat{p}_{kn} : \left(\hat{X}_n, \hat{m}_n\right) \to \left(\hat{X}_k, \hat{m}_k\right)$, $0 \le k \le n < \infty$, be another projective system with property (a). Then it is easy to show, by using the minimality of filtrations $\hat{\eta}^{(n)} = \theta^{(n)}/\theta_n$, that the projective system $\{\hat{p}_{kn}\}$ is isomorphic to $\{\hat{\pi}_{kn}\}$.

We have shown that the filtration $\theta = \{\theta_n\}_{n=0}^{\infty} = \left\{\pi_n^{-1}\hat{\eta}_n\right\}_{n=0}^{\infty}$ is determined by $\left\{\hat{\eta}^{(n)} , n \in \mathbb{Z}^+\right\}$ uniquely up to isomorphism. Since

$$\eta_0 = \omega(\eta) = \bigvee_{n=0}^{\infty} \omega_n(\eta) , \quad \omega_n(\eta) = \pi_n^{-1}\, \hat{\eta}_0^{(n)} , \quad \eta_n = \eta_0 \wedge \theta_n , \quad n \in \mathbb{Z}^+ ,$$

the filtration η in turn is determined by θ. □

1.5. Finite cofiltrations. Let $\xi = \{\xi_k\}_{k=0}^n$ be a *finite cofiltration*, i.e., a decreasing sequence of measurable partitions on a probability Lebesgue space (X, m), and let $\xi_0 = \varepsilon_X$.

Given two finite cofiltrations $\xi = \{\xi_k\}_{k=0}^n$ and $\xi' = \{\xi'_k\}_{k=0}^n$, we denote by $\mathcal{H}om\,(\xi, \xi')$ the set of all *cofiltration homomorphisms* of ξ onto ξ', which are introduced in accordance with Definition 1.2.

For any element $C \in \xi_n$ we consider the restrictions $\xi_{k,C} := \xi_k|_C$ of the partitions ξ_k on (C, m^C), and let $\xi_C = \{\xi_{k,C}\}_{k=0}^n$ be the finite cofiltration

$$\varepsilon_C = \xi_{0,C} \geq \xi_{1,C} \geq \ldots \geq \xi_{n,C} = \nu_C\,, \quad C \in \xi_n\,,$$

induced by ξ on (C, m^C).

We shall use the measurable partition $\omega(\xi) \leq \xi_n$ defined by

$$C \overset{\omega(\xi)}{\sim} C' \iff \xi_C \simeq \xi_{C'}\,, \quad C, C' \in \xi_n\,. \tag{1.10}$$

A cofiltration $\xi = \{\xi_k\}_{k=0}^n$ is called *simple* if

(1) the factor space $(X/\xi_n, m/\xi_n)$ is non-atomic,
(2) almost all elements $\left(C, m^C\right)$ of ξ_n are purely atomic spaces,
(3) the factor space $(X/\omega(\xi), m/\omega(\xi))$ is purely atomic.

If ξ is simple, then there exists a measurable partition θ_0 such that

(i) $\theta_0 \vee \xi_n = \varepsilon_X\,,\ \theta_0 \wedge \xi_n = \omega(\xi)$,
(ii) $\theta_0 \perp \xi_k \pmod{\theta_0 \wedge \xi_k}\,,\ 0 \leq k \leq n$.

This means that the partition θ_0 factorizes the cofiltration ξ in the sense of Definition 1.1 and it "gathers" all isomorphic cofiltrations $\{\xi_C\,,\ C \in \xi_n\}$.

Given such a *minimal factorizing* partition θ_0, we have a cofiltration $\theta = \{\theta_k\}_{k=0}^n$, where $\theta_k = \theta_0 \wedge \xi_k$. The factor partitions $\hat{\xi}_k = \theta_k/\theta_0$ and the factor cofiltration $\hat{\xi} = \xi/\theta_0 = \left\{\hat{\xi}_k\right\}_{k=0}^n$ are well defined on the purely atomic factor space $(\hat{X}, \hat{m}) = (X/\theta_0, m/\theta_0)$.

The natural projection $\hat{\pi} : (X, m, \xi) \to \left(\hat{X}, \hat{m}, \hat{\xi}\right)$ is a cofiltration homomorphism.

Moreover, the cofiltration $\hat{\xi}$ is *minimal*, i.e., all the cofiltrations $\hat{\xi}_{\hat{C}}\,,\ \hat{C} \in \hat{\xi}_n$, are pairwise non-isomorphic, $\omega(\hat{\xi}) = \hat{\xi}_n$.

The cofiltration $\hat{\xi}$ is determined by ξ uniquely up to an isomorphism, and it will be called the *minimal cofiltration, majorized by* ξ.

Furthermore, there exists a decomposition

$$(X, m, \xi) \simeq \left(\hat{X} \times Y, \hat{m} \otimes m_Y, \hat{\xi} \times \varepsilon_Y\right),$$

where $\xi_k = \hat{\xi}_k \times \varepsilon_Y$ for $0 \leq k \leq n$ and (Y, m_Y) is a non-atomic Lebesgue space.

Since this presentation is determined uniquely up to an isomorphism, we have

PROPOSITION 1.6. *Two finite simple cofiltrations* $\xi = \{\xi_k\}_{k=0}^n$ *and* $\xi' = \{\xi'_k\}_{k=0}^n$ *are isomorphic iff the corresponding minimal cofiltrations* $\hat{\xi} = \left\{\hat{\xi}_k\right\}_{k=0}^n$ *and* $\hat{\xi}' = \left\{\hat{\xi}'_k\right\}_{k=0}^n$ *are isomorphic.*

2. Stochastic graphs and one-sided Markov shifts

2.1. Stochastic graphs. Consider a directed graph with countable (finite or infinite) set G of edges. Denote by $G^{(0)}$ the set of all vertices of the graph and, also, by $s(g)$ the starting vertex and by $t(g)$ the terminal vertex of an edge $g \in G$. The maps

$$s : G \ni g \mapsto s(g) \in G^{(0)}\,, \quad t : G \ni g \mapsto t(g) \in G^{(0)}$$

completely determine the structure of the graph. We set

$${}_uG := \{g \in G : s(g) = u\}\,, \quad G_v := \{g \in G : t(g) = v\}\,, \quad {}_uG_v := {}_uG \cap G_v$$

for any pair $u, v \in G^{(0)}$.

Let $G^{(n)}$ denote the set of all finite G-paths $g_1 \to g_2 \to \ldots \to g_n$ of length n in G, i.e.,

$$G^{(n)} = \{g_1 g_2 \ldots g_n \in G^n : t(g_k) = s(g_{k+1})\,, \, 1 \leq k \leq n-1\}.$$

A graph G is said to be irreducible if for every pair of vertices $u, v \in G^{(0)}$ there exists $g_1 g_2 \ldots g_n \in G^{(n)}$ such that $u = s(g_1)$ and $v = t(g_n)$.

A graph G is called *stochastic* if its edges g are equipped with positive numbers $p(g)$ such that $\sum_{g \in {}_uG} p(g) = 1$ for every $u \in G^{(0)}$.

Any stochastic graph (G, p) induces a Markov chain on the state space G with the transition probabilities matrix $P = (P(g,h))_{g,h \in G}$, where

$$P(g,h) = \begin{cases} p(h), & \text{if } t(g) = s(h), \\ 0, & \text{otherwise.} \end{cases}$$

We consider only stochastic graphs, which correspond to *positively recurrent* Markov chains. This condition implies that there exist stationary probabilities $p^{(0)}(u) > 0$ on $G^{(0)}$ such that, for all vertices $u, v \in G^{(0)}$,

$$\sum_{u \in G^{(0)}} p^{(0)}(u) = 1\,, \quad \sum_{g \in G_v} p^{(0)}(s(g))\, p(g) = p^{(0)}(v)\,.$$

If, in addition, G is irreducible, the stationary probabilities $p^{(0)}(u)\,,\, u \in G^{(0)}$, on the vertices are uniquely determined by the weights $p(g)\,,\, g \in G$ on the edges.

DEFINITION 2.1. Let G and H be two stochastic graphs.

(i) A map $\phi : G \to H$ is a *graph homomorphism* if there exists a map $\phi^{(0)} : G^{(0)} \to H^{(0)}$ such that

$$s(\phi(g)) = \phi^{(0)}(s(g))\,, \quad t(\phi(g)) = \phi^{(0)}(t(g))\,, \quad g \in G\,.$$

(Note that the map $\phi^{(0)}$ is unique if it exists.)

(ii) A graph homomorphism $\phi : G \to H$ is *deterministic* if $\phi^{(0)}(G^{(0)}) = H^{(0)}$ and for every $u \in G^{(0)}$ the restriction

$$\phi\,|_{{}_uG} : {}_uG \to {}_{\phi^{(0)}(u)}H$$

of the map ϕ to ${}_uG$ is a bijection of this set onto ${}_{\phi^{(0)}(u)}H$.

(iii) A graph homomorphism is *weight preserving* if $p(\phi(g)) = p(g)$ for all $g \in G$.

In this paper we shall use the term "homomorphism" for the *weight-preserving deterministic* graph homomorphisms only.

EXAMPLE 2.2. Let G be a stochastic graph and $n > 1$. Then $G^{(n)}$ can be considered as a stochastic graph if we set $(G^{(n)})^{(0)} := G^{(n-1)}$ and

$$s(g^{(n)}) := g_1 g_2 \dots g_{n-1} , \quad t(g^{(n)}) := g_2 g_3 \dots g_n , \quad p(g^{(n)}) := p(g_n) ,$$

for $g^{(n)} = g_1 g_2 \dots g_n \in G^{(n)}$.

It is easy to see that $t : G^{(n)} \to G^{(n-1)}$ is a homomorphism of the stochastic graphs, while $s : G^{(n)} \to G^{(n-1)}$ is a graph homomorphism, which is neither weight preserving nor deterministic in general.

2.2. Normal partitions and factor graphs.

DEFINITION 2.3. A partition τ of a stochastic graph G is called *normal* if there exists a partition $\tau^{(0)}$ of $G^{(0)}$ such that

(i) $g \overset{\tau}{\sim} g' \implies s(g) \overset{\tau^{(0)}}{\sim} s(g')$, $t(g) \overset{\tau^{(0)}}{\sim} t(g')$,
(ii) for any element C of the partition τ, the restriction $s|_C : C \to s(C)$ is a bijection of C onto the element $s(C) \in \tau^{(0)}$ of the partition $\tau^{(0)}$,
(iii) $g \overset{\tau}{\sim} g' \implies p(g) = p(g')$.

Condition (ii) implies that $s(\tau) := \{s(C) \, , \, C \in \tau\}$ is a partition on $G^{(0)}$ and $\tau^{(0)} = s(\tau)$. So the partition $\tau^{(0)}$ is uniquely determined by τ.

Normal partitions are exactly the partitions induced by homomorphisms of graphs.

PROPOSITION 2.4.

(i) *Let $\phi : G \to H$ be a homomorphism, and let τ_ϕ and $\tau_\phi^{(0)}$ be defined by*

$$\tau_\phi = \phi^{-1}\varepsilon_H \, , \quad \tau_\phi^{(0)} = {\phi^{(0)}}^{-1}\varepsilon_{H^{(0)}} \, . \tag{2.1}$$

Then τ_ϕ is a normal partition on G and $(\tau_\phi)^{(0)} := s(\tau_\phi) = \tau_\phi^{(0)}$.

(ii) *Let τ be a normal partition on G with $\tau^{(0)} = s(\tau)$. Then the factor maps $\pi_\tau : G \to G/\tau$ and $\pi_{s(\tau)} : G^{(0)} \to G^{(0)}/s(\tau)$ uniquely induce a stochastic graph structure on G/τ, such that $\pi_\tau : G \to G/\tau$ is a homomorphism, and $(G/\tau)^{(0)} = G^{(0)}/\tau^{(0)}$ and $(\pi_\tau)^{(0)} = \pi_{\tau^{(0)}}$.*

(iii) *Let $\phi : G \to H$ be a homomorphism, and let $\tau = \tau_\phi$. Then the stochastic graphs G/τ and H are isomorphic, and there exists a unique isomorphism $\chi : G/\tau_\phi \to H$ such that $\phi = \chi \circ \pi_{\tau_\phi}$.*

The stochastic graph G/τ constructed in (ii) by means of a normal partition τ will be called the *factor graph* of G induced by τ. We omit the proof of Proposition 2.4, which follows from comparing Definitions 2.1 and 2.3.

PROPOSITION 2.5. *Let τ_1 and τ_2 be two normal partitions such that $s(\tau_1) = s(\tau_2)$. Then the factor graphs G/τ_1 and G/τ_2 are isomorphic.*

PROOF. Since $s(\tau_1) = s(\tau_2)$, the sets of vertices of the factor graphs G/τ_1 and G/τ_2 coincide, and we may set

$$\sigma := s(\tau_1) = s(\tau_2) \, , \quad G^{(0)}/\sigma = (G/\tau_1)^{(0)} = (G/\tau_2)^{(0)} \, .$$

Both natural projections $\pi_{\tau_1}^{(0)}$ and $\pi_{\tau_2}^{(0)}$ induced by π_{τ_1} and π_{τ_2}, respectively, coincide with $\pi_\sigma : G^{(0)} \to G^{(0)}/\sigma$.

For any pair $\tilde{u}, \tilde{v} \in G^{(0)}/\sigma$ we choose and fix a pair of vertices $u, v \in G^{(0)}$ such that $\pi_\sigma(u) = \tilde{u}$ and $\pi_\sigma(v) = \tilde{v}$. Then by setting $V := \pi_\sigma^{-1}\tilde{v}$ and ${}_uG_V := \{ {}_uG_{v'} \,,\, v' \in V\}$, we have

$$ {}_{\tilde{u}}(G/\tau_1)_{\tilde{v}} \neq \emptyset \iff {}_{\tilde{u}}(G/\tau_2)_{\tilde{v}} \neq \emptyset \iff {}_uG_V \neq \emptyset \,. $$

If ${}_uG_V \neq \emptyset$, the restriction $\pi_{\tau_i}|_{{}_uG_V}$ is a weight-preserving bijection between the set ${}_uG_V$ and ${}_{\tilde{u}}(G/\tau_i)_{\tilde{v}}$ for each $i = 1, 2$.

Thus for every pair $\tilde{u}\,,\, \tilde{v} \in G^{(0)}/\sigma$ one can choose a weight-preserving bijection $\chi_{\tilde{u}\tilde{v}}$ between ${}_{\tilde{u}}(G/\tau_1)_{\tilde{v}}$ and ${}_{\tilde{u}}(G/\tau_2)_{\tilde{v}}$. Taking into account the decompositions

$$ G/\tau_i := \bigcup_{\tilde{u}\,,\,\tilde{v} \in G^{(0)}/\sigma} {}_{\tilde{u}}(G/\tau_i)_{\tilde{v}} \,,\quad \tilde{u}\,,\, \tilde{v} \in G^{(0)}/\sigma \,,\quad i = 1\,,2\,, $$

we see that any choice of the bijections $\chi_{\tilde{u}\tilde{v}}\,,\ \tilde{u}, \tilde{v} \in G^{(0)}/\sigma$, uniquely determines the desired isomorphism $\chi : G/\tau_1 \to G/\tau_1$. □

2.3. Filtrations generated by stochastic graphs. Let G be a stochastic graph. Consider the space

$$ X_G^+ = \{x = \{g_n\}_{n=1}^\infty \in G^{\mathbb{N}} : s(g_{n+1}) = t(g_n)\,,\ n \in \mathbb{N}\} \tag{2.2} $$

of all infinite forward G-paths $g_1 \to g_2 \to g_3 \to \ldots$ in G. This space is equipped with the natural probability Markov measure m_G^+, which is uniquely defined by

$$ m_G^+\left(A_G^+(\,g_1g_2\ldots g_n\,)\right) = p^{(0)}(s(g_1))\,p(g_1)\,p(g_2)\ldots p(g_n) $$

on the cylinder sets of the form

$$ A_G^+(g_1g_2\ldots g_n) = \{x = \{h_k\}_{k=1}^\infty \in X_G^+ : h_1 = g_1, \ldots, h_n = g_n\} $$

for all $g_1g_2\ldots g_n \in G^{(n)}$.

The coordinate functions

$$ Z_n : X_G^+ \ni x = \{g_k\}_{k=1}^\infty \mapsto g_n \in G\,,\quad n \in \mathbb{N}\,, $$

produce a stationary Markov chain $Z_1, Z_2, \ldots, Z_n, \ldots$ on the path space (X_G^+, m_G^+) with the transition probabilities

$$ P(g,h) = m_G^+\{Z_{n+1} = h\,|\, Z_n = g\} = p(h)\,,\quad (g,h) \in G^{(2)}\,,\quad n \in \mathbb{N}\,. $$

The coordinate functions generate an increasing sequence of measurable partitions $\eta = \eta_G^+ = \{\eta_n\}_{n=0}^\infty$ on X_G^+, where

$$ \eta_n := \bigvee_{k=1}^{n} {Z_k}^{-1}\varepsilon_G = \left\{ A_G^+(\,g_1g_2\ldots g_n\,),\ g_1g_2\ldots g_n \in G^{(n)} \right\},\quad n \in \mathbb{N}\,, $$

and

$$ \eta_0 := \left\{ {}_uX_G^+\,,\ u \in G^{(0)} \right\}, $$

with

$$ {}_uX_G^+ = \{x = (g_1g_2\ldots) \in X_G^+ : s(g_1) = u\}\,. $$

For every fixed $u \in G^{(0)}$ we shall use also the family $\{{}_u\eta\,,\ u \in G^{(0)}\}$ of increasing sequences of measurable partitions ${}_u\eta = {}_u\eta_G = \{{}_u\eta_n\}_{n=0}^\infty$ of the space $({}_uX_G^+\,,\ {}_um_G^+)$. Here ${}_u\eta_n$ is the restriction of η_n to ${}_uX_G^+$, and ${}_um_G^+$ is the conditional probability induced by the probability m_G^+ on ${}_uX_G^+$.

We shall refer to the constructed filtration $\eta_G^+ = \{\eta_{n,G}\}_{n=0}^\infty$ as the *forward filtration* generated by the stochastic graph G and to the filtration ${}_u\eta_G^+ = \{{}_u\eta_{n,G}\}_{n=0}^\infty$ as the *forward filtration rooted at* u.

Now let G and H be two stochastic graphs and let $\phi : G \to H$ be a homomorphism. Then ϕ induces a map $\Phi_\phi^+ : X_G^+ \to X_H^+$ defined by

$$\Phi_\phi^+ : X_G^+ \ni \{g_n\}_{n=1}^\infty \mapsto \{\phi(g_n)\}_{n=1}^\infty \in X_H^+ . \tag{2.3}$$

It follows from the definition of the Markov measures m_G^+ and m_H^+ that Φ_ϕ^+ is measure preserving. Moreover, let $u \in G^{(0)}$ and $v = \phi^{(0)}(u) \in H^{(0)}$. Then $\Phi_\phi^+ \left({}_uX_G^+\right) = {}_vX_H^+$ and the restriction of Φ_ϕ^+ to ${}_uX_G^+$ is a filtration isomorphism

$$ {}_u\Phi_\phi^+ := \Phi_\phi^+ \,|_{\,{}_uX_G^+} : \left({}_uX_G^+ \,,\, {}_um_G^+ \,,\, {}_u\eta_{n,G}^+\right) \to \left({}_vX_H^+ \,,\, {}_vm_H^+ \,,\, {}_u\eta_{n,H}^+\right), \tag{2.4}$$

where ${}_u\Phi_\phi^+ \left({}_u\eta_G^+\right) = {}_v\eta_H^+$ means that ${}_u\Phi_\phi^+ \left({}_u\eta_{n,G}^+\right) = {}_v\eta_{n,H}^+$ for all n.

Notice also that

$$\phi^{(0)}(u) = \phi^{(0)}(u') \implies {}_u\,\eta_G^+ \simeq {}_{u'}\,\eta_G^+, \qquad u, u' \in G^{(0)} .$$

2.4. Cofiltrations generated by stochastic graphs. Let G be a stochastic graph. Consider the space

$$X_G^- = \left\{x = \{g_n\}_{n=-1}^{-\infty} \in G^{-\mathbb{N}} : t(g_{n-1}) = s(g_n)\,,\; n \in -\mathbb{N}\,\right\} \tag{2.5}$$

of all infinite G-paths $\ldots \to g_{-2} \to g_{-1}$.

The space is equipped with the natural probability Markov measure m_G^-, which is uniquely defined by

$$m_G^-\left(A_G^-\left(g_{-n}\ldots g_{-2}g_{-1}\right)\right) = p^{(0)}(s(g_{-n}))\,p(g_{-n})\,\ldots\,p(g_{-2})\,p(g_{-1})$$

on the cylinder sets of the form

$$A_G^-(g_{-n}\ldots g_{-2}g_{-1}) = \left\{x = \{h_k\}_{k=-1}^{-\infty} \in X_G^- : h_{-n} = g_{-n}, \ldots, h_{-1} = g_{-1}\right\}$$

for all $g_{-n}\ldots g_{-2}g_{-1} \in G^{(n)}$.

The *one-sided Markov shift* $\mathbf{T_G}$ generated by G acts on the probability space (X_G^-, m_G^-) by

$$T_G\left(\{\,g_n\,\}_{n=-1}^{-\infty}\right) = \{g_{n-1}\}_{n=-1}^{-\infty} .$$

The stationarity of the probability $p^{(0)}$ on $G^{(0)}$ provides that T_G preserves the Markov measure m_G^-. If the graph G is irreducible, the stationary distribution $p^{(0)}$ on $G^{(0)}$ is unique, and therefore the Markov measure m_G^- is uniquely determined by the graph G.

The coordinate functions

$$Z_n : X_G^- \ni x = \{g_k\}_{k=-1}^{-\infty} \mapsto g_n \in G\;, \quad n \in -\mathbb{N}\,,$$

produce a stationary Markov chain $\ldots, Z_{-3}, Z_{-2}, Z_{-1}$ on the path space $\left(X_G^-, m_G^-\right)$ with the transition probabilities

$$P\,(\,g\,,\,h\,) \;=\; m_G^-\{Z_n = h\,|\,Z_{n-1} = g\} = p(h)\;, \quad (g,h) \in G^{(2)}\;, \quad n \in -\mathbb{N}\,.$$

Let $\zeta_G = \{A^-(g)\}_{g\in G}$, where $A^-(g) = \left\{x = \{h_k\}_{k=-1}^{-\infty} \in X_G^- : h_{-1} = g\right\}$. Then $\bigvee_{n=1}^\infty T_G^{-n+1}\zeta_G = \varepsilon_{X_G^-}$, i.e., ζ_G is a one-sided generating partition for T_G.

The partition ζ_G is called the *standard one-sided Markov generator* of the one-sided Markov shift T_G.

The cofiltration $\xi_G^- = \xi(T_G) = \{\xi_n\}_{n=0}^{\infty}$ corresponding to a stochastic graph G is defined by

$$\xi_0 = \varepsilon_{X_G^-}\,, \quad \xi_n = T_G^{-n}\varepsilon_{X_G^-} = \bigvee_{k=n+1}^{\infty} T_G^{-k}\zeta_G\,, \quad n \in \mathbb{N}\,.$$

The measurable partition ξ_n is generated by the coordinate functions $Z_{-k}, k > n$.

The element $C_{\xi_n}(x)$ of ξ_n, containing a point $x = \{g_n\}_{n=-1}^{-\infty}$, has the form

$$C_{\xi_n}(x) = T_G^{-n}\left(T_G^n x\right) = \left\{x' = \{g'_k\}_{k=-1}^{-\infty} \in X_G^- : g'_k = g_k\,,\ k < -n\right\},$$

and the conditional measures on $C_{\xi_n}(x)$ are defined by

$$m^{C_{\xi_n}(x)}\left(\{x'\}\right) = p(g'_{-n})\dots p(g'_{-2})p(g'_{-1}), \quad x' = \{g'_k\}_{k=-1}^{-\infty} \in C_{\xi_n}(x)\,.$$

The Markov property of the measure m_G^- implies that the conditional measures depend only on $x'_{-n}, \dots, x'_{-1}$.

REMARK 2.6. The Markov measures m_G^+ and m_G^- as well as the corresponding stationary Markov chains $Z_1, Z_2, Z_3, \dots$ and $\dots, Z_{-3}, Z_{-2}, Z_{-1}$ on the path spaces X_G^+ and X_G^-, respectively, have the same finite-dimensional distributions

$$\mu^{(n)}(g_1 g_2 \dots g_n) := p^{(0)}(s(g_1))p(g_1)p(g_2)\dots p(g_n), \quad g_1 g_2 \dots g_n \in G^{(n)}\,. \tag{2.6}$$

Namely, the corresponding projections

$$\sigma_n^+ : \left(X_G^+, m_G^+\right) \to \left(G^{(n)}, \mu^{(n)}\right), \quad \sigma_n^- : \left(X_G^-, m_G^-\right) \to \left(G^{(n)}, \mu^{(n)}\right),$$

defined by

$$\begin{aligned} \sigma_n^+ &: X_G^+ \ni (g_1 g_2 \dots) \ \mapsto\ g_1\, g_2\, \cdots\, g_n \in G^{(n)}\,, \\ \sigma_n^- &: X_G^- \ni (\dots g_{-2} g_{-1}) \ \mapsto\ g_{-n} \cdots g_{-2} g_{-1} \in G^{(n)}\,, \end{aligned} \tag{2.7}$$

are measure space homomorphisms.

3. Minimal stochastic graphs

3.1. Minimal normal partitions.

Let G be a stochastic graph.

DEFINITION 3.1. A normal partition τ of G is called *minimal* if for any normal partition τ_1 of G, $\tau_1 \le \tau \Longrightarrow \tau_1 = \tau$.

To describe minimal normal partitions, we shall use the forward filtration η_G^+ generated by the graph G and the corresponding filtrations ${}_u\eta_G^+$ rooted at $u \in G^{(0)}$; see Section 2.3.

Let us consider a partition ω^G on $G^{(0)}$, which is defined by

$$u \overset{\omega^G}{\sim} u' \iff {}_u\eta_G^+ \simeq {}_{u'}\eta_G^+\,, \quad u\,,\, u' \in G^{(0)}\,. \tag{3.1}$$

This means that u and u' belong to the same element of ω^G iff the corresponding filtrations rooted at u and u', respectively, are isomorphic.

PROPOSITION 3.2.

(i) *There exists a normal partition τ such that $s(\tau) = \omega^G$.*
(ii) *For any normal partition τ, $s(\tau) \ge \omega^G$.*
(iii) *A normal partition τ is minimal iff $s(\tau) = \omega^G$.*

(iv) *For any pair τ_1, τ_2 of minimal normal partitions the factor graphs G/τ_1 and G/τ_1 are isomorphic.*

PROOF. (i). Let us consider $\omega^G = \{W_k\}_{k\in K}$, where K is an index set such that $|K| = |\,G^{(0)}/\omega^G|$. For each $k \in K$, we choose and fix an element $w_k \in W_k$ and also for every $u \in W_k$ we choose and fix a filtration isomorphism

$$ {}_u\Psi : \left({}_uX_G^+\,,\, {}_um_G^+\,,\, {}_u\eta_G^+\right) \to \left({}_{w_k}X_G^+\,,\, {}_{w_k}m_G^+\,,\, {}_{w_k}\eta_G^+\right)\,. $$

Using the projection

$$ \sigma_1^+ : X_G^+ \ni x = \{g_n\}_{n=1}^{\infty} \mapsto g_1 \in G $$

and the restrictions

$$ {}_u\sigma_1^+ := \sigma_1^+|_{{}_uX_G^+} :{}_u X_G^+ \mapsto \sigma_1^+\left({}_uX_G^+\right) = {}_uG\,, $$

we get a family of maps

$$ {}_u\psi : {}_uG \to {}_{w_k}G\,, \quad u \in W_k \subseteq G^{(0)}\,, \quad k \in K, $$

such that ${}_u\psi \circ {}_u\sigma_1^+ = {}_{w_k}\sigma_1^+ \circ {}_u\Psi$ for all u.

We can now define a desired partition τ of G by

$$ g \overset{\tau}{\sim} g' \iff s(g) \overset{\omega^G}{\sim} s(g')\,, \quad {}_{s(g)}\psi\,(g) = {}_{s(g')}\psi\,(g')\,. $$

For any pair $u \overset{\omega^G}{\sim} u'$ the map

$$ \Psi_{u\,,\,u'} := {}_{u'}\Psi^{-1} \circ {}_u\Psi : \left({}_uX_G^+\,,\, {}_um_G^+\right) \to \left({}_{u'}X_G^+\,,\, {}_{u'}m_G^+\right) $$

is a measure space isomorphism such that $\Psi_{u,u'}\left({}_u\eta_G^+\right) = {}_{u'}\eta_G^+$. Hence, the map $\psi_{u,u'} := {}_{u'}\psi^{-1} \circ {}_u\psi$ is a weight-preserving bijection of ${}_uG$ onto ${}_{u'}G$ such that

$$ t(\psi_{u,u'}(g)) \overset{\omega^G}{\sim} t(g)\,, \quad g \in {}_uG\,. $$

Whence the partitions τ is normal and $s(\tau) = \omega^G$. Part (i) is proved.

(ii). Let τ be a normal partition. By Proposition 2.4 there exists a homomorphism $\phi : G \to H$ such that $\tau = \tau_\phi$ and $\tau^{(0)} = s(\tau) = \tau_\phi^{(0)}$, where the partitions τ_ϕ and $\tau_\phi^{(0)}$ are defined by (2.1).

Let $u \in G^{(0)}$ and $v = \phi^{(0)}(u) \in H^{(0)}$. Then by (2.3) and (2.4) the homomorphism ϕ induces a filtration isomorphism ${}_u\Phi_\phi^+ : {}_u\eta_G^+ \to {}_v\eta_H^+$.

This implies

$$ (3.2) \qquad u \overset{\tau_\phi^{(0)}}{\sim} u' \implies {}_u\eta_G^+ \simeq {}_{u'}\eta_G^+\,, \quad u, u' \in G^{(0)}\,. $$

Comparing (3.2) and (3.1) shows that $s(\tau) = \tau_\phi^{(0)} \geq \omega^G$.

(iii). By Definition 3.1 we have

$$ \tau_1 \leq \tau_2 \implies s(\tau_1) \leq s(\tau_2) $$

and

$$ \tau_1 \leq \tau_2\,, \quad s(\tau_1) = s(\tau_2) \implies \tau_1 = \tau_2 $$

for any pair of normal partitions τ_1 and τ_2. Thus (iii) follows from (ii).

Finally, (iv) follows from (iii) and Proposition 2.5. □

3.2. Existence and uniqueness of minimal graphs.

DEFINITION 3.3.

(i) A stochastic graph H is said to be *majorized* by a stochastic graph G if there exists a homomorphism $\phi : G \to H$.

(ii) A stochastic graph M is called *minimal* if every homomorphism $\phi : M \to H$ is an isomorphism.

Let G be a stochastic graph. We shall construct a minimal stochastic graph M majorized by G as a factor graph G/τ by a minimal normal partition τ of G.

THEOREM 3.4. *For any stochastic graph G there exists a minimal stochastic graph M majorized by G. The minimal graph M is unique up to isomorphism.*

PROOF. *Existence.* For a given stochastic graph G we consider the corresponding partition ω^G defined by (3.1). By using Proposition 3.2(i), we choose a normal partition τ of G with $s(\tau) = \omega^G$. Then the factor map $\pi_\tau : G \to G/\tau$ is a homomorphism of G onto the factor graph G/τ, and the normal partition τ is minimal.

We want to show that the corresponding factor graph is minimal. Let $\phi_1 : G/_\tau \to H$ be a homomorphism of G/τ to a stochastic graph H. Then $\phi = \phi_1 \circ \pi_\tau$ is a homomorphism of G onto H such that $\tau_\phi^{(0)} \leq \tau_{1\phi}^{(0)} = \omega^G$. On the other hand $\tau_\phi^{(0)} \geq \omega^G$ by Proposition 3.2(iii) and hence $\tau_\phi^{(0)} = \omega^G = \tau_{1\phi}^{(0)}$. This means that $\phi_1^{(0)}$ is a bijection, whence ϕ_1 is an isomorphism. Thus G/τ is a minimal stochastic graph majorized by G.

Uniqueness. Let $\phi_i : G \to M_i$, $i = 1, 2$, be two homomorphisms such that the stochastic graphs M_1 and M_2 are minimal.

Considering two normal partitions τ_{ϕ_i}, $i = 1, 2$, of G, we have by Proposition 3.2(iii) that $\tau_{\phi_1}^{(0)} = \tau_{\phi_2}^{(0)} = \omega^G$. Then the factor graphs G/τ_i, $i = 1, 2$, are isomorphic by Proposition 2.5. Whence, the graphs M_i, $i = 1, 2$, are isomorphic by Proposition 2.4(iii). The proof is completed. □

Also we have proved the following consequences.

(1) A normal partition τ is minimal iff the corresponding factor graph G/τ is minimal.

(2) A stochastic graph G is minimal iff $\omega^G = \varepsilon_{G^{(0)}}$, i.e., all the filtrations ${}_u\eta_G^+$, $u \in G^{(0)}$, are pairwise non-isomorphic, i.e., the filtration η_G is minimal.

(3) Two stochastic graphs G_1 and G_2 majorize a stochastic graph H iff they majorize a minimal stochastic graph M.

4. Minimal filtrations and cofiltrations

4.1. Filtrations generated by minimal stochastic graphs. Let M be an irreducible minimal stochastic graph and let η_M^+ be the corresponding infinite filtration on the measure space (X_M^+, m_M^+).

The graph M is minimal iff $\omega^M = \varepsilon_{M^{(0)}}$, i.e., all the filtrations ${}_u\eta_M^+$, $u \in M^{(0)}$, are pairwise non-isomorphic.

We want to show that any minimal irreducible stochastic graph M is uniquely determined by its forward filtration η_M^+. More exactly

PROPOSITION 4.1. *Let M_1 and M_2 be two minimal irreducible stochastic graphs. Then M_1 and M_2 are isomorphic iff the corresponding filtrations $\eta_{M_1}^+$ and $\eta_{M_2}^+$ are isomorphic.*

PROOF. Let $\eta = \{\eta_n\}_{n=0}^{\infty}$ be a filtration on a probability Lebesgue space (X, m), and let

$$\pi_0 : (X, m) \to (X_0, m_0) := (X/\eta_0, m/\eta_0) ,$$
$$\pi_1 : (X, m) \to (X_1, m_1) := (X/\eta_1, m/\eta_1) ,$$

and let $\pi_{01} : (X_1, m_1) \to (X_0, m_0)$ be the natural projections on the corresponding factor spaces.

For any $x_0 \in X_0$ we denote by ${}_{x_0}\eta$ the restriction of η to the element $(\pi_0)^{-1}x_0$ of η_0. Also for any $x_1 \in X_1$ we denote by ${}_{x_1}\eta$ the restriction of η to the element ${\pi_1}^{-1}x_1$ of η_1.

Suppose the filtration η is isomorphic to the forward filtration η_M^+ of a minimal irreducible stochastic graph M. Then η satisfies the following conditions:

- All the filtrations ${}_{x_0}\eta$, $x_0 \in X_0$, are pairwise non-isomorphic.
- For any $x_1 \in X_1$ there exists the only $x_0 = (x_1)^* \in X_0$ such that the filtration ${}_{x_1}\eta$ is isomorphic to ${}_{x_0}\eta$.

These conditions allow one to construct a stochastic graph H by setting $H := X_1$, $H^{(0)} := X_0$ and

$$s(x_1) := \pi_{01}(x_1) , \quad t(x_1) := (x_1)^*, \quad p(x_1) := (m_1(x_1))^{-1} m_0(\pi_{01}(x_1)) .$$

In the case when $\eta = \eta_M^+$ the spaces X_0 and X_1 are naturally identified with $M^{(0)}$ and M, respectively, and the constructed graph H coincides with the graph M by this identification. Thus any minimal irreducible stochastic graph M is uniquely determined by its minimal filtration η_M^+. Whence, $\eta_{M_1}^+ \simeq \eta_{M_2}^+ \Longrightarrow M_1 \simeq M_2$.

The converse is obvious. □

REMARK 4.2. It can be proved that any minimal irreducible stochastic graph M is uniquely determined by each of the filtrations ${}_u\eta^M$, $u \in M^{(0)}$. Thus two minimal irreducible stochastic graphs M_1 and M_2 are isomorphic iff there exist $u \in M^{(0)}$ and $u' \in M'^{(0)}$ such that the filtrations ${}_u\eta_M^+$ and ${}_{u'}\eta_{M'}^+$ are isomorphic.

4.2. Finitely isomorphic cofiltrations. Let G be an irreducible stochastic graph and let $\xi(T_G) = \xi_G^- = \{\xi_{n,G}\}_{n=0}^{\infty}$ be the corresponding backward cofiltration on the probability Lebesgue space (X_G^-, m_G^-).

Recall that we assume throughout that the graph G is non-degenerate, i.e., $|{}_uG| > 1$ for some $u \in G^{(0)}$. This condition (together with irreducibility) implies that for every n the corresponding finite cofiltration $\xi_G^{(n)} = \{\xi_{k,G}\}_{k=0}^{n}$ is simple in the sense of Section 1.5.

LEMMA 4.3. *Let G and H be two irreducible stochastic graphs. If there exists a homomorphism $\phi : G \to H$, then the cofiltrations ξ_G^+ and ξ_H^+, are finitely isomorphic.*

PROOF. The homomorphism $\phi : G \to H$ induces a map $\Phi_\phi^- : X_G^- \to X_H^-$ defined by

$$\Phi_\phi^- : X_G^- \ni (\ldots g_{-2}\, g_{-1}) \mapsto (\ldots \phi(g_{-2})\, \phi(g_{-1})) \in X_H^- , \tag{4.1}$$

which is a measure space homomorphism of $\left(X_G^-, m_G^-\right)$ onto $\left(X_H^-, m_H^-\right)$ and for every n

$$\Phi_\phi^- : \left(X_G^-, m_G^-, \xi_G^{(n)}\right) \to \left(X_H^-, m_H^-, \xi_H^{(n)}\right)$$

is a cofiltration homomorphism.

The finite cofiltrations $\xi_G^{(n)}$ and $\xi_H^{(n)}$ are simple. Let

$$\hat{\pi}_H \ : \ \left(X_H^- \, , \, m_H^- \, , \, \xi_H^{(n)}\right) \ \to \ \left(\hat{X}_H^- \, , \, \hat{m}_H^- \, , \, \hat{\xi}_H^{(n)}\right)$$

be a cofiltration homomorphism such that $\hat{\xi}_H^{(n)}$ is a minimal cofiltration majorized by $\xi_H^{(n)}$. Then

$$\hat{\pi} \ := \ \hat{\pi}_H \circ \Phi_\phi^- \ : \ \left(X_G^- \, , \, m_G^- \, , \, \xi_G^{(n)}\right) \ \to \ \left(\hat{X}_H^- \, , \, \hat{m}_H^- \, , \, \hat{\xi}_H^{(n)}\right)$$

is a cofiltration homomorphism, and $\hat{\xi}_H^{(n)}$ is also a minimal cofiltration majorized by $\xi_G^{(n)}$.

Thus the finite simple cofiltrations $\xi_G^{(n)}$ on $\left(X_G^-, m_G^-\right)$ and $\left(X_H^-, m_H^-\right)$ are isomorphic by Proposition 1.6. □

COROLLARY 4.4. *Cofiltrations $\xi(G)$ and $\xi(M(G))$ are finitely isomorphic.*

Furthermore we consider the forward filtration $\eta_G^+ = \{\eta_{n,G}\}_{n=0}^\infty$ on $\left(X_G^+ \, , \, m_G^+\right)$ and its finite filtrations $\eta_G^{(n)} = \{\eta_{k,G}\}_{k=0}^n \, , \; n \in \mathbb{N}$.

LEMMA 4.5. *Let G be an irreducible stochastic graph and $n \in \mathbb{N}$. There exist a minimal cofiltration $\hat{\xi}^{(n)} = \{\hat{\xi}_k\}_{k=0}^n$ and a minimal filtration $\hat{\eta}^{(n)} = \{\hat{\eta}_k\}_{k=0}^n$ on a purely atomic measure space $(\hat{X}, \hat{m})$ such that $\hat{\xi}^{(n)}$ is majorized by $\xi_G^{(n)}$, $\hat{\eta}^{(n)}$ is majorized by $\eta_G^{(n)}$, and $\hat{\xi}_k^{(n)} = \hat{\eta}_{n-k}^{(n)}$ for every $0 \le k \le n$.*

PROOF. We use the space $\left(G^{(n)}, \mu^{(n)}\right)$ defined by (2.6) and the projections σ_n^+ and σ_n^- from $\left(X_G^+, m_G^+\right)$ and $\left(X_G^-, m_G^+\right)$ onto the space $(G^{(n)}, \mu^{(n)})$ defined by (2.7).

Let the partitions $\tilde{\eta}_k = \tilde{\xi}_{n-k} \, , \; 1 \le k \le n$, be defined on $G^{(n)}$ by

$$g_1 g_2 \cdots g_n \overset{\tilde{\eta}_k}{\sim} g_1' g_2' \cdots g_n' \quad \Longleftrightarrow \quad g_1 = g_1' \, , \, g_2 = g_2' \, , \, \ldots \, , \, g_k = g_k',$$

and let $\tilde{\eta}_0 = \tilde{\xi}_n$ be defined by

$$g_1 g_2 \cdots g_n \overset{\tilde{\eta}_0}{\sim} g_1' g_2' \cdots g_n' \quad \Longleftrightarrow \quad s(g_1) = s(g_1')$$

for $g_1 g_2 \cdots g_n \, , \, g_1' g_2' \cdots g_n' \in G^{(n)}$. Then

$$\sigma_n^+ : \left(X_G^+ \, , \, m_G^+ \, , \, \xi_G^{(n)}\right) \ \to \ \left(G^{(n)} \, , \, \mu^{(n)} \, , \, \tilde{\eta}^{(n)}\right)$$

is a filtration homomorphism, and

$$\sigma_n^- \ : \ \left(X_G^- \, , \, m_G^- \, , \, \xi_G^{(n)}\right) \ \to \ \left(G^{(n)} \, , \, \mu^{(n)} \, , \, \tilde{\xi}^{(n)}\right)$$

is a cofiltration homomorphism.

Therefore, one can find a factor map $\hat{\pi}_n : \left(G^{(n)}, \mu^{(n)}\right) \to \left(\hat{X}_n, \hat{m}_n\right)$, a minimal filtration $\hat{\eta}^{(n)} = \{\hat{\eta}_k\}_{k=0}^n$ and a minimal cofiltration $\hat{\xi}^{(n)} = \{\hat{\xi}_k\}_{k=0}^n$ on $\hat{X}_n$ such that

$$\hat{\pi}_n : \left(G^{(n)}, \mu^{(n)} \, , \, \tilde{\eta}^{(n)}\right) \ \to \ \left(\hat{X} \, , \, \hat{m} \, , \, \hat{\eta}^{(n)}\right)$$

is a filtration homomorphism,

$$\hat{\pi}_n : \left(G^{(n)}, \mu^{(n)}, \tilde{\xi}^{(n)}\right) \to \left(\hat{X}_n, \hat{m}_n, \hat{\xi}^{(n)}\right)$$

is a cofiltration homomorphism, and $\hat{\eta}_k = \hat{\xi}_{n-k}$ for all $0 \le k \le n$.

Thus the minimal $\hat{\eta}^{(n)}$ and $\hat{\xi}^{(n)}$ are majorized by $\eta_G^{(n)}$ and $\xi_G^{(n)}$ by means of homomorphisms $\hat{\pi}_n \circ \sigma_n^+$ and $\hat{\pi}_n \circ \sigma_n^-$, respectively. □

COROLLARY 4.6. $\hat{\xi}_{G_1}^{(n)} \simeq \hat{\xi}_{G_2}^{(n)} \iff \hat{\eta}_{G_1}^{(n)} \simeq \hat{\eta}_{G_2}^{(n)}$.

Let $M = M(G)$ be a minimal graph majorized by G. The backward filtration $\eta_M^+ = \{\eta_{n,M}\}_{n=0}^{\infty}$ is minimal, but the corresponding finite filtrations $\eta_M^{(n)} = \{\eta_{k,M}\}_{k=0}^{n}$ need not be minimal in general. We denote by $\hat{\eta}_M^{(n)} = \{\hat{\eta}_{k,M}\}_{k=0}^{n}$ the minimal finite filtrations majorized by $\eta_M^{(n)}$.

PROPOSITION 4.7. *Let G_1 and G_2 be two irreducible stochastic graphs and let $M_1 = M(G_1)$ and $M_2 = M(G_2)$ be the corresponding minimal graphs. Then the cofiltrations $\xi_{G_1}^-$ and $\xi_{G_2}^-$ are finitely isomorphic iff the minimal filtrations $\eta_{M_1}^+$ and $\eta_{M_2}^+$ are isomorphic.*

PROOF. By Corollary 4.4, Proposition 1.6 and Corollary 4.6 we have for any n

$$\xi_{G_1}^{(n)} \simeq \xi_{G_2}^{(n)} \iff \xi_{M_1}^{(n)} \simeq \xi_{M_2}^{(n)} \iff \hat{\xi}_{M_1}^{(n)} \simeq \hat{\xi}_{M_2}^{(n)} \iff \hat{\eta}_{M_1}^{(n)} \simeq \hat{\eta}_{M_2}^{(n)}.$$

Proposition 1.5 completes the proof. □

4.3. Classification theorems. We have now all the ingredients necessary to prove the following two theorems.

THEOREM 4.8. *Let G_1 and G_2 be two non-degenerate irreducible stochastic graphs and let $M(G_1)$ and $M(G_2)$ be minimal stochastic graphs majorized by G_1 and G_2, respectively. Then the following conditions are equivalent:*

(i) *The minimal stochastic graphs $M(G_1)$ and $M(G_2)$ are isomorphic.*
(ii) *The cofiltrations $\xi_{G_1}^-$ and $\xi_{G_2}^-$ are finitely isomorphic.*

PROOF. The cofiltrations $\xi_{G_1}^-$ and $\xi_{G_2}^-$ are finitely isomorphic iff the minimal filtrations $\eta_{M_1}^+$ and $\eta_{M_2}^+$ are isomorphic (Proposition 4.7). These filtrations, in turn, are isomorphic iff the minimal stochastic graphs $M(G_1)$ and $M(G_2)$ are isomorphic (Proposition 4.1). □

THEOREM 4.9. *Let M_1 and M_2 be two minimal irreducible stochastic graphs. Then the following conditions are equivalent:*

(i) *The stochastic graphs M_1 and M_2 are isomorphic.*
(ii) *The one-sided Markov shifts T_{M_1} and T_{M_2} are isomorphic.*
(iii) *The cofiltrations $\xi_{M_1}^-$ and $\xi_{M_2}^-$ are isomorphic.*
(iii′) *The cofiltrations $\xi_{M_1}^-$ and $\xi_{M_2}^-$ are finitely isomorphic.*
(iv) *The filtrations $\eta_{M_1}^+$ and $\eta_{M_2}^+$ are isomorphic.*
(iv′) *The filtrations $\eta_{M_1}^+$ and $\eta_{M_2}^+$ are finitely isomorphic.*

PROOF. (i) $\Longrightarrow$ (ii) $\Longrightarrow$ (iii) $\Longrightarrow$ (iii′) and (i) $\Longrightarrow$ (iv) $\Longrightarrow$ (iv′) are obvious.

(iii′) $\Longrightarrow$ (iv) by Proposition 4.7,
(iv′) $\iff$ (iv) by Proposition 1.3, and
(iv) $\Longrightarrow$ (i) by Proposition 4.1. □

References

[AMT] J. Ashley, B. Marcus, S. Tuncel, *The classification of one-sided Markov chains,* Ergod. Th. Dyn. Syst. **17** (1997), 269–295.

[FR] J. Feldman, D. J. Rudolph, *Standardness of sequences of σ-fields given by certain endomorphisms,* Fund. Math. **157** (1998), 175–189.

[H] C. Hoffman, *A zero entropy T such that the (T, Id)-endomorphism is nonstandard,* Proc. AMS **128** (2000), 183–188.

[HH] D. Heicklen, C. Hoffman, *(T, T^{-1}) is not standard,* Ergod. Th. Dyn. Syst. **18** (1998), 875–878.

[HHR] D. Heicklen, C. Hoffman, D. J. Rudolph, *Entropy and dyadic equivalence of random walks on a random scenery,* Adv. Math. **156** (2000), 157–179.

[HR] C. Hoffman, D. Rudolph, *Uniform endomorphisms which are isomorphic to a Bernoulli shift,* Ann. of Math. **156** (2002), 79–101.

[R1] B.-Z. Rubshtein, *On decreasing sequences of measurable partitions,* Sov. Math. Dokl. **13** (1972), 962–965.

[R2] B.-Z. Rubshtein, *Decreasing sequences of measurable partitions generated by endomorphisms,* Usp. Math. Nauk **28** (1973), 247–248.

[R3] B.-Z. Rubshtein, *Generating partitions of a Markov endomorphism,* Func. Anal. Appl. **8** (1974), 84–85.

[R4] B.-Z. Rubshtein, *On nonhomogeneous finitely Bernoulli sequences of measurable partitions,* Func. Anal. Appl. **10** (1976), 39–44.

[R5] B.-Z. Rubshtein, *Lacunary isomorphism of decreasing sequences of measurable partitions,* Israel J. Math. **97** (1997), 317–345.

[R6] B.-Z. Rubshtein, *On finitely Bernoulli one-sided Markov shifts and their cofiltrations,* Ergod. Th. Dyn. Syst. **19** (1999), 1527–1564.

[R7] B.-Z. Rubshtein, *On a class of one-sided Markov shifts,* preprint, `arXiv:math.DS/0406059`.

[S] A. M. Stepin, *On entropy invariants of decreasing sequences of measurable partitions,* Func. Anal. Appl. **5** (1971), 80–84.

[Ve1] A. M. Vershik, *A lacunary isomorphism theorem for monotone sequences of measurable partitions,* Func. Anal. Appl. **2** (1968), 17–21.

[Ve2] A. M. Vershik, *Decreasing sequences of measurable partitions and their applications,* Sov. Math. Dokl. **11** (1970), 1007–1011.

[Ve3] A. M. Vershik, *A continuum of pairwise non-isomorphic dyadic sequences,* Func. Anal. Appl. **5** (1971), 16–18.

[Ve4] A. M. Vershik, *Theory of decreasing sequences of measurable partitions,* St. Petersburg Math. J. **6** (1995), 705–761.

[Vi] V. G. Vinokurov, *Two non-isomorphic exact endomorphisms of the Lebesgue space with isomorphic sequences of partitions,* Random processes and related topics, vol. 1, Tashkent, 1970, pp. 43–45 (in Russian).

Department of Mathematics, Ben-Gurion University of the Negev, Beer-Sheva, 84105, Israel

E-mail address: `benzion@math.bgu.ac.il`

Amer. Math. Soc. Transl.
(2) Vol. **217**, 2006

Quotients of $\ell^\infty(\mathbb{Z},\mathbb{Z})$ and Symbolic Covers of Toral Automorphisms

Klaus Schmidt

ABSTRACT. This note gives an account of the algebraic construction of symbolic covers and representations of ergodic automorphisms of compact abelian groups. For expansive toral automorphisms this subject was initiated by A. M. Vershik.

1. Introduction

In [**V**], A. M. Vershik showed that the two-sided golden mean shift is a symbolic representation of the hyperbolic automorphism $\alpha = \left(\begin{smallmatrix} 0 & 1 \\ 1 & 1 \end{smallmatrix}\right)$ of the two-torus $\mathbb{T}^2$. The construction underlying this result was subsequently extended to arbitrary hyperbolic automorphisms of $\mathbb{T}^2$ in [**SV**], and the paper [**KV**] describes an algebraic construction of finite-to-one sofic covers of arbitrary hyperbolic toral automorphisms by using an alphabet consisting of a suitable finite set of integers in an algebraic number field associated with the automorphism.

A closely related algebraic construction of symbolic covers of expansive group automorphisms (and, more generally, of expansive $\mathbb{Z}^d$-actions by automorphisms of compact abelian groups) in [**ES1**], based on the analysis of the 'homoclinic group' of such automorphisms, was developed further in [**Sc3**], where it was shown that the two-sided beta-shift of any Pisot number β provides a finite-to-one symbolic cover of the toral automorphism defined by the companion matrix of the minimal polynomial of β (cf. also [**Si**]).

The analogous problem of finding a connection between the two-sided beta-shift of a Salem number β and the corresponding toral automorphism was one of the principal motivations for the paper [**LS2**], which investigated to what extent irreducible nonhyperbolic toral automorphisms can have symbolic representations.

2000 *Mathematics Subject Classification.* Primary 37A45, 37B10.

Key words and phrases. Ergodic nonhyperbolic toral automorphism, Markov partitions, beta-shifts.

This research was supported in part by the FWF Project P16004–N05, the American Institute of Mathematics and NSF grant DMS 0222452. I would furthermore like to express my gratitude to the Max Planck Institute for Mathematics, Bonn, and the Mathematics Department of the University of Washington, Seattle, for their hospitality while some of this work was done. Special thanks are due to Boris Solomyak for his assistance with Lemma 5.6, as well as for a number of interesting conversations.

Since such automorphisms have no nontrivial homoclinic points, any straightforward translation of the ideas in [**V**], [**KV**] or [**Sc3**] is doomed to failure. However, as was shown in [**LS2**], there exists a much weaker form of symbolic representation of such automorphisms. The key to understanding these symbolic 'representations' in the nonexpansive case lies in the study of certain quotients of the space $\ell^\infty(\mathbb{Z},\mathbb{Z})$ of all bounded two-sided integer sequences which are determined by an irreducible polynomial h with integer coefficients.

In order to explain this in a special case we assume that the polynomial h is of the form $h(u) = u^d + h_{d-1}u^{d-1} + \dots h_1 u \pm 1$ and write α_h for the automorphism of the d-dimensional torus $\mathbb{T}^d$ defined by the companion matrix of h. Let σ be the shift on $\ell^\infty(\mathbb{Z},\mathbb{Z})$ and consider the homomorphism $h(\sigma) = \sigma^d + h_{d-1}\sigma^{d-1} + \dots + h_1\sigma \pm \mathrm{Id}\colon \ell^\infty(\mathbb{Z},\mathbb{Z}) \longrightarrow \ell^\infty(\mathbb{Z},\mathbb{Z})$. If the polynomial h is hyperbolic (i.e. has no roots of absolute value 1), then the quotient group $Q^{(h)} = \ell^\infty(\mathbb{Z},\mathbb{Z})/h(\sigma)(\ell^\infty(\mathbb{Z},\mathbb{Z}))$ is naturally isomorphic to $\mathbb{T}^d$, and this isomorphism carries the automorphism of $Q^{(h)}$ induced by the shift σ to α_h (cf. [**Sc3**] and Theorem 3.1). If h is nonhyperbolic, but also noncyclotomic, then the appropriate quotient space turns out to be $Q^{(h)} = \ell^\infty(\mathbb{Z},\mathbb{Z})/(\ell^\infty(\mathbb{Z},\mathbb{Z}) \cap h(\sigma)(\ell^*(\mathbb{Z},\mathbb{Z})))$, where $\ell^*(\mathbb{Z},\mathbb{Z})$ is the space of two-sided integer sequences with at most linear growth; here $Q^{(h)}$ can be identified naturally with the quotient of $\mathbb{T}^d$ by the dense 'central' subgroup of $\mathbb{T}^d$ on which α_h acts isometrically as a rotation (cf. Theorem 4.1).

In this language the search for symbolic models of α_h translates into the search for appropriate 'symbolic covers' of the space $Q^{(h)}$ by closed, bounded, shift-invariant subsets of $\ell^\infty(\mathbb{Z},\mathbb{Z})$ (cf. Definition 2.2).

The existence and, if they exist, the combinatorial structures of such covers are an open problem (Problem 6.1). However, the more modest task of finding good 'partial' symbolic covers of $Q^{(h)}$, i.e. closed, bounded, shift-invariant subsets of $\ell^\infty(\mathbb{Z},\mathbb{Z})$ which have small nonempty intersections with 'most' classes of $\Delta^{(h)}$, seems more manageable. In Theorem 6.4 we construct symbolic partial covers $V_L^{(h)} \subset \ell^\infty(\mathbb{Z},\mathbb{Z})$ of $Q^{(h)}$ with the following properties: the topological entropy $h(\sigma_{V_L^{(h)}})$ of the restriction of σ to $V_L^{(h)}$ is equal to $h(\alpha_h)$, and for every *weakly* d-*bounded* shift-invariant probability measure ν on $V_L^{(h)}$ (Definition 6.2) there exists a ν-a.e. countable-to-one equivariant map $\psi\colon \ell^\infty(\mathbb{Z},\mathbb{Z}) \longrightarrow \mathbb{T}^d$ which sends ν to an α_h-invariant probability measure μ of equal entropy on $\mathbb{T}^d$ (the construction of α_h-invariant probability measures on $\mathbb{T}^d$ other than Lebesgue measure is of interest due to their somewhat exotic properties; cf. [**LS2**]). The existence of weakly d-bounded shift-invariant probability measures on $V_L^{(h)}$ with entropies arbitrarily close to $h(\alpha_h)$ has so far been verified only in some special cases in [**LS2**] (cf. Problem 6.8).

Most of the material in this note is based on [**Sc3**] and [**LS2**], with an extension of some of the results in [**LS2**] about companion matrices of Salem numbers to arbitrary irreducible, nonhyperbolic and ergodic automorphisms of compact connected abelian groups.

2. Equivalence relations on $\ell^\infty(\mathbb{Z},\mathbb{Z})$ defined by polynomials

Let $R = \mathbb{Z}(u^{\pm 1})$ be the ring of Laurent polynomials with integer coefficients in the variable u. We write every $f \in R$ as

$$f = \sum_{n\in\mathbb{Z}} f_n u^n \tag{2.1}$$

with $f_n \in \mathbb{Z}$ for every $n \in \mathbb{Z}$, set

$$\|f\|_1 = \sum_{n\in\mathbb{Z}} |f_n| < \infty$$

and write

$$\mathsf{S}(f) = \{n \in \mathbb{Z} : f_n \neq 0\} \tag{2.2}$$

for the *support* of f. An element $f \in R$ is *irreducible* if it cannot be written as a product $f = f_1 f_2$ with $\|f_i\|_1 > 1$ for $i = 1, 2$, and f is *hyperbolic* if it has no roots of absolute value 1.

Let

$$\ell^*(\mathbb{Z},\mathbb{Z}) = \left\{ w = (w_n) \in \mathbb{Z}^{\mathbb{Z}} : \sup_{n\in\mathbb{Z}} \frac{|w_n|}{|n|+1} < \infty \right\}$$
$$\supset \ell^\infty(\mathbb{Z},\mathbb{Z}) = \left\{ w = (w_n) \in \mathbb{Z}^{\mathbb{Z}} : \|w\|_\infty = \sup_{n\in\mathbb{Z}} |w_n| < \infty \right\}.$$

Both $\ell^*(\mathbb{Z},\mathbb{Z})$ and $\ell^\infty(\mathbb{Z},\mathbb{Z})$ will be furnished with the topology of coordinatewise convergence. We write $\sigma\colon \ell^*(\mathbb{Z},\mathbb{Z}) \longrightarrow \ell^*(\mathbb{Z},\mathbb{Z})$ for the *shift*, defined by

$$(\sigma w)_n = w_{n+1} \tag{2.3}$$

for every $w = (w_n) \in \ell^*(\mathbb{Z},\mathbb{Z})$.

For the remainder of this discussion we consider an irreducible polynomial

$$h = \sum_{n=0}^{d} h_n u^n \in R \text{ with } h_0 > 0,\ \ d > 0 \text{ and } h_d \neq 0, \tag{2.4}$$

and we define a group homomorphism $h(\sigma)\colon \ell^*(\mathbb{Z},\mathbb{Z}) \longrightarrow \ell^*(\mathbb{Z},\mathbb{Z})$ by

$$h(\sigma) = \sum_{n\in\mathbb{Z}} h_n \sigma^n. \tag{2.5}$$

The *Mahler measure* of h is given by

$$\mathsf{M}(h) = \exp\left(\int_0^1 \log|h(e^{2\pi i t})|\, dt \right) = |h_d| \cdot \prod_{\{\gamma\in\mathbb{C} : h(\gamma)=0\}} \max\{|\gamma|, 1\} \tag{2.6}$$

(cf. [**LSW**, (3-2)]). According to Kronecker's theorem [**K**], $\mathsf{M}(h) = 1$ if and only if h is cyclotomic (i.e. if and only if h divides the polynomial $u^n - 1$ for some $n \geq 1$).

Consider the shift-invariant subgroup

$$\ell_h^\infty(\mathbb{Z},\mathbb{Z}) = \ell^\infty(\mathbb{Z},\mathbb{Z}) \cap h(\sigma)(\ell^*(\mathbb{Z},\mathbb{Z})) \tag{2.7}$$

of $\ell^\infty(\mathbb{Z},\mathbb{Z})$, where $h(\sigma)$ is defined in (2.5). We write

$$Q^{(h)} = \ell^\infty(\mathbb{Z},\mathbb{Z})/\ell_h^\infty(\mathbb{Z},\mathbb{Z}) \tag{2.8}$$

for the corresponding quotient group and

$$q^{(h)}\colon \ell^\infty(\mathbb{Z},\mathbb{Z}) \longrightarrow Q^{(h)} \tag{2.9}$$

for the quotient map. It will be convenient to write

$$\Delta^{(h)} = \left\{ (w,w') \in \ell^\infty(\mathbb{Z},\mathbb{Z}) \times \ell^\infty(\mathbb{Z},\mathbb{Z}) : w - w' \in \ell_h^\infty(\mathbb{Z},\mathbb{Z}) \right\} \tag{2.10}$$

for the equivalence relation arising from the partition of $\ell^\infty(\mathbb{Z},\mathbb{Z})$ into cosets of $\ell_h^\infty(\mathbb{Z},\mathbb{Z})$ and to denote by

$$\Delta^{(h)}(w) = w + \ell_h^\infty(\mathbb{Z},\mathbb{Z}) \tag{2.11}$$

the $\Delta^{(h)}$-equivalence class of $w \in \ell^\infty(\mathbb{Z},\mathbb{Z})$. In this notation $Q^{(h)}$ is simply the space of equivalence classes of $\Delta^{(h)}$.

REMARKS 2.1.

(1) It will be clear from (4.3)–(4.4) that, if w is a two-sided integer sequence with polynomial growth such that $h(\sigma)w$ is bounded, then w has at most linear growth and therefore lies in $\ell^*(\mathbb{Z},\mathbb{Z})$. We could thus have defined $\Delta^{(h)}$ equivalently as the set of all pairs $(w,w') \in \ell^\infty(\mathbb{Z},\mathbb{Z}) \times \ell^\infty(\mathbb{Z},\mathbb{Z})$ whose difference is of the form $h(\sigma)w$ for some integer sequence w of polynomial (or, indeed, sub-exponential) growth.

(2) If the polynomial h is hyperbolic, Theorem 3.1 will show that $\ell_h^\infty(\mathbb{Z},\mathbb{Z}) = h(\sigma)(\ell^\infty(\mathbb{Z},\mathbb{Z}))$, so that

$$Q^{(h)} = \ell^\infty(\mathbb{Z},\mathbb{Z})/h(\sigma)(\ell^\infty(\mathbb{Z},\mathbb{Z})). \tag{2.12}$$

DEFINITION 2.2. A closed, bounded, shift-invariant set $V \subset \ell^\infty(\mathbb{Z},\mathbb{Z})$ is a *symbolic partial cover* of $Q^{(h)}$. A symbolic partial cover $V \subset \ell^\infty(\mathbb{Z},\mathbb{Z})$ is a *symbolic cover* of $Q^{(h)}$ if

$$q^{(h)}(V) = Q^{(h)}. \tag{2.13}$$

V is a (partial) *finite-to-one* (or *countable-to-one*) symbolic cover of $Q^{(h)}$ if $V \cap \Delta^{(h)}(w)$ is finite (or countable) for every $w \in \ell^\infty(\mathbb{Z},\mathbb{Z})$.

The main reason for trying to understand the quotient group $Q^{(h)}$ and to find covers of it is that these objects are intimately connected with symbolic covers — in the sense of [**KV**], [**LS2**], [**Sc3**] and [**V**] — of the irreducible ergodic automorphism α_h of the compact connected abelian group X_h in (3.2)–(3.3) determined by the polynomial h and that any symbolic cover of $Q^{(h)}$ defines a kind of symbolic cover of α_h (cf. Theorems 3.1 and 4.1). Even symbolic partial covers can be useful for constructing invariant probability measures of, for example, nonhyperbolic ergodic toral automorphisms (cf. [**LS2**]).

3. Expansive group automorphisms and quotients of ℓ^∞

Let $h \in R$ be an irreducible, noncyclotomic, and not necessarily hyperbolic, polynomial of the form (2.4). The following discussion describes the connection between the quotient space $Q^{(h)}$ in (2.8) and a certain irreducible automorphism α_h of a compact connected abelian group X_h (where *irreducible* means that every closed α_h-invariant subgroup $Y \subsetneq X_h$ is finite). Background and details of can be found in [**ES1**], [**LS2**], [**Sc1**], [**Sc2**], and [**Sc3**].

We write $\tau \colon \mathbb{T}^\mathbb{Z} \longrightarrow \mathbb{T}^\mathbb{Z}$ for the shift

$$(\tau x)_n = x_{n+1}, \quad x = (x_n) \in \mathbb{T}^\mathbb{Z}, \tag{3.1}$$

and define a closed, shift-invariant subgroup $X_h \subset \mathbb{T}^\mathbb{Z}$ by

$$X_h = \Big\{ x = (x_n) \in \mathbb{T}^\mathbb{Z} : \sum_{i=0}^{d} h_i x_{n+i} = 0 \ (\mathrm{mod}\, 1) \quad \text{for every } n \in \mathbb{Z} \Big\}. \tag{3.2}$$

The restriction

$$\alpha_h = \tau_{X_h} \tag{3.3}$$

of τ to X_h is a continuous, irreducible and ergodic automorphism of the compact abelian group X_h with entropy $\log \mathsf{M}(h)$. Furthermore, α_h is expansive if and only if h is hyperbolic (cf. [**Sc2**, Theorem 7.1 and Propositions 7.2–7.3]).

In the case where $h_0 = |h_d| = 1$ in (2.4) and X_h is therefore isomorphic to $\mathbb{T}^d$, the automorphism α_h in (3.3) is algebraically conjugate to the companion matrix

$$M_h = \begin{bmatrix} 0 & 1 & 0 & \dots & 0 & 0 \\ 0 & 0 & 1 & \dots & 0 & 0 \\ \vdots & & \vdots & \ddots & \vdots & \\ & & & & & 0 \\ 0 & 0 & 0 & \dots & 0 & 1 \\ -h_0 & -h_1 & -h_2 & \dots & -h_{d-2} & -h_{d-1} \end{bmatrix}, \tag{3.4}$$

of h, acting on $\mathbb{T}^d$ from the left, where the isomorphism between X_h and $\mathbb{T}^d$ is the coordinate projection

$$x \mapsto \begin{bmatrix} x_0 \\ x_1 \\ \vdots \\ x_{d-1} \end{bmatrix}.$$

More generally, every irreducible ergodic automorphism α of a compact connected abelian group X is finitely equivalent to such an automorphism α_h for an appropriate polynomial $h \in R$ (cf. [**Sc1**], [**Sc2**] or [**LS1**]–[**LS2**]).

We return to our polynomial h in (2.4), extend the shift σ on $\ell^*(\mathbb{Z},\mathbb{Z}) \subset \mathbb{Z}^\mathbb{Z}$ linearly to an automorphism $\bar\sigma$ of

$$\ell^*(\mathbb{Z},\mathbb{R}) = \left\{ w = (w_n) \in \mathbb{R}^\mathbb{Z} : \sup_{n\in\mathbb{Z}} \frac{|w_n|}{|n|+1} < \infty \right\}$$

and define

$$h(\bar\sigma) = \sum_{n=0}^{d} h_n \bar\sigma^n \colon \ell^*(\mathbb{Z},\mathbb{R}) \longrightarrow \ell^*(\mathbb{Z},\mathbb{R}) \tag{3.5}$$

as in (2.5). Denote by Ω_h the set of roots of h and set

$$\begin{aligned} \Omega_h^- &= \{\omega \in \Omega_h : |\omega| < 1\}, \ \ \Omega_h^{(0)} = \{\omega \in \Omega_h : |\omega| = 1\}, \\ \Omega_h^+ &= \{\omega \in \Omega_h : |\omega| > 1\}. \end{aligned} \tag{3.6}$$

According to [**LS2**, (2.15)], the kernel

$$W_h^{(0)} = \ker h(\bar\sigma) \tag{3.7}$$

is the linear span of the vectors $\{w^{(1)}(\omega), w^{(2)}(\omega) : \omega \in \Omega_h^{(0)}\}$ with

$$w^{(1)}(\omega)_n = \mathrm{Re}(\omega^n), \qquad w^{(2)}(\omega)_n = \mathrm{Im}(\omega^n) \tag{3.8}$$

for every $n \in \mathbb{Z}$ and $\omega \in \Omega_h^{(0)}$, where Re and Im denote the real and imaginary parts.

Let $\rho\colon \ell^*(\mathbb{Z},\mathbb{R}) \longrightarrow \mathbb{T}^\mathbb{Z}$ be the map

$$\rho(w)_n = w_n \ (\mathrm{mod}\ 1), \ \ w = (w_n) \in \ell^*(\mathbb{Z},\mathbb{R}). \tag{3.9}$$

Then

$$\rho \circ \bar\sigma = \tau \circ \rho,$$

and the set

$$W_h = h(\bar\sigma)^{-1}(\ell^\infty(\mathbb{Z},\mathbb{Z})) \subset \rho^{-1}(X_h) \subset \ell^*(\mathbb{Z},\mathbb{R}) \tag{3.10}$$

is a closed and shift-invariant subgroup which contains both $\ell^\infty(\mathbb{Z},\mathbb{Z})$ and $W_h^{(0)}$ (cf. (3.16) and (4.5)).

Following [**LS2**] we write

$$\frac{1}{h(u)} = \frac{1}{h_d} \sum_{\omega\in\Omega_h} \frac{b_\omega}{u-\omega}$$

for the partial fraction decomposition of $1/h$ with $b_\omega \in \mathbb{C}$ for every $\omega \in \Omega_h$ and define elements $w^{\Delta_\pm}$ and w^{Δ_0} in $\ell^\infty(\mathbb{Z},\mathbb{R})$ by

$$(3.11)\qquad \begin{aligned} w_n^{\Delta_+} &= \begin{cases} \frac{1}{h_d}\cdot\sum_{\omega\in\Omega_h^-} b_\omega\omega^{n-1} & \text{if } n\ge 1,\\ \frac{1}{h_d}\cdot\sum_{\omega\in\Omega_h^{(0)}\cup\Omega_h^+} -b_\omega\omega^{n-1} & \text{if } n\le 0,\end{cases}\\ w_n^{\Delta_-} &= \begin{cases} \frac{1}{h_d}\cdot\sum_{\omega\in\Omega_h^-\cup\Omega_h^{(0)}} b_\omega\omega_i^{n-1} & \text{if } n\ge 1,\\ \frac{1}{h_d}\cdot\sum_{\omega\in\Omega_h^+} -b_\omega\omega^{n-1} & \text{if } n\le 0,\end{cases}\\ w_n^{\Delta_0} &= \frac{1}{h_d}\cdot\sum_{\omega\in\Omega_h^{(0)}} b_\omega\omega^{n-1} \quad \text{for every } n\in\mathbb{Z}.\end{aligned}$$

Then

$$(3.12)\qquad \begin{gathered} w^{\Delta_0}\in W_h^{(0)}, \qquad w^{\Delta_+}+w^{\Delta_0}=w^{\Delta_-}\in W_h,\\ h(\bar\sigma)(w^{\Delta_+})_n = h(\bar\sigma)(w^{\Delta_-})_n = v_n^{\Delta} =: \begin{cases}1 & \text{if } n=0,\\ 0 & \text{if } 0\ne n\in\mathbb{Z},\end{cases}\end{gathered}$$

and

$$(3.13)\qquad \begin{gathered} x^{\Delta_\pm}=\rho(w^{\Delta_\pm})\in X_h,\ x^{\Delta_0}=\rho(w^{\Delta_0})\in X_h^{(0)},\ x^{\Delta_+}+x^{\Delta_0}=x^{\Delta_-},\\ \lim_{n\to\infty} w_n^{\Delta_+} = \lim_{n\to\infty} w_{-n}^{\Delta_-}=0 \quad \text{exponentially fast.}\end{gathered}$$

Now suppose that h is hyperbolic. Then $w^{\Delta_0}=0$ and we set

$$(3.14)\qquad w^{\Delta}=w^{\Delta_+}=w^{\Delta_-},\ x^{\Delta}=\rho(w^{\Delta}),$$

and define group homomorphisms

$$(3.15)\qquad \bar\xi\colon \ell^\infty(\mathbb{Z},\mathbb{Z})\longrightarrow W_h, \qquad \xi=\rho\circ\bar\xi\colon \ell^\infty(\mathbb{Z},\mathbb{Z})\longrightarrow X_h$$

by

$$\bar\xi(v)=\sum_{n\in\mathbb{Z}} v_n\bar\sigma^{-n}w^{\Delta}, \qquad \xi(v)=\sum_{n\in\mathbb{Z}} v_n\alpha_h^{-n}x^{\Delta}$$

for every $v\in\ell^\infty(\mathbb{Z},\mathbb{Z})$. Then

$$h(\bar\sigma)\circ\bar\xi(w)=w$$

for every $w\in\ell^\infty(\mathbb{Z},\mathbb{Z})$, and since $\ker h(\bar\sigma)=\{0\}$, we conclude that

$$(3.16)\qquad \begin{gathered} W_h = h(\bar\sigma)^{-1}(\ell^\infty(\mathbb{Z},\mathbb{Z})) = \bar\xi(\ell^\infty(\mathbb{Z},\mathbb{Z}))\subset\ell^\infty(\mathbb{Z},\mathbb{R}),\\ h(\bar\sigma)(W_h)=\ell^\infty(\mathbb{Z},\mathbb{Z}),\\ \xi(\ell^\infty(\mathbb{Z},\mathbb{Z}))=X_h,\\ \ker\xi=\{w\in\ell^\infty(\mathbb{Z},\mathbb{Z}):\xi(w)=0\}=h(\sigma)(\ell^\infty(\mathbb{Z},\mathbb{Z})).\end{gathered}$$

Furthermore, ξ is equivariant in the sense that

$$(3.17)\qquad \xi\circ\sigma=\alpha_h\circ\xi$$

and induces an isomorphism

$$(3.18)\qquad \xi'\colon \ell^\infty(\mathbb{Z},\mathbb{Z})/h(\sigma)(\ell^\infty(\mathbb{Z},\mathbb{Z}))\longrightarrow X_h,$$

which is again equivariant in the obvious sense. We obtain the following result.

THEOREM 3.1 ([**Sc3**]). *Let $h \in R$ be a nonconstant irreducible hyperbolic polynomial, and let α_h be the irreducible expansive automorphism of the compact abelian group X_h in (3.2)–(3.3). If $h(\sigma)\colon \ell^*(\mathbb{Z},\mathbb{Z}) \longrightarrow \ell^*(\mathbb{Z},\mathbb{Z})$ is the homomorphism (2.5) and $Q^{(h)}$ is the quotient group (2.8), then*

$$Q^{(h)} = \ell^\infty(\mathbb{Z},\mathbb{Z})/h(\sigma)(\ell^\infty(\mathbb{Z},\mathbb{Z})),$$

and the equivariant map $\xi\colon \ell^\infty(\mathbb{Z},\mathbb{Z}) \longrightarrow X_h$ in (3.15) has kernel

$$\ker \xi = h(\sigma)(\ell^\infty(\mathbb{Z},\mathbb{Z}))$$

and thus induces an equivariant bijection

$$\xi'\colon Q^{(h)} \longrightarrow X_h.$$

Finally, if $V \subset \ell^\infty(\mathbb{Z},\mathbb{Z})$ is a symbolic cover of $Q^{(h)}$ in the sense of Definition 2.2, then $\xi(V) = X_h$.

PROOF. The first inclusion in (3.16) shows that every $v \in \ell^\infty(\mathbb{Z},\mathbb{Z})$ with $h(\sigma)v \in \ell^\infty(\mathbb{Z},\mathbb{Z})$ must itself lie in $\ell^\infty(\mathbb{Z},\mathbb{Z})$, and all other statements follow from (3.16)–(3.18). □

4. Nonexpansive group automorphisms and quotients of ℓ^∞

In this section we assume that h is nonhyperbolic (but still noncyclotomic) and that $\Omega_h^{(0)}$ is therefore nonempty and $W_h^{(0)} = \ker h(\bar\sigma) \neq \{0\}$ (cf. (3.6)–(3.7)). Galois theory implies that the restriction of ρ to $W_h^{(0)}$ is injective, and the *central subgroup*

$$X_h^{(0)} = \rho(W_h^{(0)}), \tag{4.1}$$

on which α_h acts isometrically, is dense in X_h by irreducibility.

We define group homomorphisms $\bar\xi^*\colon \ell^\infty(\mathbb{Z},\mathbb{Z}) \longrightarrow W_h$ and $\xi^*\colon \ell^\infty(\mathbb{Z},\mathbb{Z}) \longrightarrow X_h$ by setting

$$\begin{aligned} \bar\xi^*(v) &= \sum_{n\ge 0} v_n \bar\sigma^{-n}(w^{\Delta_-}) + \sum_{n<0} v_n \bar\sigma^{-n}(w^{\Delta_+}), \\ \xi^*(v) &= \rho \circ \bar\xi^*(v), \end{aligned} \tag{4.2}$$

for every $v = (v_n) \in \ell^\infty(\mathbb{Z},\mathbb{Z})$. Since the coordinates $w_n^{\Delta_+}$ and $w_{-n}^{\Delta_-}$ decay exponentially as $n \to \infty$ and $W_h \subset \ell^*(\mathbb{Z},\mathbb{R})$ is closed, $\bar\xi^*$ is well defined by (3.12). The second equation in (3.12) shows that

$$h(\bar\sigma) \circ \bar\xi^*(v) = v \tag{4.3}$$

for every $v \in \ell^\infty(\mathbb{Z},\mathbb{Z})$ and hence that

$$\bar\xi^* \circ h(\bar\sigma)w - w \in W_h^{(0)} = \ker h(\bar\sigma) \tag{4.4}$$

for every $w \in W_h$. From (3.10) and (4.4) we see that

$$W_h = h(\bar\sigma)^{-1}(\ell^\infty(\mathbb{Z},\mathbb{Z})) = \bar\xi^* \circ h(\bar\sigma)(W_h) + W_h^{(0)} \tag{4.5}$$

(cf. (3.10)) and that

$$\bar\xi^* \circ h(\bar\sigma)(W_h \cap \ell^\infty(\mathbb{Z},\mathbb{R})) \subset W_h \cap \ell^\infty(\mathbb{Z},\mathbb{R}). \tag{4.6}$$

The map $\bar{\xi}^*\colon \ell^\infty(\mathbb{Z},\mathbb{Z}) \longrightarrow W_h$ is obviously not shift-equivariant. Indeed,

$$
\begin{aligned}
\mathsf{d}(n,v) &= \bar{\sigma}^n \circ \bar{\xi}^*(v) - \bar{\xi}^* \circ \sigma^n(v) \\
&= \begin{cases} \sum_{j=0}^{n-1} v_j \bar{\sigma}^{n-j} w^{\Delta_0} & \text{if } n > 0, \\ 0 & \text{if } n = 0, \\ -\sum_{j=1}^{n} v_{-j} \bar{\sigma}^{j-n} w^{\Delta_0} & \text{if } n < 0 \end{cases}
\end{aligned} \tag{4.7}
$$

for every $n \in \mathbb{Z}$ and $v \in \ell^\infty(\mathbb{Z},\mathbb{Z})$, and the resulting map

$$
\mathsf{d}\colon \mathbb{Z} \times \ell^\infty(\mathbb{Z},\mathbb{Z}) \longrightarrow W_h^{(0)} \tag{4.8}
$$

satisfies the cocycle equation

$$
\mathsf{d}(m, \sigma^n v) + \bar{\sigma}^m \mathsf{d}(n,v) = \mathsf{d}(m+n, v) \tag{4.9}
$$

for every $m, n \in \mathbb{Z}$ and $v \in \ell^\infty(\mathbb{Z},\mathbb{Z})$. We put

$$
\tilde{Y} = \ell^\infty(\mathbb{Z},\mathbb{Z}) \times W_h^{(0)} \tag{4.10}
$$

and consider the continuous surjective maps $\tilde{\sigma}\colon \tilde{Y} \longrightarrow \tilde{Y}$, $\bar{\vartheta}\colon \tilde{Y} \longrightarrow W_h$ and $\vartheta\colon \tilde{Y} \longrightarrow X_h$ defined by

$$
\begin{aligned}
\tilde{\sigma}(v,w) &= (\sigma v, \bar{\sigma} w + \mathsf{d}(1,v)), \\
\bar{\vartheta}(v,w) &= \bar{\xi}^*(v) + w, \\
\vartheta(v,w) &= \rho \circ \bar{\vartheta}(v,w)
\end{aligned} \tag{4.11}
$$

for every $(v,w) \in \tilde{Y} = \ell^\infty(\mathbb{Z},\mathbb{Z}) \times W_h^{(0)}$. The map $\tilde{\sigma}$ is obviously a homeomorphism, and

$$
\bar{\vartheta} \circ \tilde{\sigma} = \bar{\sigma} \circ \bar{\vartheta}, \quad \vartheta \circ \tilde{\sigma} = \alpha_h \circ \vartheta. \tag{4.12}
$$

Since the restriction of ρ to $W_h^{(0)}$ is injective and $W_h^{(0)} \cap \ell^*(\mathbb{Z},\mathbb{Z}) = \{0\}$,

$$
\vartheta(v,w) + X_h^{(0)} = \vartheta(v',w') + X_h^{(0)} \text{ if and only if } v - v' \in \ell_h^\infty(\mathbb{Z},\mathbb{Z}) \tag{4.13}
$$

for all $(v,w), (v',w') \in \tilde{Y}$. We obtain the following result.

THEOREM 4.1 ([**LS2**]). *Let $h \in R$ be an irreducible nonhyperbolic polynomial which is not cyclotomic, and let α_h be the irreducible, ergodic and nonexpansive automorphism of the compact abelian group X_h in* (3.2)–(3.3). *Let $h(\sigma)\colon \ell^*(\mathbb{Z},\mathbb{Z}) \longrightarrow \ell^*(\mathbb{Z},\mathbb{Z})$ be the homomorphism* (2.5), *and let $\ell_h^\infty(\mathbb{Z},\mathbb{Z}) \subset \ell^\infty(\mathbb{Z},\mathbb{Z})$ be the subgroup defined in* (2.7).

(1) *If $\xi^*\colon \ell^\infty(\mathbb{Z},\mathbb{Z}) \longrightarrow X_h$ is the nonequivariant group homomorphism* (4.2), *then*

$$
\xi^{*-1}(X_h^{(0)}) = \ell_h^\infty(\mathbb{Z},\mathbb{Z}),
$$

and ξ^* induces an (equivariant) bijection

$$
\xi^{*\prime}\colon Q^{(h)} \longrightarrow X_h/X_h^{(0)}.
$$

(2) *If $V \subset \ell^\infty(\mathbb{Z},\mathbb{Z})$ is a symbolic cover of $Q^{(h)}$ in the sense of Definition* 2.2, *then $\xi^*(V) + X_h^{(0)} = X_h$.*

(3) *If $\vartheta\colon \tilde{Y} = \ell^\infty(\mathbb{Z},\mathbb{Z}) \times W_h^{(0)} \longrightarrow X_h$ is the equivariant map defined in* (4.11)–(4.12), *then two points $(v,w), (v',w') \in \tilde{Y}$ are mapped by ϑ to the same coset of $X_h^{(0)} \subset X_h$ if and only if $v - v' \in \ell_h^\infty(\mathbb{Z},\mathbb{Z})$.*

REMARK 4.2. It is easy to find symbolic covers of $Q^{(h)}$ (irrespective of whether h is hyperbolic or not): if $Y_h = W_h \cap [0,1)^{\mathbb{Z}}$, then $\rho(Y_h) = X_h$, and $Z_h = \overline{h(\bar{\sigma})(Y_h)} \subset \ell^\infty(\mathbb{Z},\mathbb{Z})$ is a closed, bounded, shift-invariant subset with $\bar{\xi}^*(Z_h) + W_h^{(0)} \supset Y_h$ by (4.4). Hence $\xi^*(Z_h) + X_h^{(0)} = X_h$, and Theorem 4.1 shows that Z_h must intersect every coset of $\ell_h^\infty(\mathbb{Z},\mathbb{Z})$ in $\ell^\infty(\mathbb{Z},\mathbb{Z})$.

The following corollary of Theorem 4.1 suggests that the search for 'small' symbolic covers of $Q^{(h)}$ for nonhyperbolic h may be considerably more difficult than in the hyperbolic case (cf. Corollary 5.4 and Theorem 5.5).

COROLLARY 4.3. *Let $h \in R$ be an irreducible nonhyperbolic polynomial which is not cyclotomic. Then $Q^{(h)}$ has no finite-to-one symbolic cover (cf. Definition* 2.2).

PROOF. We are claiming that there is no closed, bounded, shift-invariant set $V \subset \ell^\infty(\mathbb{Z},\mathbb{Z})$ which intersects every coset of $\ell_h^\infty(\mathbb{Z},\mathbb{Z})$ in a nonempty finite set.

Suppose that such a set V exists. Since $\ell^*(\mathbb{Z},\mathbb{Z})$ is sigma-compact in the topology of pointwise convergence, $h(\sigma)(\ell^*(\mathbb{Z},\mathbb{Z})) \subset \ell^*(\mathbb{Z},\mathbb{Z})$ is again sigma-compact. It follows that the equivalence relation

$$\Delta_V^{(h)} = \Delta^{(h)} \cap (V \times V)$$

is a Borel subset of $V \times V$ such that $\pi_1^{-1}(w) \cap \Delta_V^{(h)}$ is finite and nonempty for every $w \in V$ (here $\pi_1 \colon V \times V \longrightarrow V$ is the first coordinate projection). In particular, $\pi_1(B) \subset V$ is Borel for every Borel set $B \subset \Delta_V^{(h)}$.

Since V must be uncountable and any two uncountable Borel sets are Borel isomorphic, there exists a Borel isomorphism $\phi \colon V \longrightarrow [0,1]$, and we use ϕ to pull the order on $[0,1]$ back to an order $\prec$ on V.

Let $E \subset \Delta_V^{(h)}$ be the uniquely determined Borel set with the following properties:

(i) the restriction of π_1 to E is a bijection of E and V,
(ii) for every $w \in V$, the unique element $(w, w') \in E$ satisfies that $w' \preceq w''$ for every $w'' \in \Delta^{(h)}(w) \cap V$.

The set E is the graph of a Borel map $\psi \colon V \longrightarrow V$ with the property that $\psi^{-1}(\{w\})$ is finite for every $w \in V$. Hence $B = \psi(E)$ is a Borel set, and our construction guarantees that $|B \cap \Delta^{(h)}(w)| = 1$ for every $w \in \ell^\infty(\mathbb{Z},\mathbb{Z})$.

The continuous map $\xi^* \colon \ell^\infty(\mathbb{Z},\mathbb{Z}) \longrightarrow X_h$ in (4.2) is injective on B, and hence $\xi^*(B) \subset X_h$ is a Borel set which intersects each coset of the dense subgroup $X_h^{(0)}$ in a single point (cf. Theorem 4.1). Since this is impossible, we have proved the corollary by contradiction. □

5. Construction of symbolic partial covers of $Q^{(h)}$

We write $\ell^1(\mathbb{Z},\mathbb{Z}) \subset \ell^\infty(\mathbb{Z},\mathbb{Z}) \subset \ell^*(\mathbb{Z},\mathbb{Z})$ for the set of all sequences with only finitely many nonzero terms. By viewing every $f = \sum_{n\in\mathbb{Z}} f_n u^n \in R$ as the element $(f_n) \in \ell^1(\mathbb{Z},\mathbb{Z})$, we identify R with $\ell^1(\mathbb{Z},\mathbb{Z})$.

Let $h \in R$ be an irreducible, nonconstant and noncyclotomic polynomial of the form (2.4). We define an equivalence relation $\Delta_1^{(h)}$ on $\ell^\infty(\mathbb{Z},\mathbb{Z})$ by

$$\Delta_1^{(h)} = \{(w,w') \in \ell^\infty(\mathbb{Z},\mathbb{Z}) \times \ell^\infty(\mathbb{Z},\mathbb{Z}) : w - w' \in h(\sigma)(\ell^1(\mathbb{Z},\mathbb{Z}))\} \tag{5.1}$$

(cf. (2.5)) and write

$$\Delta_1^{(h)}(w) = \{w' \in \ell^\infty(\mathbb{Z},\mathbb{Z}) : (w,w') \in \Delta_1^{(h)}\} \tag{5.2}$$

for the equivalence class of $w \in \ell^\infty(\mathbb{Z},\mathbb{Z})$.

We introduce a lexicographic order $\prec$ on the ring R by setting $0 \prec f$ if and only if $f_m > 0$ for the *smallest* $m \in \mathbb{Z}$ with $f_m \neq 0$ (cf. (2.1)) and by saying that $f \prec f'$ whenever $0 \prec f' - f$. The order $\prec$ on R induces a lexicographic order (again denoted by $\prec$) on each equivalence class of $\Delta_1^{(h)}$: if $(v,v') \in \Delta_1^{(h)}$, then $v - v' \in h(\sigma)(\ell^1(\mathbb{Z},\mathbb{Z}))$, and $v \prec v'$ if and only if $v - v' = h(\sigma)f$ for some $f \prec 0$.

Let $V \subset \ell^\infty(\mathbb{Z},\mathbb{Z})$ be a closed, bounded, shift-invariant subset. We put $P = \{f \in R : 0 \prec f\}$ and set

$$\begin{aligned} V^{(h)} &= \bigcap_{f\in P} (V \smallsetminus (V - h(\sigma)f)) = V \smallsetminus \bigcup_{f\in P} (V - h(\sigma)f) \\ &= \{w \in V : w' \preceq w \text{ for every } w' \in V \cap \Delta^{(h)}(w)\}. \end{aligned} \tag{5.3}$$

THEOREM 5.1. *Let $h \in R$ be an irreducible, nonconstant and noncyclotomic polynomial of the form (2.4), L a positive integer,*

$$V_L = \{0,\dots,L-1\}^{\mathbb{Z}}, \tag{5.4}$$

and let $V_L^{(h)} \subset V_L$ be the subset defined by (5.3). Then $V_L^{(h)}$ is closed, shift-invariant and has the following properties.

1. $|V_L^{(h)} \cap \Delta_1^{(h)}(w)| \le 1$ *for every* $w \in \ell^\infty(\mathbb{Z},\mathbb{Z})$.
2. *If h is hyperbolic, then $V_L^{(h)}$ is a partial finite-to-one symbolic cover of $Q^{(h)}$.*
3. *If $\sigma_{V_L^{(h)}}$ is the restriction to $V_L^{(h)}$ of the shift σ on $\ell^\infty(\mathbb{Z},\mathbb{Z})$ in (2.3), then its topological entropy satisfies that $h(\sigma_{V_L^{(h)}}) \le \log \mathsf{M}(h)$ (cf. (2.6)).*

We start the proof of Theorem 5.1 with a lemma.

LEMMA 5.2. *The polynomial h in (2.4) is hyperbolic if and only if there exists a constant $b > 0$ with*

$$\|h(\sigma)w\|_\infty \ge b \cdot \|w\|_\infty \text{ for every } w \in \ell^\infty(\mathbb{Z},\mathbb{Z}), \tag{5.5}$$

where $h(\sigma)$ is defined in (2.5).

PROOF. Since h is noncyclotomic, Galois theory implies that $\ker h(\sigma) = \{0\}$ (cf. [**LS2**, (2.15)]).

Suppose that h is hyperbolic. We define w^Δ by (3.14) and conclude from (3.13) there exist constants $\gamma \in (0,1)$ and $C > 0$ such that $|w_n^\Delta| \le C\cdot\gamma^{|n|}$ for every $n \in \mathbb{Z}$.

The shift-equivariant map $\bar\xi\colon \ell^\infty(\mathbb{Z},\mathbb{Z}) \longrightarrow \ell^\infty(\mathbb{Z},\mathbb{R})$ in (3.15) satisfies that $\|\bar\xi(v)\|_\infty \le 2C \cdot \|v\|_\infty \cdot \sum_{n\ge0}\gamma^n$ and $h(\sigma)\circ\bar\xi(w) = \bar\xi\circ h(\sigma)(w) = w$ for every $w \in \ell^\infty(\mathbb{Z},\mathbb{Z})$. Hence

$$\|w\|_\infty = \|\bar\xi(h(\sigma)w)\|_\infty \le 2C\cdot\|h(\sigma)w\|_\infty \cdot \sum_{n\ge0}\gamma^n$$

for every $w \in \ell^\infty(\mathbb{Z},\mathbb{Z})$, which proves the existence of a lower bound $b > 0$ in (5.5).

If h is noncyclotomic and nonhyperbolic, we choose $\theta \in \Omega_h^{(0)}$ (cf. (3.6)) and define, for every integer $j \geq 0$, a point $\omega^{(j)} = (\omega_n^{(j)}) \in \ell^\infty(\mathbb{Z}, \mathbb{R})$ by setting

$$\omega_n^{(j)} = \begin{cases} \theta^n + \theta^{-n} & \text{if } n \geq j, \\ 0 & \text{otherwise.} \end{cases}$$

Then

$$(h(\bar{\sigma})\omega^{(j)})_n = 0$$

for $n < j - d$ and $n \geq j$, and $\|h(\bar{\sigma})\omega^{(j)}\|_\infty \leq 2 \cdot \|h\|_1$.

For every $t \in \mathbb{R}$ we denote by $\lceil t \rceil$ the smallest integer $\geq t$, and we set, for every $M \geq 1$ and $n \in \mathbb{Z}$,

$$\tilde{\omega}^{(M)} = \sum_{j=0}^{M} \omega^{(3jd)}, \qquad w_n^{(M)} = \lceil \tilde{\omega}_n^{(M)} \rceil.$$

The resulting sequence $(w^{(M)}, M \geq 1)$ in $\ell^\infty(\mathbb{Z}, \mathbb{Z})$ satisfies that $\|w^{(M)}\|_\infty = M \cdot \|w^{(1)}\|_\infty$ and $\|h(\sigma)w^{(M)}\|_\infty \leq 3 \cdot \sum_{i=0}^{d} |h_i|$ for every $M \geq 1$. This proves that there is no $b > 0$ satisfying (5.5). □

For the proof of Theorem 5.1 as well as that of Theorem 5.3 below we recall some facts from [**Sc1**] (cf. also [**C**] or [**W**]). We fix $\gamma \in \Omega_h$ (cf. (3.6)), denote by $K = \mathbb{Q}(\gamma)$ the algebraic number field generated by γ, and write $P^{(K)}$, $P_f^{(K)}$ and $P_\infty^{(K)}$ for the sets of places (or equivalence classes of valuations), finite places and infinite places of K. For every place v of K we denote by K_v the *v-adic completion* of K and consider the valuation $|\cdot|_v \in v$ defined by

$$|a|_v \cdot \lambda_v(C) = \lambda_v(a \cdot C) \tag{5.6}$$

for every $a \in K$, where λ_v is a Haar measure on the additive group of the locally compact, metrizable field K_v and where $C \subset K_v$ is a compact neighbourhood of 0. We write

$$\iota_v \colon K \longrightarrow K_v$$

for the embedding of K in K_v and set

$$R_v = \{a \in K : |a|_v \leq 1\}, \ \bar{R}_v = \{a \in K_v : |a|_v \leq 1\}. \tag{5.7}$$

Let

$$S = P_\infty^{(K)} \cup \{v \in P_f^{(K)} : |\gamma|_v \neq 1\}. \tag{5.8}$$

The set

$$K_S = \prod_{v \in S} K_v \tag{5.9}$$

is a locally compact abelian group with respect to coordinatewise addition, and we write

$$\iota \colon K \longrightarrow K_S \tag{5.10}$$

for the diagonal embedding $a \mapsto \iota(a) = (\iota_v(a), v \in S)$, $a \in K$. If

$$R_S = \bigcap_{v \in P^{(K)} \smallsetminus S} R_v \tag{5.11}$$

is the ring of S-integers in K, then $\iota(R_S)$ is a discrete co-compact subgroup of K_S.

For every subset $F \subset \mathbb{Z}$ we write

$$\pi_F : \ell^\infty(\mathbb{Z}, \mathbb{Z}) \longrightarrow \mathbb{Z}^F \tag{5.12}$$

for the projection onto the coordinates in F.

PROOF OF THEOREM 5.1. For every $f \in P$ we set

$$V(f) = V_L \setminus (V_L - h(\sigma)f). \tag{5.13}$$

Since $h(\sigma)f$ has only finitely many nonzero coordinates, $V(f) \subset V_L$ is closed and open, and hence $V_L^{(h)}$ is a closed — and obviously shift-invariant — subset of V_L.

If $V_L^{(h)}$ meets some equivalence class of $\Delta^{(h)}$ in more than one point, then there exist $w \in V_L^{(h)}$ and $f \in P$ such that $w' = w + h(\sigma)f \in V_L^{(h)}$. Hence $w' \in V_L \cap (V_L - h(\sigma)f)$, which contradicts the definition of $V_L^{(h)}$.

In order to prove (2), we set, for $N_1 < N_2$,

$$\begin{aligned} B(N_1, N_2, K) - \{w \in \ell^*(\mathbb{Z}, \mathbb{Z}) : |w|_n \le K \\ \text{for } N_1 \le n \le N_1 + d \text{ and } N_2 \le n \le N_2 + d\}, \end{aligned} \tag{5.14}$$

where d appears in (2.4), and claim that

$$\left|\pi_{\{N_1,\dots,N_2\}}\left((w + h(\sigma)(B(N_1, N_2, K))) \cap V_L^{(h)}\right)\right| \le (2K+1)^{2d+2} \tag{5.15}$$

for every $w \in V_L^{(h)}$, $K \ge 0$ and $N_1 < N_2$ (cf. (5.12)).

Indeed, if (5.15) does not hold for some $w \in V_L^{(h)}$, $K \ge 0$ and $N_1 < N_2$, then we can find elements $y, y' \in B(N_1, N_2, K)$ with the following properties:

(i) $y_r = y'_r$ for $r = N_1, \dots, N_1 + d$ and $r = N_2, \dots, N_2 + d$,
(ii) $(y_{N_1+d+1}, \dots, y_{N_2-1}) \ne (y'_{N_1+d+1}, \dots, y'_{N_2-1})$,
(iii) $v = w + h(\sigma)y \in V_L^{(h)}$, $v' = w + h(\sigma)y' \in V_L^{(h)}$.

We define a Laurent polynomial $f \in R$ by

$$f_r = \begin{cases} y_r - y'_r & \text{for } r = N_1 + d + 1, \dots, N_2 - 1, \\ 0 & \text{otherwise} \end{cases}$$

(cf. (2.1)) and observe that the points $v - h(\sigma)f$ and $v' + h(\sigma)f$ both lie in $V_L^{(h)}$. By comparing this with (5.3), we obtain a contradiction in one of the two cases. This proves (5.15).

If h is hyperbolic, then Lemma 5.2 implies that there exists a $K \ge 0$ such that

$$\{v \in \ell^\infty(\mathbb{Z}, \mathbb{Z}) : h(\sigma)v \in (V_L^{(h)} - V_L^{(h)})\} \subset \{-K, \dots, K\}^{\mathbb{Z}}.$$

Hence

$$V_L^{(h)} \cap \left(w + h(\sigma)(B(N_1, N_2, K))\right) = V_L^{(h)} \cap \Delta^{(h)}(w)$$

for every $w \in V_L^{(h)}$ and $N_1 < N_2$, and (3) follows from (5.15).

For every $w \in \ell^\infty(\mathbb{Z}, \mathbb{Z})$ and $r, s \in \mathbb{Z}$ with $r \le s$ we set

$$w_r^s(\gamma) = \sum_{i=r}^{s} w_i \gamma^i. \tag{5.16}$$

Let

$$\tilde{h} = u^d \cdot \sum_{n \in \mathbb{Z}} h_n u^{-n} \in R \tag{5.17}$$

be the *reversal* of h, and let $N \geq 1$. If $w, w' \in V_L^{(h)}$ and $w_0^{N-1}(\gamma) = w_0'^{N-1}(\gamma)$, then the Laurent polynomials $f(u) = \sum_{k=0}^{N-1} w_k u^k$ and $f'(u) = \sum_{k=0}^{N-1} w'_k u^k$ differ by a multiple $h(\sigma)f$ of $\tilde{h}$ which we may assume to lie in $P\tilde{h}$ (by interchanging w and w', if necessary). This implies that the point $w' + h(\sigma)f$ lies in V_L, and hence that $w' \in V_L \cap (V_L - h(\sigma)f)$. As in the preceding paragraph we obtain a contradiction to our hypothesis that $w' \in V_L^{(h)}$. This shows that $w_0^{N-1}(\gamma) \neq w_0'^{N-1}(\gamma)$ whenever $w, w' \in V_L^{(h)}$ and $\pi_{\{0,\dots,N-1\}}(w) \neq \pi_{\{0,\dots,N-1\}}(w')$ (cf. (5.16)).

For every $N \geq 1$, the set $\iota(\{w_0^{N-1}(\gamma) : w \in V_L^{(h)}\}) \subset K_S$ is contained in the set

$$F(N) = \prod_{v \in P_\infty^{(K)}} \{a \in K_v : |a|_v \leq LN \cdot \max(1, |\gamma|_v^N)\} \cdot \prod_{v \in S \smallsetminus P_\infty^{(K)}} (\bar{R}_v \cup \gamma^N \bar{R}_v).$$

We fix a Haar measure λ on K_S with $\lambda(\prod_{v\in S} \bar{R}_v) = 1$ (cf. (5.7)). As N varies, the same calculation as in [**LW**] shows that

$$\begin{aligned}\lambda(F(N)) \leq & \prod_{\{v \in P_\infty^{(K)} : K_v = \mathbb{R}\}} LN \cdot \max(1, |\gamma|_v^N) \\ & \cdot \prod_{\{v \in P_\infty^{(K)} : K_v = \mathbb{C}\}} L^2N^2 \cdot \max(1, |\gamma|_v^N) \cdot \prod_{v \in S \smallsetminus P_\infty^{(K)}} \max(1, |\gamma|_v^N) \\ & \leq \mathsf{M}(\tilde{h})^N \cdot (LN)^{\mathsf{r}+2\mathsf{s}} = \mathsf{M}(h)^N \cdot (LN)^{\mathsf{r}+2\mathsf{s}},\end{aligned}$$

where r and s denote the numbers of real and complex places of K (i.e. the number of $v \in P_\infty^{(K)}$ with $K_v = \mathbb{R}$ and $K_v = \mathbb{C}$, respectively). As $\iota(R_S)$ is discrete and co-compact in K_S, there exists a constant $C > 0$ such that, for every $N \geq 1$,

$$\left|\{w_0^{N-1}(\gamma) : w \in V_L^{(h)}\}\right| = \left|\pi_{\{0,\dots,N-1\}}(V_L^{(h)})\right| \leq C \cdot \mathsf{M}(h)^N \cdot (LN)^{\mathsf{r}+2\mathsf{s}},$$

where $|\cdot|$ denotes cardinality. This implies that

$$h(\sigma_{V_L^{(h)}}) = \lim_{N\to\infty} \frac{1}{N} \log \left|\pi_{\{0,\dots,N-1\}}(V_L^{(h)})\right| \leq \log \mathsf{M}(h). \qquad \square$$

THEOREM 5.3. *Let $h \in R$ be an irreducible, nonconstant and noncyclotomic polynomial of the form* (2.4), $L \geq 1$, *and let $V_L^{(h)} \subset V_L$ be the closed, shift-invariant subset defined in* (5.3). *If L is sufficiently large, then $h(\sigma_{V_L^{(h)}}) = \log \mathsf{M}(h)$.*

COROLLARY 5.4. *Suppose that the polynomial h is hyperbolic. If L is sufficiently large, then $V_L^{(h)}$ is a finite-to-one symbolic cover of $Q^{(h)}$.*

PROOF OF COROLLARY 5.4, GIVEN THEOREM 5.3. If L is sufficiently large, then $h(\sigma_{V_L^{(h)}}) = \log \mathsf{M}(h)$ by Theorem 5.3. By Theorem 5.1(2), $V_L^{(h)}$ is a partial finite-to-one symbolic cover of $Q^{(h)}$, which implies that the restriction to $V_L^{(h)}$ of the equivariant group homomorphism $\xi \colon \ell^\infty(\mathbb{Z}, \mathbb{Z}) \longrightarrow X_h$ in (3.15) is finite-to-one (cf. Theorem 3.1). In particular, $Y = \xi(V_L^{(h)})$ is a closed, α_h-invariant subset of X_h such that the restriction of α_h to Y has topological entropy $\log \mathsf{M}(h) = h(\alpha_h)$, and the uniqueness of the measure of maximal entropy of α_h implies that $\lambda_{X_h}(Y) = 1$. Hence $Y = X_h$ and $V_L^{(h)}$ is a symbolic cover of $Q^{(h)}$. $\square$

Under the hypotheses of Corollary 5.4 we can even find almost one-to-one symbolic covers of $Q^{(h)}$:

THEOREM 5.5. *Under the hypotheses of Theorem 3.1 there exists a closed, bounded, shift-invariant subset $V^* \subset \ell^\infty(\mathbb{Z},\mathbb{Z})$ with the following properties.*

(1) *V^* is a sofic shift.*
(2) *V^* is a finite-to-one symbolic cover of $Q^{(h)}$ (cf. Definition 2.2).*
(3) *$\lambda_{X_h}(\{x \in X_h : |\xi^{-1}(\{x\}) \cap V^*| = 1\}) = 1$, where λ_{X_h} is the normalized Haar measure of X_h.*

PROOF. This is [**Sc3**, Theorem 5.1]. □

For the proof of Theorem 5.3 we put, in the notation of (5.8),

$$S^- = \{v \in S : |\gamma|_v < 1\}, \quad K_{S^-} = \prod_{v \in S^-} K_v, \tag{5.18}$$

and denote by $\iota^- : K \longrightarrow K_{S^-}$ the diagonal embedding of K in K_{S^-} (cf. (5.10)).

We set $\mathbb{N} = \{0, 1, 2, \dots\}$ and write $\ell^\infty(\mathbb{N},\mathbb{Z})$ for the space of one-sided bounded integer sequences, furnished with the topology of coordinatewise convergence, the maximum norm $\|\cdot\|_\infty$ and the one-sided shift σ_+ defined as in (2.3).

LEMMA 5.6. *Let $\phi\colon \ell^\infty(\mathbb{N},\mathbb{Z}) \longrightarrow K_{S^-}$ be defined by*

$$\phi(w) = \sum_{k \geq 0} w_k \cdot \iota^-(\gamma^k)$$

for every $w = (w_k) \in \ell^\infty(\mathbb{N},\mathbb{Z})$. Then there exists an integer $L \geq 1$ such that $\phi(B_L^+)$ has nonempty interior in K_{S^-}, where

$$B_L^+ = \{-L, \dots, L\}^{\mathbb{N}}.$$

PROOF. The basic idea for the proof of this lemma is due to Boris Solomyak (personal communication). We set $R(\gamma) = \mathbb{Z}[\gamma^{\pm 1}] = \{f(\gamma) : f \in R\} \subset R_S$, where $\gamma \in \Omega_h$ was chosen for the proof of Theorem 5.1. As discussed in [**ES2**] or [**Sc1**], $R_S/R(\gamma)$ is finite, and $R(\gamma)^- = \iota^-(R(\gamma))$ is therefore dense in K_{S^-}.

We denote by $M_\gamma\colon K_{S^-} \longrightarrow K_{S^-}$ diagonal multiplication by γ on K_{S^-}. The set $Q = \prod_{v \in S^-} \bar{R}_v$ is a compact neighbourhood of $0 \in K_{S^-}$. As $R(\gamma)^-$ is dense in K_{S^-} and $M_\gamma Q \subset Q$ has nonempty interior, $\bigcup_{c \in R(\gamma)^-}(c + M_\gamma Q) \supset Q$, and the compactness of Q implies that there exists a finite set $F = \{f^{(1)}, \dots, f^{(l)}\} \subset R$ with

$$Q \subset \bigcup_{i=1}^{l} \bigl(\iota^-(f^{(i)}(\gamma)) + M_\gamma Q\bigr). \tag{5.19}$$

We fix $a \in Q$. By (5.19) there exists a $t_0(a) \in \{1, \dots, l\}$ with

$$a \in \iota^-(f^{(t_0(a))}(\gamma)) + M_\gamma Q,$$

and by repeating this argument, we find a sequence

$$(t_0(a), t_1(a), t_2(a), \dots) \in \{1, \dots, l\}^{\mathbb{N}}$$

with

$$a \in \sum_{i=0}^{s} M_\gamma^i [\iota^-(f^{(t_i(a))}(\gamma))] + M_\gamma^{s+1} Q \tag{5.20}$$

for every $s \geq 0$. As M_γ is a strict contraction on K_{S^-}, this implies that

$$a = \sum_{i=0}^{\infty} M_\gamma^i[\iota^-(f^{(t_i(a))}(\gamma))]. \tag{5.21}$$

There exist integers $J, J' \geq 0$ with $\mathsf{S}(f^{(i)}) \subset \{-J, \dots, J\}$ and $\|f^{(i)}\|_\infty \leq J'$ for every $i = 1, \dots, l$ (cf. (2.2)). We set $L = (2J+1)J'$ and obtain from (5.21) that there exists a sequence $w = (w_k) \in B_L^+$ with $a = \sum_{i \geq 0} w_k \cdot \iota^-(\gamma^{i-J})$. As $a \in Q$ was arbitrary, this implies that

$$\phi(B_L^+) \supset M_\gamma^J Q,$$

which proves our claim. □

LEMMA 5.7. *Let $L \geq 1$ and B_L^+ be chosen as in Lemma 5.6. Then there exists a $c > 0$ such that*

$$\left|\left\{\sum_{i=0}^{s-1} w_i \cdot \iota^-(\gamma^i) : w = (w_i) \in B_L^+\right\}\right| \geq c \cdot \mathsf{M}(h)^s \tag{5.22}$$

for every $s \geq 1$, where $|\cdot|$ denotes cardinality.

PROOF. According to Lemma 5.6, $\phi(B_L^+) \subset K_{S^-}$ has nonempty interior and hence positive Haar measure $\lambda_{K_{S^-}}(\phi(B_L^+))$. The inclusion (5.20) implies that

$$\begin{aligned} Q &\subset \bigcup_{(j_0,\dots,j_{s-1}) \in \{1,\dots,l\}^s} \sum_{i=0}^{s-1} M_\gamma^i(\iota^-(f^{(j_i)}(\gamma))) + M_\gamma^s Q \\ &\subset \bigcup_{w \in B_L^+} \sum_{i=0}^{s-1+2J} w_i \cdot \iota^-(\gamma^{i-J}) + M_\gamma^s Q \end{aligned}$$

for every $s \geq 1$. Since we need at least $\lambda_{K_{S^-}}(Q)/\lambda_{K_{S^-}}(M_\gamma^s Q) = \mathsf{M}(h)^s$ distinct translates of $M_\gamma^s Q$ to cover Q, this implies that

$$\left|\left\{\sum_{i=0}^{s-1+2J} w_i \cdot \iota^-(\gamma^{i-J}) : w \in B_L^+\right\}\right| \geq \mathsf{M}(h)^s$$

for every $s \geq 1$, which proves (5.22). □

LEMMA 5.8. *There exist an integer $L \geq 1$ and a $c > 0$ such that*

$$\left|\left\{\sum_{i=0}^{s-1} w_i \cdot \iota^-(\gamma^i) : w = (w_i) \in V_L^+\right\}\right| \geq c \cdot \mathsf{M}(h)^s \tag{5.23}$$

for every $s \geq 1$, where

$$V_L^+ = \{0, \dots, L-1\}^{\mathbb{N}}.$$

PROOF. Let $L \geq 1$ be the integer appearing in Lemma 5.7. We set $L' = 2L+1$, $\bar{w} = (L, L, L, \dots)$ and $\bar{a} = \phi(\bar{w})$. Then $V_{L'}^+ = B_L^+ + \bar{w}$ and $\phi(V_{L'}^+) = \phi(B_L^+) + \bar{a}$. Equation (5.23) follows from (5.22) with L' replacing L. □

PROOF OF THEOREM 5.3. For every quadruple of integers $s_1 \leq r_1 < r_2 \leq s_2$ we set

$$(5.24)\qquad \begin{aligned} V^{(s_1,s_2)} &= \bigcap_{\{f\in P:\, \mathsf{S}(h(\sigma)f)\subset\{s_1,\dots,s_2\}\}} (V_L \smallsetminus (V_L + h(\sigma)f)),\\ V^{(s_1,s_2)}_{(r_1,r_2)} &= \pi_{\{r_1,\dots,r_2\}}(V^{(s_1,s_2)}) \subset \{0,\dots,L-1\}^{\{r_1,\dots,r_2\}},\\ N^{(s_1,s_2)}_{(r_1,r_2)} &= \bigl|V^{(s_1,s_2)}_{(r_1,r_2)}\bigr|, \end{aligned}$$

where $\mathsf{S}(h(\sigma)f)$ is the support of $h(\sigma)f$ (cf. (5.3)). Clearly,

$$\{w^{s_2}_{s_1}(\gamma) : w \in V^{(s_1,s_2)}\} = \{w^{s_2}_{s_1}(\gamma) : w \in V_L\},$$

and

$$\begin{aligned} N^{(s_1,s_2)}_{(s_1,s_2)} &= |\{w^{s_2}_{s_1}(\gamma) : w \in V^{(s_1,s_2)}\}| = |\{w^{s_2}_{s_1}(\gamma) : w \in V_L\}|\\ &= \left|\left\{\sum_{i=s_1}^{s_2} w_i \cdot \iota^-(\gamma^{i-s_1}) : w \in V\right\}\right| \geq c\cdot \mathsf{M}(h)^{s_2-s_1+1} \end{aligned}$$

by (5.23). For $s_1' \leq s_1 \leq r_1 < r_2 \leq s_2 \leq s_2'$,

$$(5.25)\qquad V^{(s_1,s_2)}_{(r_1,r_2)} \supset V^{(s_1',s_2')}_{(r_1,r_2)} \text{ and hence } N^{(s_1,s_2)}_{(r_1,r_2)} \geq N^{(s_1',s_2')}_{(r_1,r_2)}.$$

We fix $M \geq 1$. From (5.25) it is clear that

$$N^{(-lM,(l+1)M-1)}_{(kM,(k+1)M-1)} \geq N^{(-(l+j)M,(l+j+1)M-1)}_{((k+i)M,(k+i+1)M-1)}$$

for every $j \geq 1$, $i = -j,\dots,j$, $l \geq 1$ and $k = -l,\dots,l$. For fixed $l \geq 1$,

$$\begin{aligned} c\cdot\mathsf{M}(h)^{(2l+2j+1)M} &= c\cdot\mathsf{M}(\tilde{h})^{(2l+2j+1)M} \leq N^{(-(l+j)M,(l+j+1)M-1)}_{(-(l+j)M,(l+j+1)M-1)}\\ &\leq \prod_{k=-l-j}^{l+j} N^{(-(l+j)M,(l+j+1)M-1)}_{(kM,(k+1)M-1)}\\ &\leq \prod_{k=-l-j}^{-j-1} N^{(-(l+j)M,(l+j+1)M-1)}_{(kM,(k+1)M-1)} \cdot \prod_{k=-j}^{j} N^{(-(l+j)M,(l+j+1)M-1)}_{(kM,(k+1)M-1)}\\ &\qquad\cdot \prod_{k=j+1}^{l+j} N^{(-(l+j)M,(l+j+1)M-1)}_{(kM,(k+1)M-1)}\\ &\leq \left(\prod_{k=-l}^{-1} N^{(-lM,(l+1)M-1)}_{(kM,(k+1)M-1)}\right)\cdot\left[N^{(-lM,(l+1)M-1)}_{(0,M-1)}\right]^{2j+1}\\ &\qquad\cdot\left(\prod_{k=1}^{l} N^{(-lM,(l+1)M-1)}_{(kM,(k+1)M-1)}\right)\\ &= \left(\prod_{k=-l}^{l} N^{(-lM,(l+1)M-1)}_{(kM,(k+1)M-1)}\right)\cdot\left[N^{(-lM,(l+1)M-1)}_{(0,M-1)}\right]^{2j} \end{aligned}$$

for every $j \geq 1$, and by letting $j \to \infty$, we conclude that

$$N^{(-lM,(l+1)M-1)}_{(0,M-1)} \geq \mathsf{M}(h)^M.$$

As $l \to \infty$, $V^{(-lM,(l+1)M-1)}_{(0,M-1)}$ decreases to $\pi_{\{0,\dots,M-1\}}(V_L^{(h)})$, and hence

$$\lim_{l\to\infty} N^{(-lM,(l+1)M-1)}_{(0,M-1)} = \bigl|\pi_{\{0,\dots,M-1\}}(V_L^{(h)})\bigr| \geq \mathsf{M}(h)^M$$

for every $M \geq 1$. By varying M, we see that $h(\sigma_{V_L^{(h)}}) \geq \log \mathsf{M}(h)$, and the reverse inequality follows from Theorem 5.1. □

REMARK 5.9. If the polynomial h is hyperbolic, the shift space $V_L^{(h)}$ in (5.3) is sofic for every $L \geq 1$ (cf. [**ES1**] and [**Sc3**]). If h is irreducible and nonexpansive, the combinatorial structure of $V_L^{(h)}$ is not well understood.

6. Invariant measures on $X_h/X_h^{(0)}$

Remark 4.2 shows that we can always find symbolic covers of $Q^{(h)}$, and for hyperbolic h these covers can even be chosen to be finite-to-one. However, if h is nonhyperbolic, Corollary 4.3 raises a number of questions.

PROBLEMS 6.1.

(1) Let $h \in R$ be irreducible and noncyclotomic. Is

$$\inf_V h(\sigma_V) = h(\alpha_h) = \log \mathsf{M}(h),$$

where the infimum is taken over all symbolic covers V of $Q^{(h)}$, and where $h(\sigma_V)$ is the topological entropy of the restriction of σ to V? For hyperbolic h the answer to this question is 'yes' (cf. Theorem 5.5).

(2) Does there always exist a symbolic cover V of $Q^{(h)}$ with $h(\sigma_V) = h(\alpha_h) = \log \mathsf{M}(h)$?

(3) Does there always exist a countable-to-one symbolic cover V of $Q^{(h)}$ (i.e. for which $\Delta^{(h)}(w) \cap V$ is countable for every $w \in V$)? A positive answer to this question would also solve (2).

(4) If there exist countable-to-one (or other nice) symbolic covers of $Q^{(h)}$, can one also find such covers with a relatively simple combinatorial structure? In light of Section 7 the best one could probably hope for is covers which are factors of countable-state shifts of finite type with well-behaved factor maps (such as, for example, beta-shifts).

In the possible absence of good symbolic covers one can try to construct partial covers V of $Q^{(h)}$ which are 'large' in the sense that $h(\sigma_V) = h(\alpha_h) = \log \mathsf{M}(h)$ and which are 'small' in the sense that, for certain natural measures on V, the factor map $q^{(h)} \colon \ell^\infty(\mathbb{Z},\mathbb{Z}) \longrightarrow Q^{(h)}$ is countable-to-one a.e. As was shown in [**LS2**], these ideas can be used to construct invariant probability measures for irreducible nonexpansive group automorphisms.

Let $\mathsf{d} \colon \mathbb{Z} \times \ell^\infty(\mathbb{Z},\mathbb{Z}) \longrightarrow W_h^{(0)}$ be the cocycle describing the nonequivariance of $\bar{\xi}^*$ in (4.7)–(4.9).

DEFINITION 6.2. A shift-invariant probability measure ν on $\ell^\infty(\mathbb{Z},\mathbb{Z})$ is *weakly* d*-bounded* if there exists, for every $\varepsilon > 0$, a compact subset $C_\varepsilon \subset W_h^{(0)}$ such that

$$\nu(\{v \in \ell^\infty(\mathbb{Z},\mathbb{Z}) : \mathsf{d}(k,v) \in C_\varepsilon\}) > 1 - \varepsilon \text{ for every } k \in \mathbb{Z}. \tag{6.1}$$

THEOREM 6.3. *Let $h \in R_1$ be an irreducible nonhyperbolic polynomial which is not cyclotomic, α_h the ergodic and nonexpansive automorphism of the compact connected abelian group X_h defined in (3.2)–(3.3), and let $\tilde{\sigma} \colon \tilde{Y} \longrightarrow \tilde{Y}$ be defined by (4.10)–(4.11). For every σ-invariant probability measure ν on $\ell^\infty(\mathbb{Z},\mathbb{Z})$ the following conditions are equivalent.*

(1) *ν is weakly d-bounded.*

(2) *There exists a $\tilde{\sigma}$-invariant probability measure $\tilde{\nu}$ on $\tilde{Y}$ with $\tilde{\pi}_*\tilde{\nu} = \nu$, where $\tilde{\pi} \colon \tilde{Y} \longrightarrow \ell^\infty(\mathbb{Z},\mathbb{Z})$ is the first coordinate projection.*

(3) *There exists a Borel map* $\mathsf{b}\colon \ell^\infty(\mathbb{Z},\mathbb{Z}) \longrightarrow W_h^{(0)}$ *with*

$$\mathsf{d}(1,v) = \mathsf{b}(\sigma v) - \bar{\sigma}\mathsf{b}(v) \text{ for } \nu\text{-a.e. } v \in \ell^\infty(\mathbb{Z},\mathbb{Z}). \tag{6.2}$$

If ν *satisfies these equivalent conditions, then the Borel maps* $\bar{\xi}_{\mathsf{b}}^*\colon \ell^\infty(\mathbb{Z},\mathbb{Z}) \longrightarrow \ell^*(\mathbb{Z},\mathbb{R})$ *and* $\xi_{\mathsf{b}}^*\colon \ell^\infty(\mathbb{Z},\mathbb{Z}) \longrightarrow X_h$, *defined by*

$$\bar{\xi}_{\mathsf{b}}^*(v) = \bar{\xi}^*(v) + \mathsf{b}(v) \text{ and } \xi_{\mathsf{b}}^*(v) = \xi^*(v) + \rho \circ \mathsf{b}(v) \tag{6.3}$$

for every $v \in \ell^\infty(\mathbb{Z},\mathbb{Z})$, *have the property that*

$$\begin{aligned} &\xi_{\mathsf{b}}^*(v) - \xi^*(v) \in X_h^{(0)} \text{ for every } v \in \ell^\infty(\mathbb{Z},\mathbb{Z}), \\ &\bar{\xi}_{\mathsf{b}}^* \circ \sigma = \bar{\sigma} \circ \bar{\xi}_{\mathsf{b}}^* \text{ and } \xi_{\mathsf{b}}^* \circ \sigma = \alpha_h \circ \xi_{\mathsf{b}}^* \ \nu\text{-a.e.}, \end{aligned} \tag{6.4}$$

and the probability measure

$$\mu = (\xi_{\mathsf{b}}^*)_*\nu \tag{6.5}$$

on X_h *is* α_h*-invariant.*

PROOF. This is [**LS2**, Theorem 4.13]. □

THEOREM 6.4. *Let* $L \geq 1$, *and let* $V_L^{(h)} \subset V_L = \{0,\dots,L-1\}^{\mathbb{Z}} \subset \ell^\infty(\mathbb{Z},\mathbb{Z})$ *be defined by* (5.3). *If* ν *is a weakly* d*-bounded shift-invariant probability measure on* $V_L^{(h)}$, *and if* $\xi_{\mathsf{b}}^*\colon \ell^\infty(\mathbb{Z},\mathbb{Z}) \longrightarrow X_h$ *is the* ν*-a.e. equivariant map* (6.3), *then the* α_h*-invariant probability measure* $\mu = (\xi_{\mathsf{b}}^*)_*\nu$ *on* X_h *is singular with respect to Haar measure and satisfies that* $h_\nu(\sigma) = h_\mu(\alpha_h)$.

For the proof of Theorem 6.4 we need several lemmas. Let $\mathbf{R} = \Delta^{(h)} \cap (V_L^{(h)} \times V_L^{(h)})$ be the equivalence relation induced by $\Delta^{(h)}$ on $V_L^{(h)}$. Exactly as in the proof of Corollary 4.3 we see that $\mathbf{R}$ is a $\bar{\sigma}\times\bar{\sigma}$-invariant Borel set in $V_L^{(h)} \times V_L^{(h)}$.

LEMMA 6.5. *Let* $Y \subset V_L^{(h)}$ *be a shift-invariant Borel set with* $\nu(Y) = 1$ *such that* (6.2) *holds for every* $v \in Y$, *and let*

$$\begin{aligned} Y(M) &= \{y \in Y : \|\mathsf{b}(y)\|_\infty \leq M\}, \\ L(M) &= \{y \in \ell^*(\mathbb{Z},\mathbb{Z}) : \|y - \bar{\xi}^* \circ h(\sigma)(y)\|_\infty \leq M\}, \\ \mathbf{R}(M,w) &= \bigl(w + h(\sigma)(L(M))\bigr) \cap V_L^{(h)} \subset \mathbf{R}(w) \end{aligned}$$

for every $M \geq 1$ *and* $w \in V_L^{(h)}$ *(cf.* (4.3)*). Then there exists a constant* $M_1 > 0$ *such that*

$$\begin{aligned} &\bigl|\pi_{\{0,\dots,n\}}\bigl(\mathbf{R}(K,w) \cap Y(M) \cap \sigma^{-n}(Y(M))\bigr)\bigr| \\ &\qquad\qquad \leq (2M_1\beta + 4M + 2K + 1)^{2d+2}, \\ &\bigl|\pi_{\{-n,\dots,0\}}\bigl(\mathbf{R}(K,w) \cap Y(M) \cap \sigma^{n}(Y(M))\bigr)\bigr| \\ &\qquad\qquad \leq (2M_1\beta + 4M + 2K + 1)^{2d+2}, \end{aligned} \tag{6.6}$$

for every $K, M \geq 1$, $w \in V_L^{(h)}$ *and* $n \geq 1$.

PROOF. By (4.2) there exists a constant $M_1 > 0$ such that

$$\max_{j=0,\dots,d} |\bar{\xi}^*(w)_j| \leq M_1 \cdot \|w\|_\infty$$

for every $w \in \ell^\infty(\mathbb{Z},\mathbb{Z})$. As $(\bar{\sigma}^*)^n \circ \bar{\xi}^*(w) = \bar{\xi}^* \circ \sigma^n(w) + \mathsf{b}(\sigma^n w) - \bar{\sigma}^n \mathsf{b}(w)$ for every $n \in \mathbb{Z}$,

$$\max_{j=0,\dots,d} |\bar{\xi}^*(w)_{n+j}| \le M_1\beta + 2M$$

for every $M \ge 1$, $n \in \mathbb{Z}$ and $w \in Y(M) \cap \sigma^{-n}(Y(M))$. We fix $w \in V_L^{(h)}$ and obtain that, for every $v \in \mathbf{R}(K,w) \cap Y(M) \cap \sigma^{-n}(Y(M))$,

$$\max_{j=0,\dots,d} |\bar{\xi}^*(v)_j| \le M_1\beta, \qquad \max_{j=0,\dots,d} |\bar{\xi}^*(v)_{n+j}| \le M_1\beta + 2M,$$

and that there exists a unique $y \in \ell^*(\mathbb{Z},\mathbb{Z})$ with $v = w + h(\sigma)(y)$ and $\|y - \bar{\xi}^* \circ h(\sigma)(y)\|_\infty \le K$.

If v' is a second element in $\mathbf{R}(K,w) \cap Y(M) \cap \sigma^{-n}(Y(M))$ with $v' = w + h(\sigma)(y')$ for some $y' \in \ell^*(\mathbb{Z},\mathbb{Z})$, then $\|y' - \bar{\xi}^* \circ h(\sigma)(y')\|_\infty \le K$, and hence

$$\max_{j=0,\dots,d} |y_j - y'_j| \le 2M_1\beta + 2K \quad \text{and} \quad \max_{j=0,\dots,d} |y_{n+j} - y'_{n+j}| \le 2M_1\beta + 4M + 2K.$$

The first inequality in (6.6) now follows from (5.15), and the proof of the second one is analogous. □

LEMMA 6.6. *For ν-a.e. $w \in Y$, the intersection $\Delta^{(h)}(w) \cap Y$ is countable, and the map $\xi^*_{\mathsf{b}} \colon V_L^{(h)} \longrightarrow X_h$ is countable-to-one ν-a.e.*

PROOF. In the notation of Lemma 6.5 we set $Y'(1) = Y(1)$ and $Y'(M) = Y(M) \smallsetminus Y(M-1)$ for every $M \ge 2$, and we define a map $q \colon Y \longrightarrow \mathbb{R}$ by setting $q(w) = 2^{-M}$ if $w \in Y'(M)$, $M \ge 1$. We fix an everywhere positive, Borel measurable and σ-invariant version $p = E_\nu(q|\mathcal{S}_Y^\sigma)$ of the conditional expectation of q, given the sigma-algebra $\mathcal{S}_Y^\sigma$ of σ-invariant Borel subsets of Y. After decreasing Y by a σ-invariant ν-null set, if necessary, we also assume that

$$\lim_{n\to\infty} \frac{1}{n} \cdot \sum_{j=1}^{n} q(\sigma^n w) = \lim_{n\to\infty} \frac{1}{n} \cdot \sum_{j=1}^{n} q(\sigma^{-n} w) = p(w)$$

for every $w \in Y$.

We claim that

$$\sum_{v \in \mathbf{R}(K,w) \cap Y} q(v)p(v)^2 = \sup_{\substack{F \subset \mathbf{R}(K,w) \cap Y \\ F \text{ is finite}}} \sum_{v \in F} q(v)p(v)^2 < \infty \tag{6.7}$$

for every $w \in Y$ and $K \ge 1$.

Indeed, if $F \subset \mathbf{R}(K,w) \cap Y$ is finite, then

$$\begin{aligned}
\sum_{v \in F} q(v)p(v)^2 &= \lim_{n\to\infty} \sum_{v\in F} q(v) \cdot \frac{1}{n^2} \cdot \Bigl(\sum_{j=1}^{n} q(\sigma^{-j}v)\Bigr) \cdot \Bigl(\sum_{j'=1}^{n} q(\sigma^{j'}v)\Bigr) \\
&= \lim_{n\to\infty} \frac{1}{n^2} \cdot \sum_{j=1}^{n} \sum_{j'=1}^{n} \sum_{M\geq 1} \sum_{M'\geq 1} \sum_{M''\geq 1} 2^{-M-M'-M''} \\
&\qquad \cdot \bigl|\pi_{\{-j,\dots,j'\}}(F \cap \sigma^j(Y'(M)) \cap Y'(M'') \cap \sigma^{-j'}(Y'(M')))\bigr| \\
&\leq \lim_{n\to\infty} \frac{1}{n^2} \cdot \sum_{j=1}^{n} \sum_{j'=1}^{n} \sum_{M\geq 1} \sum_{M'\geq 1} \sum_{M''\geq 1} 2^{-M-M'-M''} \\
&\qquad \cdot \bigl|\pi_{\{-j,\dots,0\}}(\mathbf{R}(K,w) \cap \sigma^j(Y'(M)) \cap Y'(M''))\bigr| \\
&\qquad \cdot \bigl|\pi_{\{0,\dots,j'\}}(\mathbf{R}(K,w) \cap \sigma^{-j'}(Y'(M')) \cap Y'(M''))\bigr| \\
&\leq \sum_{M\geq 1} \sum_{M'\geq 1} \sum_{M''\geq 1} 2^{\;M \quad M' \quad M''} \\
&\qquad \cdot \bigl[(2M_1\beta + 2K + 4\max\{M, M''\} + 1)^{2d+2}\bigr] \\
&\qquad \cdot \bigl[(2M_1\beta + 2K + 4\max\{M', M''\} + 1)^{2d+2}\bigr] < \infty
\end{aligned}$$

by (5.15), which proves (6.7).

Since the maps $p, q \colon Y \longrightarrow \mathbb{R}$ are everywhere positive, (6.7) implies that the sets $\mathbf{R}(K,w) \cap Y$ and $\mathbf{R}(w) \cap Y = \bigcup_{K\geq 1} \mathbf{R}(K,w) \cap Y$ are countable for every $w \in Y$ and that the equivariant Borel map $\xi^*_{\mathsf{b}} \colon Y \longrightarrow X$ in (6.3) is therefore countable-to-one. □

PROOF OF THEOREM 6.4. By Lemma 6.6 there exists a shift-invariant Borel set $Y \subset V_L^{(h)}$ with $\nu(Y) = 1$ such that the Borel map $\xi^*_{\mathsf{b}} \colon Y \longrightarrow X_h$ in (6.3) is countable-to-one. Since countable-to-one factor maps do not decrease entropy, $h_\nu(\sigma) = h_\mu(\alpha_h)$. Furthermore, the Borel set $Z = \xi^*_{\mathsf{b}}(Y) \subset X_h$ is α_h-invariant with $\mu(Z) = 1$ and intersects each coset of $X_h^{(0)}$ in a countable set. Hence $\lambda_{X_h}(Z) = 0$, which proves that λ_{X_h} and μ are mutually singular. □

COROLLARY 6.7. *Let $L \geq 1$, and let $V_L^{(h)} \subset V_L = \{0, \dots, L-1\}^{\mathbb{Z}} \subset \ell^\infty(\mathbb{Z}, \mathbb{Z})$ be defined by* (5.3). *If ν is a weakly* d*-bounded shift-invariant probability measure on $V_L^{(h)}$, then the map $\xi^\# \colon \ell^\infty(\mathbb{Z},\mathbb{Z}) \longrightarrow X_h/X_h^{(0)}$ induced by the group homomorphism $\xi^* \colon \ell^\infty(\mathbb{Z},\mathbb{Z}) \longrightarrow X_h$ has the following properties.*

(1) *The probability space $(X_h/X_h^{(0)}, \mathcal{B}_{X_h/X_h^{(0)}}, \xi^\#_* \nu)$ is standard, where by $\mathcal{B}_{X_h/X_h^{(0)}}$ we denote the Borel field of $X_h/X_h^{(0)}$.*

(2) *If $\alpha_h^\#$ is the group automorphism of $X_h/X_h^{(0)}$ induced by α_h, then $\xi^\#_*\nu$ is $\alpha^\#$-invariant and $h_{\xi^\#_*\nu}(\alpha^\#) = h_\nu(\sigma)$.*

PROOF. By [**LS1**, Proposition 4.17] there exists a solution b' of (6.2) and an α_h-invariant Borel set $Z \subset X_h$ which intersects each coset of $X_h^{(0)}$ in at most one point, such that $(\xi^*_{\mathsf{b}'})_*\nu(Z) = 1$. This implies all our assertions. □

PROBLEM 6.8. Theorems 5.3 and 6.4 raise the following question: is

$$\sup_\nu h_\nu(V_L^{(h)}) = h(\sigma_{V_L^{(h)}}), \tag{6.8}$$

where the supremum in (6.8) is taken over all weakly bounded shift-invariant probability measures on $V_L^{(h)}$?

A shift-invariant probability measure ν on $V_L^{(h)}$ is *bounded* if there exists a compact subset $C \subset W_h^{(0)}$ such that $\mathsf{d}(n,w) \in C$ for every $n \in \mathbb{Z}$ and ν-a.e. $w \in V_L^{(h)}$. The following conditions can be shown to be equivalent.

(1) $\sup_{\nu \text{ bounded}} h_\nu(V_L^{(h)}) = h(\sigma_{V_L^{(h)}})$.

(2) $\sup_{N\ge 1} h(\sigma_{V_L^{(h)}\cap h(\sigma)(B_N)}) = h(\sigma_{V_L^{(h)}})$, where $B_N = \{-N,\dots,N\}^{\mathbb{Z}}$ for every $N \ge 0$.

If β is a Salem number with minimal polynomial $h \in R$, then $h = \tilde{h}$ (cf. (5.17)), and the positive answer to these equivalent questions follows from Proposition 7.2 below and [**LS2**, Theorem 7.1]. In the general case this question is still open.

7. Some examples

EXAMPLE 7.1 (Beta-shifts). Suppose that h has a single root $\gamma < 1$, that all other roots of h have absolute values ≥ 1, and that $h_0 = 1$ (i.e. that the inverse $\beta = \gamma^{-1}$ of γ is either an integer, a Pisot number or a Salem number). The S^- in (5.18) consists of a single real place.

Following [**P**], we consider the map

$$T_\beta x = \beta x \pmod 1 \tag{7.1}$$

from the unit interval $I = [0,1]$ to itself and define, for every $x \in I$, the *beta-expansion* $\omega_\beta(x) = (\omega_\beta(x)_n)$ of x by setting

$$\omega_\beta(x)_n = \beta T_\beta^n x - T_\beta^{n+1} x \tag{7.2}$$

for every $n \ge 0$. Note that $\omega_\beta(x)_n \in \{0,\dots,\lceil\beta-1\rceil\}$ for every $n \ge 1$, where $\lceil\beta-1\rceil$ is the smallest integer $\ge \beta - 1$ and that

$$x = \sum_{n\ge 0} \omega_\beta(x)_n \beta^{-n-1} \tag{7.3}$$

for every $x \in I$. We set

$$\omega_\beta^*(1) = \sup_{x\in[0,1)} \omega_\beta(x), \tag{7.4}$$

where the supremum is taken with respect to the lexicographic order $\prec$ on the space $\ell^\infty(\mathbb{N},\mathbb{Z})$ and observe that

$$1 = \sum_{n\ge 0} \omega_\beta^*(1)_n \beta^{-n-1}. \tag{7.5}$$

Recall that $\sigma_+^k \omega_\beta^*(1) \ne (0,0,0,\dots)$ and

$$\sigma_+^k \omega_\beta^*(1) \preceq \omega_\beta^*(1) \tag{7.6}$$

for every $k \ge 1$ (cf. [**P**]). The restriction of σ_+ to the closed, shift-invariant set

$$V_\beta^+ = \{v \in \ell^\infty(\mathbb{N},\mathbb{N}) : \sigma_+^n v \preceq \omega_\beta^*(1) \text{ for every } n \ge 0\} \subset \{0,\dots,\lceil\beta-1\rceil\}^{\mathbb{N}}$$

is called the *β-shift*. If we set

$$\eta_\beta(v) = \sum_{n\ge 0} v_n \beta^{-n-1} \tag{7.7}$$

for every $v \in V_\beta^+$, then $\eta_\beta\colon V_\beta^+ \longrightarrow [0,1]$ is surjective and at most two-to-one. Furthermore, if $v, v' \in V_\beta^+$ satisfy that $\eta_\beta(v) = \eta_\beta(v')$, and if $v \prec v'$, then there exists an integer $k \ge 0$ such that $v_n = v'_n$ for $n < k$, $v'_k = v_k + 1$, and $v'_n = 0$, $v_n = \omega_\beta^*(1)_{n-k-1}$ for $n > k$. Finally, if $v, v' \in V_\beta^+$, then $v \preceq v'$ if and only if $\eta_\beta(v) \le \eta_\beta(v')$ (cf. [**P**]).

In order to define the two-sided beta-shift space $V_\beta \subset \{0, \dots, \lceil \beta - 1 \rceil\}^{\mathbb{Z}}$, we set $v^+ = (v_0, v_1, v_2, \dots) \in \ell^\infty(\mathbb{N}, \mathbb{Z})$ for every $v = (v_n) \in \ell^\infty(\mathbb{Z}, \mathbb{Z})$ and put

$$V_\beta = \{v \in \ell^\infty(\mathbb{Z}, \mathbb{Z}) : (\sigma^n v)^+ \in V_\beta^+ \text{ for every } n \in \mathbb{Z}\}. \tag{7.8}$$

From the description of the potential non-injectiveness of the map η_β in (7.7) it is clear that V_β intersects every equivalence class of $\Delta_1^{(h)}$ in at most one point (cf. (5.1)).

PROPOSITION 7.2. *Suppose that the polynomial h in (2.4) satisfies that $h_0 = 1$ and that h has a root $\gamma < 1$ and all other roots of h have absolute values ≥ 1. Put $\beta = \gamma^{-1}$ and denote by $V_\beta \subset \{0, \dots, \lceil \beta - 1 \rceil\}^{\mathbb{Z}}$ the two-sided beta-shift (7.8). If $L > \beta$, then $V_L^{(h)} \supset V_\beta$ and $h(\sigma_{V_L^{(h)}}) = h(\sigma_{V_\beta}) = \log \beta$ (cf. (5.3)).*

PROOF. In order to verify that $V_L^{(h)} \supset V_\beta$, we argue by contradiction and assume that there exists a $v \in V_\beta \smallsetminus V_L^{(h)}$. Then (5.3) and (5.13) show that there exists an $f \in P$ with $v \in V_L \cap (V_L - h(\sigma)f)$, i.e. that $v' = v + h(\sigma)f \in V_L$ for some $f \in P$. We set $n = \min \mathrm{S}(f)$ (cf. (2.2)) and assume without loss of generality that $n = 0$ (by shifting v and f, if necessary). According to (7.7)–(7.8), $\eta_\beta(v^+) = \eta_\beta(v'^+)$. As $v'_0 \ge v_0 + 1$, we conclude that $1 \le v'_0 - v_0 + \sum_{n \ge 0} v'_n \beta^{-n} = \sum_{n \ge 1} v_n \beta^{-n}$. Since $v \in V_\beta$, it follows that $v'_0 = v_0 + 1$ and that $v_n = \omega_\beta^*(1)_{n-1}$ and $v'_n = 0$ for every $n \ge 1$. This is clearly impossible, since v and v' differ in only finitely many coordinates.

The last identity follows from Theorem 5.3, since $h(\sigma_{V_\beta}) = \log \beta$. □

Take, for example, the polynomial $h(u) = 1 - u - u^2$ with roots $\gamma = \frac{2}{1+\sqrt{5}} < 1$ and $\gamma' = -1/\gamma$. If $V_2 = \{0,1\}^{\mathbb{Z}}$ (cf. (5.4)), then Proposition 7.2 shows that $V_2^{(h)} \supset V_\beta$, where $\beta = 1/\gamma = \frac{1+\sqrt{5}}{2}$, and $h(\sigma_{V_2^{(h)}}) = h(\sigma_{V_\beta}) = \log \beta$. One can check that every point $w \in V_2^{(h)} \smallsetminus V_\beta$ is either of the form

$$w_k = 1 \text{ for every } k \in \mathbb{Z}$$

or that there exists an $l \in \mathbb{Z}$ with

$$w_k = 1 \text{ for every } k < l,\ w_l = 0 \text{ and } (w_{l+1}, w_{l+2}, w_{l+3}, \dots) \in V_\beta^+.$$

We also have that $V_3^{(h)} \supset V_\beta$ and $h(\sigma_{V_3^{(h)}}) = \log \beta$, but neither of the spaces $V_2^{(h)}$ and $V_3^{(h)}$ contains the other:

$$(\dots, 1, 1, 1, \dots) \in V_2^{(h)} \smallsetminus V_3^{(h)},$$

whereas

$$(\dots, 2, 2, 2, \dots) \in V_3^{(h)} \smallsetminus V_2^{(h)}.$$

In this example, V_β is a shift of finite type and the spaces $V_L^{(h)}$, $L \ge 2$, are sofic by Remark 5.9.

EXAMPLE 7.3 (The polynomial $h(u) = 5 - 6u + 5u^2$). The roots of h are of the form $\gamma = \frac{3}{5} + i \cdot \frac{4}{5}$, $\bar{\gamma} = \frac{3}{5} - i \cdot \frac{4}{5}$ with absolute values equal to 1, and the set S in (5.8) consists of a single infinite complex place v_∞ (corresponding to γ and $\bar{\gamma}$) and two copies of the finite place 5 with $K_5 = \mathbb{Q}_5$, the 5-adic rationals (note that h has two roots $\gamma_1, \gamma_2 \in \mathbb{Q}_5$ with $|\gamma_1|_5 = 1/5$ and $|\gamma_2|_5 = 5$, where $|\cdot|_5$ is the 5-adic valuation). We write v_5 for the place of $\mathbb{Q}(\gamma)$ corresponding to γ_1 and obtain that $S^- = \{v_5\}$ and $K_{S^-} = \mathbb{Q}_5$ (cf. (5.18)).

Since every $t \in \mathbb{Z}_5 = \bar{R}_{v_5}$ can be expressed uniquely as

$$t = \sum_{n \geq 0} a_n \gamma_1^n$$

with $a_n \in \{0, 1, 2, 3, 4\}$ for every $n \geq 0$, we have that $V_5^{(h)} = V_5$, and the proofs of Lemma 5.7 and Theorem 5.3 show that $h(\sigma_{V_5^{(h)}}) = \log \mathsf{M}(h) = \log 5$ (cf. (2.6)). More generally, if $L \geq 5$, and if $V_L = \{0, \ldots, L-1\}^{\mathbb{Z}}$, then the same argument as in Proposition 7.2 shows that $V_L^{(h)} \supset V_5^{(h)}$, and Theorem 5.1 guarantees that $h(\sigma_{V_L^{(h)}}) = h(\sigma_{V_5^{(h)}}) = \log 5$.

EXAMPLE 7.4 (Reversing polynomials). Let $h \in R$ be of the form (2.4), and let $g = \operatorname{sgn}(h_d)\tilde{h}$, where sgn stands for sign (cf. (5.17)). Then $\mathsf{M}(h) = \mathsf{M}(g)$ by (2.6), but the spaces $V_L^{(h)}$ and $V_L^{(g)}$ may not be reversals of each other (due to the possible sign-change involved in the definition of g).

For example, if $h(u) = 1 - u - u^2$ is the polynomial appearing at the end of Example 7.1, then $g(u) = 1 + u - u^2$, and $V_2^{(g)}$ is the set of all sequences in $V_2^{(h)}$, reversed and with zeros and ones interchanged.

Similarly, if $h(u) = 1 - u^2 - u^3$, then h has a single small root $\gamma = 0.75487\cdots < 1$ and two complex conjugate roots with absolute values > 1. If $\beta = \gamma^{-1} = 1.32472\cdots$, then $V_\beta \subset V_2$.

Examples 7.1 and 7.3–7.4 had the property that $|S^-| = 1$. Here is an example with $|S^-| \geq 2$ and $|S \smallsetminus S^-| \geq 2$.

EXAMPLE 7.5 (The polynomial $h(u) = 1 - u^2 - u^4$). The two roots of h of absolute value < 1 are given by $\gamma = \pm\sqrt{\frac{2}{1+\sqrt{5}}}$, and S^- consists of the two places corresponding to these roots.

Let $\beta = \frac{1+\sqrt{5}}{2}$, and let V_β be the corresponding two-sided beta-shift space consisting of all sequences $(v_n) \in \{0,1\}^{\mathbb{Z}}$ with $v_n v_{n+1} = 0$ for every $n \in \mathbb{Z}$. One can check as in Example 7.1 that $Y \subset V_2^{(h)}$, where Y is the shift of finite type determined by the condition that $y_n y_{n+2} = 0$ for every $n \in \mathbb{Z}$. Note that Y consists of two interspersed copies of V_β and that $h(\sigma_Y) = h(\sigma_{V_\beta}) = \log \beta = \mathsf{M}(h) = h(\sigma_{V_2^{(h)}})$.

References

[C] J. W. S. Cassels, *Local Fields*, Cambridge University Press, Cambridge, 1986.

[ES1] M. Einsiedler, K. Schmidt, *Markov partitions and homoclinic points of algebraic $\mathbb{Z}^d$-actions*, Proc. Steklov Inst. Math. **216** (1997), 259–279.

[ES2] M. Einsiedler, K. Schmidt, *Irreducibility, homoclinic points and adjoint actions of algebraic $\mathbf{Z}^d$-actions of rank one*, Nonlinear Phenomena and Complex Systems, A. Maass, S. Martinez and J. San Martin (eds.), Kluwer Academic Publishers, Dordrecht, 2002, pp. 95–124.

[K] L. Kronecker, *Zwei Sätze über Gleichungen mit ganzzahligen Coefficienten*, J. Reine Angew. Math. **53** (1857), 173–175.

[KV] R. Kenyon, A. Vershik, *Arithmetic construction of sofic partitions of hyperbolic toral automorphisms*, Ergod. Th. Dynam. Sys. **18** (1998), 357–372.

[LS1] E. Lindenstrauss, K. Schmidt, *Invariant measures of nonexpansive group automorphisms*, Israel J. Math. **144** (2004), 29–60.

[LS2] E. Lindenstrauss, K. Schmidt, *Symbolic representations of nonexpansive group automorphisms*, preprint.

[LSW] D. Lind, K. Schmidt, T. Ward, *Mahler measure and entropy for commuting automorphisms of compact groups*, Invent. Math. **101** (1990), 593–629.

[LW] D. Lind, T. Ward, *Automorphisms of solenoids and p-adic entropy*, Ergod. Th. Dynam. Sys. **8** (1988), 411–419.

[P] W. Parry, *On the β-expansions of real numbers*, Acta Math. **11** (1960), 401–416.

[Sc1] K. Schmidt, *Automorphisms of compact abelian groups and affine varieties*, Proc. London Math. Soc. **61** (1990), 480–496.

[Sc2] K. Schmidt, *Dynamical systems of algebraic origin*, Birkhäuser Verlag, Basel-Berlin-Boston, 1995.

[Sc3] K. Schmidt, *Algebraic coding of expansive group automorphisms and two-sided beta-shifts*, Monatsh. Math. **129** (2000), 37–61.

[Si] N. Sidorov, *Bijective and general arithmetic codings for Pisot toral automorphisms*, J. Dynam. Cont. Sys. **7** (2001), 447–472.

[SV] N. Sidorov, A. Vershik, *Bijective arithmetic codings of the 2-torus, and binary quadratic forms*, J. Dynam. Cont. Sys. **4** (1998), 365–400.

[V] A. M. Vershik, *Arithmetic isomorphism of hyperbolic toral automorphisms and sofic shifts*, Funktsional. Anal. i Prilozhen. **26** (1992), 22–27.

[W] A. Weil, *Basic Number Theory*, Springer Verlag, Berlin-Heidelberg-New York, 1974.

Mathematics Institute, University of Vienna, Nordbergstrasse 15, A-1090 Vienna, Austria

and

Erwin Schrödinger Institute for Mathematical Physics, Boltzmanngasse 9, A-1090 Vienna, Austria

E-mail address: klaus.schmidt@univie.ac.at